Medical Virology 9

Medical Virology 9

Edited by
Luis M. de la Maza
and
Ellena M. Peterson
University of California, Irvine Medical Center
Orange, California

SPRINGER SCIENCE+BUSINESS MEDIA, LLC

&

Library of Congress Catalog Card Number 89-657524

ISBN 978-1-4684-5858-9   ISBN 978-1-4684-5856-5 (eBook)
DOI 10.1007/978-1-4684-5856-5
ISSN 1043-1837

Proceedings of the 1989 International Symposium on Medical Virology,
held October 16–18, 1989, in San Francisco, California

© 1990 by Springer Science+Business Media New York
Originally published by Plenum Press New York in 1990
Softcover reprint of the hardcover 1st edition 1990

FOREWORD

It would have been difficult at the beginning of the 80's to have predicted that by the end of the decade, Medical Virology would have become one of the most important topics in the area of both basic and clinical research. Although we were expecting a progressive increase in awareness of the role played by viruses in different diseases, we did not expect the outbreak of a fatal disease that was going to shake the roots of our society. The appearance of the human immunodeficiency virus (HIV-1) in the early 80's, has prompted a unique research impetus in the area of Medical Virology. The knowledge that we are gaining in our attempt to understand the biology of HIV-1 and the immunological response to this virus should not only help us control the spread of this virus, but should also help us to better understand other viral infections. Let us hope that during the 1990's we can learn how to control HIV-1 infections so that by the end of the decade, no more human lives succumb to an infection with this virus.

Luis M. de la Maza   Irvine, California
Ellena M. Peterson   March, 1990

ACKNOWLEDGEMENTS

We would like to thank all the speakers that came to San Francisco and shared their knowledge during the lectures and for writing the chapters in this book. We also want to acknowledge the participants that provided lively discussions, and to Dr. Edwin H. Lennette, who chaired one of the sessions. We would also like to acknowledge Dr. Thomas Cesario for chairing a session, however, his arrival in San Francisco was deterred at Orange County Airport by the announcement of seismic activity in the Bay Area. We do, however, acknowledge his efforts.

We wish to thank all the staff of the Division of Medical Microbiology at the University of California Irvine Medical Center for continuously supporting the Symposium, and a special recognition should go to Marie Pezzlo and Sandra Aarnaes who helped us to coordinate this meeting. We also want to express our appreciation to Penny Welter and Lillian Dalgleish for their administrative support. We are grateful to Gregory Safford and Melanie Yelaty from Plenum Publishing Corporation for their encouragement and support in the publication of this book.

This year, we want to express our gratitude to the staff of the Park Fifty-Five Hotel for the help and support they provided to all the participants of the Symposium following the earthquake of October 17, 1989.

It would have been impossible to organize the Symposium without the financial support provided by the following organizations:

> Beckman Instruments, Inc.
> Bio-Rad
> Cambridge BioScience
> Diamedix Corp.
> Genentech, Inc.
> GIBCO/BRL Research Products
> ICN Pharmaceuticals, Inc.
> Merck Sharp and Dohme
> Microbiology Reference Laboratory (MRL)
> Microbix Biosystems, Inc.
> Organon Teknika
> Pharmacia LKB Biotechnology Inc.
> Sandoz Research Institute
> Smith Kline & French Laboratories
> Stellar Bio Systems, Inc.
> Sterling Research Group
> Syva Co.

Triton Biosciences Inc.
The Upjohn Company
ViroMed Laboratories, Inc.
Virion (U.S.), Inc.
Whittaker Bioproducts, Inc.
Wyeth-Ayerst Research

CONTENTS

Evaluation of Immunoassays for Electron Microscopy   1
   FRANCES W. DOANE, NAN ANDERSON, FRANCIS LEE,
      KATHRYN PEGG-FEIGE and JOHN HOPLEY

Monoclonal Time-Resolved Fluoroimmunoassay: Sensitive Systems
      for the Rapid Diagnosis of Respiratory Virus Infections 17
   JOHN C. HIERHOLZER, LARRY J. ANDERSON and PEKKA E.
   HALONEN

The Use of the Polymerase Chain Reaction in the Detection, Quantifica-
      tion and Characterization of Human Retroviruses 47
   BERNARD J. POIESZ, GARTH D. EHRLICH, BRUCE C. BYRNE,
      KEITH WELLS, SHIRLEY KWOK and JOHN SNINSKY

ROUND TABLE: POLYMERASE CHAIN REACTION (PCR) 77

Clinical Serological and Intestinal Immune Responses to Rotavirus
      Infection of Humans 85
   RUTH BISHOP, JENNIFER LUND, ELIZABETH CIPRIANI,
      LEANNE UNICOMB and GRAEME BARNES

Medical Virology of Small Round Gastroenteritis Viruses 111
   NEIL R. BLACKLOW

Rabies - New Challenges by an Ancient Foe 129
   RICHARD W. EMMONS

HTLV-I Infections 147
   YORIO HINUMA

Human Herpesvirus 6: Basic Biology and Clinical Associations 163
   JOHN A. STEWART

Strains of Respiratory Syncytial Virus: Implications for Vaccine
      Development 187
   LARRY J. ANDERSON

Plans for Human Trials of a Vaccine Against Epstein-Barr Virus
      Infection 207
   M.A. EPSTEIN

Advances in the Treatment of HIV-1 Infections 217
   MARTIN S. HIRSCH

ABSTRACTS 237

CONTRIBUTORS 273

AUTHOR INDEX 275

SUBJECT INDEX 277

EVALUATION OF IMMUNOASSAYS FOR ELECTRON MICROSCOPY

Frances W. Doane, Nan Anderson, Francis Lee, Kathryn
Pegg-Feige and John Hopley

Department of Microbiology
University of Toronto
Toronto, Ontario, Canada

INTRODUCTION

In recent years immunoelectron microscopy (IEM) has become one of the several immunoassays available to virologists for detection and identification of viruses. Like other immunological assays, the reliability depends on careful attention to test conditions (e.g. antigen/antibody concentration, incubation time, temperature, pH) and to a regard for incorporation of a range of appropriate controls. This chapter will discuss several IEM techniques that can be applied to fluid specimens commonly encountered in diagnostic virology. Consideration will be given to some of the variables that affect the efficiency of these tests, and to the advantages and disadvantages of IEM relative to other immunoassays. Additional practical details are available from several published sources (Doane, 1974, 1986, 1987, 1988a, 1988b; Doane and Anderson, 1977, 1987; Katz and Kohn, 1984; Kjeldsberg, 1986).

HISTORY

The prospect of using electron microscopy (EM) to examine antigen-antibody complexes was first reported in 1941 by Anderson and Stanley in the U.S.A. and by von Ardenne et al. in Germany. Both groups observed that visible aggregates were formed when tobacco mosaic virus (TMV) was incubated with homologous antibody. Virologists were slow to adopt this technique which has been termed "immune electron microscopy" (Almeida and Waterson, 1969) or "immunoelectron microscopy" (Kelen et al. 1971). With the introduction of the negative staining technique, however, (Brenner and Horne, 1959) the potential usefulness of IEM in the study of virus-antibody interaction gradually became apparent (Almeida et al. 1963; Bayer and Mannweiler, 1963; Kleczkowski, 1961; Lafferty and Oertelis, 1961, 1963). It also became apparent that IEM could be used to identify different viral antigens (Hummeler et al. 1961; Watson and Wildy, 1963), to identify elusive or fastidious viruses (Best et al. 1967; Kapikian et al. 1972a, 1972b, 1973; Paver et al. 1973), and to increase by several hundred-fold the sensitivity of detection of viruses by EM (Doane, 1974; Anderson and Doane, 1973). Recent improvements of IEM methodology have included the introduction of colloidal gold as a specific marker; this has resulted in enhanced EM visualization of immune complexes and permits the detection of soluble viral antigens (Doane, 1987, 1988b; Kjeldsberg, 1986), and enables quantitation of low levels of viral antibody (Hopley and Doane, 1985; Vreeswijk et al. 1988).

MATERIALS

Viral Antigens

IEM can be used on crude clinical specimens such as fecal samples, on virus-infected cell culture lysates, or on purified virus preparations. When the interpretation of the test depends on the presence or absence of viral antibody aggregates, it is preferable to work with at least a partially purified specimen.

Antibodies

Either monoclonal or polyclonal antibodies can be used, although the majority of reports to date have employed polyclonal antibodies. Unfractionated serum is suitable in most instances, but where fine resolution (immunological or ultrastructural) is required, sera should be clarified by centrifugation for 1 hr at 100,000 x g to remove protein aggregates, or should be purified to obtain the globulin fraction. Antisera should be inactivated at 56°C for 30 min to avoid complications arising from the presence of complement (Almeida and Waterson, 1969).

The concentration of antibody used, in relation to the amount of antigen, affects not only the sensitivity of the assay but also the appearance of the resultant immune complexes (Figure 1). Whereas higher concentrations may increase the sensitivity, and enlarge the immune complexes thereby making them easier to find by EM, too high a concentration may lead to antibody coating of individual virus particles and a reduction of antibody-antigen aggregates (Lee, 1977). In addition, higher antibody concentrations may produce non-specific cross reactions and increased background deposits of serum protein that tend to interfere with visualization of immune complexes. Ideally, a box titration on all antibody preparations to be used as IEM probes should be performed against positive and negative virus controls to obtain an "IEM endpoint" - viz. the highest dilution of antibody that produces a positive reading. Between 5-10 times that concentration should be used in subsequent tests.

Negative Stains

A variety of negative stains are suitable for IEM, but phosphotungstic acid (PTA) remains the universal favorite. We use 2% PTA in Millipore filtered distilled water, adjusted to pH 6.5 with 1N KOH.

Specimen Grids

Copper grids of 300-mesh size coated with Formvar or parlodion offer a stable support for negatively stained specimens. Maximum stability of the plastic support film is achieved by coating it with a thin layer of evaporated carbon prior to use.

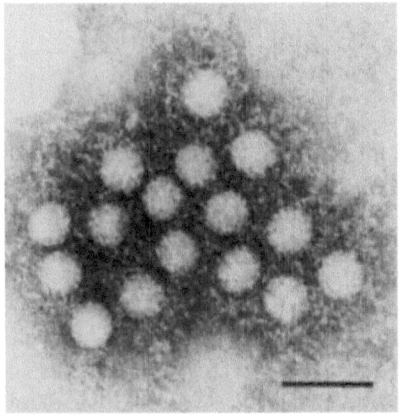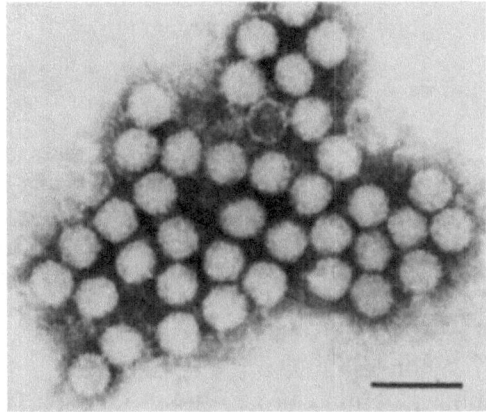

Figure 1. The effect of antibody concentration on the morphology of the immune complex. Left: high antibody concentration. Right: low antibody concentration. Bars equal 50 nm.

IMMUNOELECTRON MICROSCOPY METHODS

When the first two methods described below are used for serotyping, a virus-antibody complex is the indicator of a positive reaction, whereas the presence of predominantly single particles indicates a negative result. Unfortunately, many viruses tend to clump naturally, especially in fecal specimens (Narang and Codd, 1981); thus, one must always be cautious in interpreting the significance of aggregates in IEM assays, paying close attention to the accompanying controls (e.g. test sample exposed to antibodies of different specificities). Differentiation is facilitated by using an antibody concentration that produces a visible halo around the immune complexes. It is also possible to enhance or "decorate" the complex by the addition of anti-species antibody or protein A (Milne and Luisoni, 1977).

Although IEM is generally used to detect a specific viral antigen, it can also be adapted as an assay for viral antibody. Kapikian et al. (1976) measured the diameter of the antibody halo to determine the relative concentration of antibody in acute and convalescent phase sera. A major limitation is that the immune complexes produced at high dilutions of serum may be devoid of a visible halo. A more sensitive and readily detectable method is provided by decorating with an anti-species antibody or protein A labelled with a gold marker (Hopley and Doane, 1985).

Direct Immunoelectron Microscopy (DIEM) Method

In the original method described by Almeida and Waterson (1969), the virus preparation is mixed with an equal volume of antiserum, incubated at 37°C for 1 hr., then placed at 4°C overnight. Immune complexes are sedimented by centrifugation at 40,000 x g for 1 hr. The pellet is resuspended in distilled water, added to a grid, and negatively stained. We have found it possible to shorten the method, without an appreciable loss in sensitivity, by eliminating the overnight and centrifugation steps (Lee, 1977). After the initial incubation at 37°C, the mixture is processed by the agar diffusion method (Anderson and Doane, 1972), in which a drop of mixture is allowed to air-dry on a coated EM specimen grid placed on agar.

Serum in Agar (SIA) Method

This is a modification of the agar diffusion method, incorporating the viral antiserum in the agar itself (Figure 2) (Anderson and Doane, 1973). As the specimen on the grid dries, homologous antibody diffuses through the agar to form immune complexes on the EM grid support film. The method has been used to increase the sensitivity of virus detection by EM (Anderson and Doane, 1973), and for serotyping enteroviruses (Anderson and Doane, 1973; Lamontagne et al. 1980; Petrovicova and Juck, 1977).

When screening a specimen for a variety of different viruses, human gamma globulin can be used in the agar (Berthiaume et al. 1981) This permits an initial family identification of a virus, or mixture of viruses, on the basis of morphology. Further identification can be performed by SIA using individual or pooled antisera.

The concentration of antiserum used for screening should be considerably higher than that used for serotyping. We routinely use immune serum globulin at a final dilution (in molten agar) of 1/50; pools of enterovirus antisera are also prepared to give a final dilution of 1/50 per serum (Doane 1986; Doane and Anderson, 1987).

A modification of this technique has been described by Furui (1986). The sample is placed directly on the agar surface, and a specimen grid coated with protein A is floated on the sample, coated side down. After 60 min, the grid is removed, washed and negatively stained.

Solid Phase Immunoelectron Microscopy (SPIEM) Method

This method is also known as immunosorbent electron microscopy (ISEM) (Katz and Kohn, 1984). Introduced by Derrick in 1973 for "trapping" plant viruses (Figure 3), it involves the coating of the specimen support film with viral antibodies. An even greater trapping effect can be achieved by the application of *Staphylococcus aureus* protein A to the film (SPIEM-SPA), prior to the antibody (Table 1) (Pegg-Feige, 1983; Pegg-Feige and Doane, 1983).

SPIEM provides a relatively simple and sensitive IEM method for detecting viruses. It has been successfully applied to rotaviruses (Katz et al. 1980; Kjeldsberg and Mortensson-Egnund, 1982; Nicolaieff et al. 1980; Obert et al. 1981; Rubenstein and Miller, 1982), adenoviruses (van Rij et al. 1982), enteroviruses (Pegg-Feige, 1983; Pegg-Feige and Doane, 1983, 1984), and hepatitis A virus (Kjeldsberg and Siebke, 1985), and Sindbis virus (Katz and Straussman, 1984).

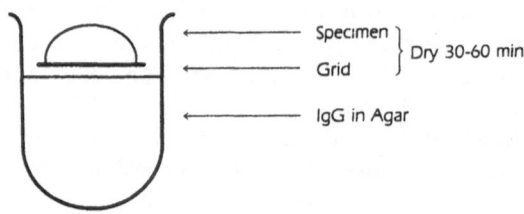

Figure 2. Serum-in-agar method.

TABLE 1. Sensitivity of Different EM Methods

	DEM	SPIEM	SPIEM-SPA	SIA	AIRFUGE
Poliovirus	6×10^{5}[a]	3×10^4	9×10^3	6×10^3	6×10^3

[a] Minimum detectable virus concentration ($TCID_{50}$/ml).

As discussed by Katz and Kohn (1984) in their comprehensive review of this technique, there are several variables that affect the efficiency of virus trapping; these include such factors as the nature of the specimen support film, the concentration and pH of the reactants, the duration and temperature of incubation steps. We have found that maximum trapping efficiency is obtained with parlodion-carbon films pretreated with ultraviolet light, or parlodion-carbon and Formvar-carbon films pretreated with glow discharge ionization (Table 2) (Pegg-Feige, 1983; Pegg-Feige and Doane, 1983, 1984). In our SPIEM protocol, pretreated grids are floated on drops of reactants at 25°C according to the following protocol: 25 µl of 0.1 mg/ml protein A, 10 min; 3 consecutive drops of Tris buffer, 1-2 min; 25 µl of diluted antiserum, 10 min; Tris buffer rinse, 1-2 min; 25 µl of test sample, 30 min; Tris buffer rinse, 1-2 min; 2% PTA pH 7.0. Each reference antiserum must be tested to determine the optimum working dilution necessary to achieve maximum virus trapping.

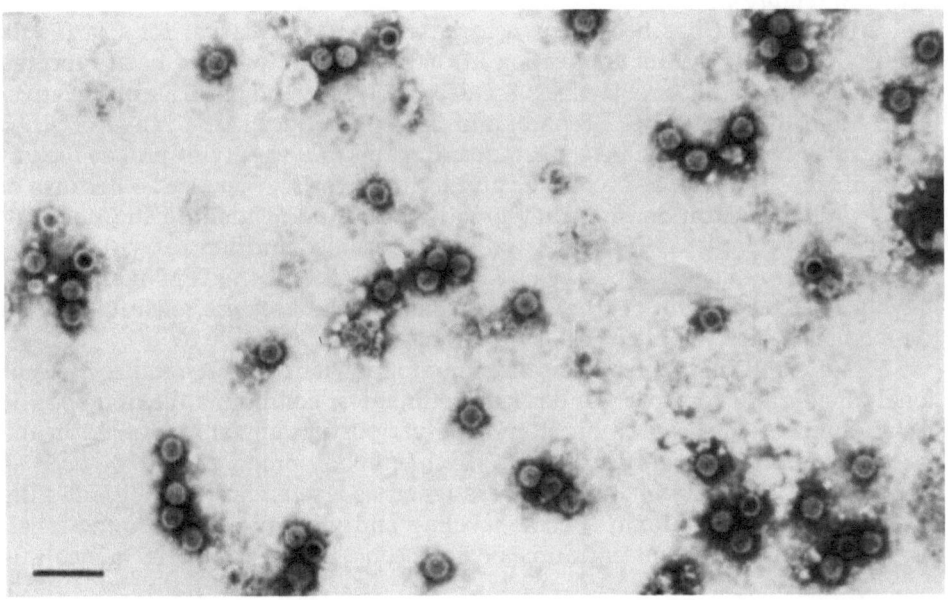

Figure 3. Rotavirus trapped on a specimen grid coated with protein A and rotavirus antiserum. Bar equals 200 nm.

TABLE 2. Effect on SPIEM Trapping Efficiency of Support Film Pretreatment

	Treatment		
Film	None	UV Light[b]	Glow Discharge[c]
Parlodion-carbon	71[a]	91	84
Formvar-carbon	9	8	74

[a] Average virus particles per grid square
[b] UV pretreatment: 1700 mw/cm^2 for 30 min.
[c] Glow discharge: 5-10 sec pink ionization of argon gas in a sputter coater.

SPIEM has also been used to serotype rotaviruses (Gerna et al. 1984, 1985, 1988) and enteric adenoviruses (Wood and Bailey, 1987) directly from fecal specimens. Individual grids were coated with a single type-specific rotavirus antiserum, and by a comparison of the number of virus particles trapped to each grid, rapid serotyping of human rotaviruses directly from stool specimens could be achieved.

Immunogold Method

The IEM methods described above rely on the presence of antibody-bound virus particles in the form of either virus-antibody aggregates or virus particles trapped to an antibody-coated grid. By using an electron-dense marker to tag the antibody probe, soluble viral antigen as well as virus particles can be detected by EM (Figure 4).

Colloidal gold has been shown to be an excellent marker for EM (Faulk and Taylor, 1971), and its application in diagnostic virology has been reported by several authors. (For review, see Kjeldsberg, 1986). The particles are relatively easy and inexpensive to prepare, and can be produced in a range of sizes. They can be conjugated via electrostatic attraction to a variety of probes including antibodies, protein A, protein G, and lectins (Horisberger, 1981). Because of their extreme electron-density, they greatly facilitate EM identification of antigen-antibody complexes. In IEM studies on partially purified rotavirus (Table 3) and rotavirus in fecal specimens (Table 4) we found the IEM method employing labelled protein A (PAG IEM) exhibited the greatest sensitivity of all EM methods tested (Hopley 1985; Hopley and Doane, 1985).

Despite the high sensitivity of immunogold labeling, it requires exceptional attention to controls. Under sub-optimal test conditions the non-specific background labeling can be high, and the various parameters involved in the assay should be thoroughly evaluated. Each antibody preparation to be used as a probe must be titrated before being assigned as a reference, to determine the dilution at which maximum specific labeling and minimum background labeling are obtained; the concentration of gold conjugate will also influence this ratio.

Several variations of the immunogold technique have been described. These include mixing the reactants directly in solution (Hopley and Doane, 1985; Stannard et al. 1982); adding each reactant by floating a specimen grid

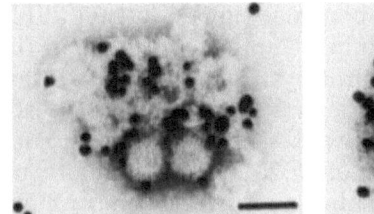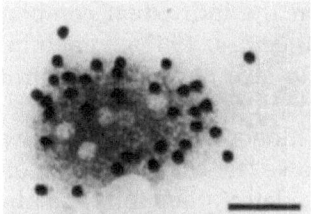

Figure 4. A colloidal gold marker greatly increases the visibility of virus-antibody aggregates that might otherwise be difficult to identify. Left: rotavirus. Right: enterovirus. Bars equal 100 nm.

TABLE 3. Sensitivity of Different EM Methods

	DEM	DIEM	PAG IEM (VP)[b]	PAG IEM (SA)[c]
Rotavirus	4×10^{5}[a]	2×10^{4}	1×10^{4}	2×10^{3}

[a] Minimum detectable virus concentration ($TCID_{50}$/ml)
[b] PAG IEM (VP): Endpoint read on basis of gold-labled virus particles.
[c] PAG IEM (SA): Endpoint read on basis of gold label.

TABLE 4. Comparison of Sensitivity of DIEM and PAG IEM for Detection of Rotavirus in Stool Suspensions

Stool Specimen	Highest reciprocal titre	
	DIEM[a]	PAG IEM[b]
HSC1	4,000	16,000
HSC2	2,000	16,000
HSC4	200	8,000
HSC5	1,000	8,000
HSC6	1,000	8,000
6749	200	4,000
5843	100	4,000
5706	2,000	4,000

[a] Based on presence of virus-antibody aggregates.
[b] Based on presence of gold label.

sequentially on the individual components (similar to the SPIEM) (Lin, 1984; Pares and Whitecross, 1982); combining with the SIA method (Doane, 1988b). A gold particle size of 5 nm constitutes a finer probe than larger particles, reducing steric hinderance. This may be appropriate in studies on the interaction of monoclonal antibodies with specific epitopes. For routine diagnostic IEM, however, a particle size of 15-20 nm has two advantages: (1) it is much easier to detect, therefore scanning a grid can be performed at a magnification of as low as 3,000x, whereas approximately 14,000x magnification is needed to detect 5 nm particles; (2) the signal-to-noise-ratio is much higher with 15-20 nm particles than with 5 nm particles (Hopley, 1985).

COMPARISON WITH OTHER IMMUNOASSAYS

IEM offers diagnostic virologists a multifunctional tool; because of its high sensitivity, it can be used to detect small amounts of viral antigen or antibody; as was first demonstrated by Best et al. (1967) with rubella virus and Bayer et al. (1968) with HBsAg, IEM can be used to reveal the identity of elusive viruses; in conjunction with antibody probes, it can be used to serotype viruses directly from clinical specimens or cell culture lysates, and to determine the location of specific epitopes on virus particles.

Various studies have shown that IEM is at least as sensitive as ELISA and RIA (Kjeldsberg and Mortensson-Egnund, 1982; Kjeldsberg and Siebke 1985; Morinet et al. 1984; Obert et al. 1981; Svensson et al. 1983). With weakly positive specimens it provides an advantage over ELISA in that the direct visualization of virus particles obviates the necessity for a confirmatory test (Obert et al. 1981).

The data in Table 1 indicate that a similar sensitivity can be achieved with the Airfuge ultracentrifuge, without the need for antibodies. With its impressive ability to concentrate viruses in small volumes, some diagnostic laboratories use this instrument routinely on many of their clinical specimens (Hammond et al. 1981).

It appears that, in general, there is little difference in the sensitivity of virus detection that can be obtained with DIEM, SIA and SPIEM; depending on the virus and test conditions, the improvement over DEM can be as high as 1,000-fold (El-Ghorr et al. 1988). The decision as to which of the three IEM methods to choose is determined to a large extent by the nature of the specimen and the information being sought. We prefer to use DIEM in combination with the agar diffusion method when only small volumes of specimen or reference antibody are available. The SIA method, incorporating gamma globulin in the agar within the microtitre cups, is well suited to the broad screening of specimens for multiple viruses. Once a virus family has been identified on the basis of morphology, subsequent serotyping can be performed by the same system, using pooled or individual antisera. A major advantage of the SIA method is the ability to store prepared microtitre plates for long periods of time at 4°C. Because of the flexible nature of the plastic plate, pairs of cups can easily be cut away from the stored plate and set into a rubber holder in preparation for the test.

Both the DIEM and SIA methods produce virus-antibody aggregates in a positive reaction. At low virus concentrations the aggregates may be in isolated areas on the EM grid and many of the grid squares may be devoid of any virus, requiring a longer examination time (Morinet et al. 1984). This phenomenon is avoided by using SPIEM, where trapped virus tends to be more uniformly distributed over the supporting film. Furthermore, virus morphology is usually more evident, as virus particles are not surrounded by a halo of antibody molecules.

The solid phase system has other advantages over DIEM and SIA methods. The antibody-coated grids can be stored for several weeks at 4°C without serious loss of trapping efficiency (drops below 50% after 4 weeks) (Pegg-Feige 1973). They can be dispatched to centers with no EM facilities, and returned later for examination. They are especially useful when dealing with crude samples such as fecal suspensions. Subsequent washing will remove background debris without removal of virus.

Even greater sensitivity can be achieved by the addition of a colloidal gold label. Provided optimum test parameters have been established and ample controls are included in every test, the immunogold method is a valuable diagnostic tool. Its full potential will undoubtedly come with the increasing availability of type specific monoclonal antibodies. Factors that remain to be established include the advantage of direct vs indirect labeling, and the relative sensitivities of antimurine antibody, protein A and protein G in the indirect immunogold assays employing monoclonal antibodies.

In assessing the value of techniques in diagnostic virology it is always necessary to consider not only the financial cost of supplies and equipment needed, but also the cost in terms of time required to perform these techniques. On this basis, ELISA is superior to IEM for screening large numbers of specimens. Using microtitre plates, automatic pipetters and readers, many dozens of specimens can be processed coincidentally by ELISA. Although multiple specimens can be prepared in parallel by IEM, each grid must be examined individually in the EM. We estimate that a single operator could process approximately 15-20 samples by IEM during an average working day (Doane, 1988b). But IEM can be applied more readily to selected specimens, is faster to process than ELISA, produces results more rapidly and provides comparably greater sensitivity. And, with the introduction of monoclonal antibodies to the diagnostic armamentarium, its full potential remains to be realized.

REFERENCES

Almeida J, Cinader B, Howatson A (1963) The structure of antigen-antibody complexes. A study by electron microscopy. J Exp Med 118:327-340.
Almeida JD, Waterson AP (1969) The morphology of virus-antibody interaction. Adv Virus Res 15:307-338.
Anderson N, Doane FW (1972) Agar diffusion method for negative staining of microbial suspensions in salt solutions. Appl Microbiol 24:495-496.
Anderson N, Doane FW (1973) Specific identification of enteroviruses by immuno-electron microscopy using a serum-in-agar method. Can J Microbiol 19:585-589.
Anderson TF, Stanley WM (1941) A study by means of the electron microscope of the reaction between tobacco mosaic virus and its antiserum. J Biol Chem 139:339-344.
Bayer ME, Mannweiler E (1963) Antigen-antibody reactions in influenza virus as seen in the electron microscope. Arch Gesamte Virusforsch 13:541-547.
Bayer ME, Blumberg BS, Werner B (1968) Particles associated with Australia antigen in the sera of patients with leukemia, Down's syndrome and hepatitis. Nature 218:1057-1059.

Berthiaume L, Alain R, McLaughlin B, Payment P (1981) Rapid detection of human viruses in faeces by a simple and routine immune electron microscopy technique. J Gen Virol 55:223-227.

Best JM, Bantvala JE, Almeida JD, Waterson AP (1967) Morphological characteristics of rubella virus. Lancet 2:237-239.

Brenner S ,Horne RW (1959) A negative staining method for high resolution microscopy of viruses. Biochem Biophys Acta 34:103-110.

Derrick KS (1973) Quantitative assay for plant viruses using serological specific electron microscopy. Virology 56:652-653.

Doane FW (1974) Identification of viruses by immunoelectron microscopy. In: Kurstak E, Morisset R (eds) Viral Immunodiagnosis. Academic Press, New York. pp 237-255.

Doane FW (1986) Electron microscopy and immunoelectron microscopy. In: Specter S, Lancz GJ (eds) Clinical Virology Manual. Elsevier, New York. pp 71-88.

Doane FW (1987) Immunoelectron microscopy in diagnostic virology. Ultrastruct Pathol 11:681-685.

Doane FW (1988a) Immunoelectron microscopy and its role in diagnostic virology. Clin Immunol News 9:159-162.

Doane FW (1988b) Electron Microscopy. In: Lennette EH, Halonen P, Murphy FA (eds) Laboratory Diagnosis of Infectious Diseases. Principles and Practice. Springer-Verlag, New York. pp 121-131.

Doane FW, Anderson N (1977) Electron and immunoelectron microscopic procedures for diagnosis of viral infections. In Kurstak E, Kurstak C (eds) Comparative Diagnosis of Viral Diseases, Vol II, part B. Academic Press, New York. pp 505-539.

Doane FW, Anderson N (1987) Electron Microscopy in Diagnostic Virology: A Practical Guide and Atlas. Cambridge University Press, New York.

El-Ghorr AA, Snodgrass DR, Scott FMM (1988) Evaluation of an immunogold electron microscopy technique for detecting bovine coronavirus. J Virol Methods 19:215-224.

Faulk WP, Taylor GM (1971) An immunocolloidal method for the electron microscope. Immunochemistry 8:1081-1083.

Furui S (1986) Use of protein A in the serum-in-agar diffusion method in immune electron microscopy for detection of virus particles in cell culture. Microbiol Immunol 30:1023-1035

Gerna G, Passarani N, Battaglia M, Percivalle E (1984) Rapid serotyping of human rotavirus strains by solid-phase immune electron microscopy. J Clin Microbiol 19:273-278.

Gerna G, Passarani N, Sarasini A, Battaglia M (1985) Characterization of serotypes of human rotavirus strains by solid-phase immune electron microscopy. J Infect Dis 152:1143-1151.

Gerna G, Sarasini A, Coulson BS, Parea M, Torsellini M, Arbustini E, Battaglia M (1988) Comparative sensitivities of solid-phase immune electron microscopy and enzyme-linked immunosorbent assay for serotyping of human rotavirus strains with neutralizing monoclonal antibodies. J Clin Microbiol 26:1383-1387.

Hammond GW, Hazelton PR, Chuang I, Klisko, B. (1981) Improved detection of viruses by electron microscopy after direct ultracentrifuge preparation of specimens. J Clin Microbiol 14:220-221.

Hopley, JFA (1985) Protein A-gold immunoelectron microscopy studies on the simian rotavirus SA11. M.Sc thesis, University of Toronto.

Hopley JFA, Doane FW (1985) Development of a sensitive protein A-gold immunoelectron microscopy method for detecting viral antigens in fluid specimens. J Virol Methods 12:135-147.

Horisberger M (1981) Colloidal gold. A cytochemical marker for light and fluorescent microscopy and for transmission and scanning electronmicroscopy. In: Johari O (ed) Scanning Electron Microscopy. AMF O'Hara, S.E.M., Inc., Chicago. vol II, pp 9-31.

Hummeler K, Anderson TF, Brown RA (1961) Identification of poliovirus particles of different antigenicity by specific agglutination as seen in the electron microscope. Virology 16:84-90.

Kapikian A, Almeida J, Stott E (1972a) Immune electron microscopy of rhinoviruses. J Virol 10:142-146.

Kapikian AZ, Wyatt RG, Dolin R, Thornhill TS, Kalica AR, Chanock RM (1972b) Visualization by immune electron microscopy of a 27 nm particle associated with acute infectious non-bacterial gastroenteritis. J Virol 10:1075-1081.

Kapikian A, James H, Kelly S, Vaughn A (1973) Detection of coronavirus 692 by immune electron microscopy. Infect Immun 7:111-116.

Kapikian AZ, Dienstag JL, Purcell RH (1976) Immune electron microscopy as a method for the detection, identification, and characterization of agents not cultivable in an *in vitro* system. In: Rose NR, Friedman H (eds) Manual of Clinical Immunology. Amer Soc Microbiol, Washington D.C. pp 467-480.

Katz D, Straussman Y, Shahar A, Kohn (1980) Solid-phase immune electron microscopy (SPIEM) for rapid viral diagnosis. J Immunol Methods 38:171-174.

Katz D, Kohn A (1984) Immunosorbent electron microscopy for detection of viruses. Adv Virus Res 29:169-194

Katz D, Straussman Y (1984) Evaluation of immunoadsorbent electron microscopic techniques for detection of Sindbis virus. J Virol Methods 8:243-254.

Kelen AE, Hathaway AE, McLeod DA (1971) Rapid detection of Australia/SH antigen and antibody by a simple and sensitive technique of immunoelectron microscopy. Can J Microbiol 17:993-1000.

Kjeldsberg E (1985) Specific labelling of human rotaviruses and adenoviruses with gold-IgG complexes. J Virol Methods 12:47-57.

Kjeldsberg E (1986) Immunonegative stain techniques for electron microscopic detection of viruses in human faeces. Ultrastr Pathol 10:553-570.

Kjeldsberg E, Mortensson-Egnund K (1982) Comparison of solid-phase immune electron microscopy, direct electron microscopy and enzyme-linked immunosorbent assay for detection of rotaviruses in faecal samples. J Virol Methods 4:45-53.

Kjeldsberg E, Siebke JC (1985) Use of immunosorbent electron microscopy for detection of rota- and hepatitis A virus in sucrose solutions. J Virol Methods 12:161-167.

Kleczkowski A (1961) Serological behaviour of tobacco mosaic virus and its protein fragments. Immunol. 4:130-141.

Lafferty KJ, Oertelis SJ (1961) Attachment of antibody to influenza virus. Nature 192:764-765.

Lafferty KJ, Oertelis S (1963) The interaction between virus and antibody. III. Examination of virus-antibody complexes with the electron microscope. Virology 21:91-99.

Lamontagne L, Marsolais G, Marois P, Assaf R (1980) Diagnosis of rotavirus, adenovirus, and herpesvirus infections by immune electron microscopy using a serum-in-agar diffusion method. Can J Microbiol 26:261-264.

Lee FK (1977) A study of enterovirus-antibody interaction in immunoelectron microscopy. M.Sc thesis, University of Toronto.

Lin N (1984) Gold IgG complexes improve the detection and identification of viruses in leaf dip preparations. J Virol Methods 8:181-190.

Milne RG, Luisoni E (1977) Rapid immune electron microscopy of virus preparations. In: Maramorosch K, Kaprowski H (eds) Methods in Virology. Academic Press, New York. vol 6, pp 265-281.

Morinet F, Ferchal F, Colimon R, Pérol Y (1984) Comparison of six methods for detecting human rotavirus in stools. Eur J Clin Microbiol 3:136-140.

Narang HK, Codd AA (1981) Frequency of pre-clumped virus in routine fecal specimens from patients with acute nonbacterial gastroenteritis. J Clin Microbiol 13:982-988.

Nicolaieff A, Obert G, Van Regenmortel MHV (1980) Detection of rotavirus by serological trapping on antibody-coated electron microscope grids. J Clin Microbiol 12:101-104.

Obert G, Gloekler R, Burckard J, Van Regenmortel MHV (1981) Comparison of immunosorbent electron microscopy, enzyme immunoassay and counterimmunoelectrophoresis for detection of human rotavirus in stools. J Virol Methods 3:99-107.

Pares RD, Whitecross MI (1982) Gold-labelled antibody decoration (GLAD) in the diagnosis of plant viruses by immune-electron microscopy. J Immunol Methods 51:23-28.

Paver W, Ashley C, Caul E, Clarke S (1973) A small virus in human faeces. Lancet 1:237-240.

Pegg-Feige (1983) Solid phase immunoelectron microscopy (SPIEM) in virus identification. M.Sc thesis. University of Toronto.

Pegg-Feige K, Doane FW (1983) Effect of specimen support film in solid phase immunoelectron microscopy. J Virol Methods 7:315-319.

Pegg-Feige K, Doane FW (1984) Solid-phase immunoelectron microscopy for rapid diagnosis of enteroviruses. Proc of the 42nd Annual Meeting of the Electron Microscopy Society of America. pp 226-227.

Petrovicova A, Juck AS (1977) Serotyping of coxsackieviruses by immune electron microscopy. Acta Virol 21:165-167.

Rubenstein AS, Miller MF (1982) Comparison of an enzyme immunoassay with electron microscopic procedures for detecting rotavirus. J Clin Microbiol 15:938-944.

Stannard LM, Lennon M, Hodgkiss M, Smuts H (1982) An electron microscopic demonstration of immune complexes of hepatitis B e-antigen using colloidal gold as a marker. J Med Virol 9:165-175.

Svensson L, Grandien M, Pettersson C-A (1983) Comparison of solid-phase immune electron microscopy by use of protein A with direct electron microscopy and enzyme-linked immunosorbent assay for detection of rotavirus in stool. J Clin Microbiol 18:1244-1249.

van Rij G, Klepper L, Peperkamp E, Schaap GJP (1982) Immune electron microscopy and a cultural test in the diagnosis of adenovirus ocular infection. Brit J Ophthamol 66:317-319.

von Ardenne M., Friedrich-Freska H, Schramm G (1941) Elektronmikroskopische untersuchung der pracipitinreaktion von tabakmosaikvirus mit kaninchenantiserum. Arch Gesamte Virusforsch 2:80-86.

Vreeswijk J, Folkers E, Wagenaar F, Kapsenberg JG (1988) The use of colloidal gold immunoelectron microscopy to diagnose varicella-zoster virus (VZV) infections by rapid discrimination between VZV, HSV-1 and HSV-2. J Virol Methods 22:255-271.

Watson DH, Wildy P (1963) Some serological properties of herpes virus particles studied with the electron microscope. Virology 21:100-111.

Wood DJ, Bailey AS (1987) Detection of adenovirus types 40 and 41 in stool specimens by immune electron microscopy. J Med Virol 21:191-199.

DISCUSSION

Riepenhoff-Talty M (Children's Hospital, Buffalo, NY):

How do you account for the two-log differences in detection between echo and coxsackie?

Doane F (University of Toronto, Toronto, Ontario, Canada):

Do you mean with respect to direct electron microscopy? I think it depends on the day; if you had taken another strain of echo and another strain of coxsackie, you may not have seen that difference. We weren't trying to compare one virus with the other, we were trying to compare those individual viruses under different conditions. It depends entirely on how many virus particles were being produced from the host cells by one virus as compared to the other.

Stewart J (Centers for Disease Control, Atlanta, GA):

When you were looking at trying to identify different viruses by the gold method, obviously if you had scattered particles with just one dot, you would not be really sure if that was specific. How many dots do you need; two, three? How do you evaluate that?

Doane F:

You have to compare the number of gold particles in the test grids versus those in the controls. On the basis of our experience with rotaviruses, we decided that it had to be more than a 14% difference in the total number of particles. If it's a very clear positive, you don't usually have to count the gold dots in your control. But if there is any question about it, then you really should count.

Stewart J:

That is in the positive specimen where you actually have a complex of particles and have several gold dots around it. As long as you don't see similar structures in your control, then you're in very good shape, I guess.

Doane F:

It's important that controls be run against negative or heterologous sera. You're treading on thin ice when you're identifying on the basis of labeled soluble antigen rather than labeled virus particles.

Merz P (Institute for Basic Research, New York, NY):

Do you have any idea what your particle to infectivity ratio is?

Doane F:

That varies with every single enterovirus preparation (especially with the enteroviruses). As you probably know, there is a great deal of variation. With enteroviruses, it can be 1,000:1 for one particular type and 100:1 for another. It depends on the kind of cells they have been grown in. For example, if a cell culture isolate is from a particularly susceptible cell line, you may have a very low particle to infectivity ratio.

Merz P:

I was actually more curious about clinical specimens.

Doane F:

We have not tried to analyze that. Very often these studies have been carried out on rotavirus, and because they are so difficult to culture, it's not that easy to determine the ratio. So, I can't answer that.

Stewart J:

In terms of the antibody in the agar, how important is that concentration? Do you have to do block titrations with this?

Doane F:

Most of our work with the serum-in-agar method has been with enterovirus and we always do block titrations to begin with. Once you've determined the optimum serum dilution for a particular reference serum, the titration doesn't have to be repeated.

Al-Nakib W (Kuwait University, Kuwait)

Have you tried the gold method with regard to respiratory viruses detection in nasal specimens, for example?

Doane F:

We have not tried it, but there should be no reason why it can't be used. However, over the years we have found that enveloped viruses are very very difficult to work with when it comes to immunoelectron microscopy. I would think that the gold technique for increasing your sensitivity of detection would work very well. But insofar as producing an immune complex and depending on that as your positive, you may run into difficulties because a lot of these enveloped viruses will aggregate even in the absence of antibody.

Al-Nakib W:

I feel this is an area that is very important. Say you take influenza, we do have antiviral chemotherapy available anyway, and this is a procedure that lends itself useful in terms of rapid diagnosis. You don't need to do it on many specimens. Basically, you're going to have a few patients, say in hospitals, that you need to have a rapid diagnosis so you can make a decision as to whether to treat or not, and I think that this is an area that's very important.

Doane F:

I was thinking of RSV and parainfluenza, but certainly IEM has been used on influenza and it's been found to work very well. But I am not aware of the use of colloidal gold labeling with influenza assays. Something to try.

Bishop R (Royal Children's Hospital, Melbourne, Australia):

In regard to the serum-in-agar technique, you have emphasized how important it is to determine the concentration before adding it to the agar. You also said that when you don't know what you're looking for, you sometimes add gamma globulin to that agar. How do you determine in such a mixture what concentration to use?

Doane F:

We don't. Because when I mentioned that it's important to titrate, I really meant with respect to typing. With typing, of course, you want to be certain that your immune complex is a specific reaction. When you're using the serum-in-agar method for simply trapping a virus, it doesn't matter whether it's too concentrated or not. You just want to be able to find an aggregate and then look inside the aggregate and see what virus it is you've captured. So, pre-titration is essential if you're going to be doing typing tests, but if you're simply using a virus detection system, we just use immune serum globulin undiluted, or at a final concentration, in the agar, of 1/50.

Bishop R:

Do you mind telling me what glow discharge is?

Doane F:

It's a procedure that is performed in a sputter coating unit. Ionization of argon is carried out over the support film and that changes the charge on the formvar or the parlodian, making it more hydrophilic. The best thing is just to use parlodian, of course, and then you don't have to worry about pre-treating the film.

Oshiro L (California Department of Health Services, Berkeley, CA):

I have a couple of technical questions. First of all, how long are your grids treated with UV light?

Doane F:

They are pre-treated with 1,7000 mW/cm^2 of UV for 30 minutes, just prior to use.

Oshiro L:

The other question is, someone in England suggested that the commercial grade or the type of carbon might also change your sensitivity. Have you had any experience with that?

Doane F:

No, we haven't.

Oshiro L:

What type of carbon do you use?

Doane F:

We have always used Union Carbide spectroscopic electrodes.

MONOCLONAL TIME-RESOLVED FLUOROIMMUNOASSAY: SENSITIVE SYSTEMS FOR THE RAPID DIAGNOSIS OF RESPIRATORY VIRUS INFECTIONS

John C. Hierholzer[1], Larry J. Anderson[1] and Pekka E. Halonen[2]

[1]Respiratory and Enteric Viruses Branch
Division of Viral Diseases/CID
Centers for Disease Control
Atlanta, Georgia 30333, USA
[2]Department of Virology
University of Turku
Kiinamyllynkatu 13, SF-20520
Turku 52, Finland

INTRODUCTION

The respiratory viruses afflict everyone, causing an average of three respiratory illnesses per person per year and millions of lost work days per year, at an enormous economic cost. There are over 200 viruses that cause these respiratory illnesses, and this fact alone has discouraged vaccine development and made specific identification of virus infections difficult. However, it remains important to identify respiratory viruses, because antiviral therapy and control and prevention measures are available for some respiratory viruses. In the event of an outbreak, rapid identification of the causative agent may allow epidemiologic control measures to be put into action more expediently.

This idea is certainly not new. Efforts at rapid identification of respiratory viruses began twenty-two years ago with immunofluorescence (IFA) (Gardner and McQuillan, 1968; Gray et al. 1968). At that time, the reagents were rather crude, and the fluorescein-labeled antisera were not well characterized. Still, direct or indirect IFA was very helpful for identifying respiratory viruses in cell cultures, as well as for detecting viral antigens in nasal and eye secretions. Since that time, other tests, such as immunoelectrophoresis, counter-electrophoresis, and single radial hemolysis, came and went. They all used specific antisera, but their level of sensitivity was too low to be useful in a diagnostic laboratory.

Then, in the late 1970's, enzyme immunoassays (EIA) were introduced (Harmon et al. 1979; Hornsleth et al. 1981; Johansson et al. 1980; Sarkkinen et al. 1980, 1981a, 1981b; Vesikari et al. 1981). Their potential for rapid diagnosis was immediately recognized, and since about 1984, high-quality reagents for both IFA and EIA have become commercially available. The availability of good reagents, plus the inherent versatility of IFA and EIA, have stimulated a new era of interest in the field of rapid diagnosis (Richman et al. 1984). Further, the increased capability in IFA and EIA spawned the time-resolved fluoroimmunoassay and its application to rapid virus diagnosis.

The time-resolved fluoroimmunoassay (TR-FIA) was first developed at the University of Turku, Finland in the early 1980's (Soini, 1985; Soini and Hemmila, 1979; Soini and Kojola, 1983). The idea was that fluorescence could be automated if a fluorophore were found that had a decay time much longer than the background decay time. For example, common proteins like human serum albumin, cytochrome c, and hemoglobin have fluorescence decay times of 3 to 4 ns, while fluorescein isothiocyanate, the usual label in FA, has a decay time of 4.5 ns. Clearly, the nonspecific background and the fluorescein label have similar decay times which cannot be separated to yield specific fluorescence.

Rare earth metals, or lanthanides, are known to have long decay times, in the order of one-thousand to one-million ns, several orders of magnitude greater than fluorescein (Soini and Hemmila, 1979). This would provide a usable separation between background and specific fluorescence. Further, the Stoke's shift, or difference between the excitation wavelength and the emission wavelength, is very large with lanthanide compounds. So, the Turku biochemists settled on trivalent europium (Eu^{3+}) coupled with EDTA as a suitable label which could be conjugated to purified proteins. From there, the test resembles a standard capture EIA test. The capture antibody is added to wells as purified IgG, then the antigen and Eu^{3+}-labeled detector antibody are added. The test is then concluded by quantifying the label in a single-photon-counting fluorometer rather than in an enzyme/substrate/color system as in EIA. The fluorometer is really the heart of the system, because it shoots xenon light pulses through the sample at a low wavelength, waits a prescribed time for the background fluorescence to decay, and then counts the specific fluorescence as photons of light emitted at a much higher wavelength. Thus, the TR-FIA is basically a sensitive detection system based on metal chelate chemistry and time-lapse fluorometry which can be readily applied to viral diagnostics (Dahlen et al. 1988; Halonen et al. 1983, 1985; Kuo et al. 1985; Lovgren et al. 1985; Siitari et al. 1983).

In preliminary studies with polyclonal antibodies, the TR-FIA appeared to be more sensitive for detecting respiratory antigens than IFA, RIA, or EIA tests (Halonen et al. 1983, 1985). Subsequent studies with monoclonal antibodies (MAbs) revealed even greater sensitivities for detecting influenza virus (Walls et al. 1986), adenovirus (Hierholzer et al. 1987), respiratory syncytial virus (RSV) and parainfluenza (PI) viruses (Hierholzer et al. 1989) in clinical specimens. In this chapter, we will review our comparisons of all-monoclonal TR-FIA tests with similarly-constructed, all-monoclonal, biotin/avidin EIA and with polyclonal capture EIA tests to determine which test had the greatest sensitivity for detecting these pathogens directly in clinical specimens. We will also include recent data on the human coronaviruses (strains 229E and OC43) and the agents of acute hemorrhagic conjunctivitis (coxsackievirus A24 variant and enterovirus 70) which have been subjected to the same comparisons.

TECHNIQUES

Virus and Antisera

Prototype strains and multiple isolates of each virus type were obtained from our reference virus collection. Isolates were initially recovered from various specimens by passage in appropriate cell cultures and identified by standard procedures (Anderson et al. 1985, 1986; Halonen et al. 1985; Hierholzer, 1973; Hierholzer and Pallansch, 1989; Hierholzer and Tannock, 1986a, 1986b; Hierholzer et al. 1969, 1975, 1988a, 1988b).

Polyclonal rabbit antiserum used in some EIA formats was prepared against purified Ad2 hexons (Hierholzer et al. 1984; Sarkkinen et al. 1981a); other rabbit antisera were prepared against purified parainfluenzaviruses, coronaviruses, and enteroviruses (Halonen et al. 1985; Hierholzer and Pallansch, 1989; Sarkkinen et al. 1981a). Polyclonal horse antisera to RSV and the parainfluenza viruses were prepared against crude virus supernatants. The production of adenovirus hexon MAb, RSV monoclonals, parainfluenza virus monoclonals, and enterovirus 70 MAb have been previously described (Anderson et al. 1984, 1985, 1986; Hierholzer et al. 1984, 1989). All were used as mouse ascitic fluids which were tested by IFA and EIA for specificity and reactivity. The selected ones were further evaluated by testing against a large panel of homotypic and heterologous virus strains. All antibodies and other reagents were titrated to optimal endpoints in checkerboard fashion, so that each test system used all reactants at their optimal dilution to give the highest signal-to-background ratio.

Specimens

Nasopharyngeal aspirates (NPA) and other specimens positive for respiratory viruses were obtained from children with pneumonia or bronchiolitis in the United States, Brazil, and Finland (Halonen et al. 1985; Meurman et al. 1983, 1984; Ruuskanen et al. 1984; Sarkkinen et al. 1981a, 1981b; Virtanen et al. 1983). The Turku specimens were stored at -70°C as a 1:5 dilution in PBS containing 20% fetal calf serum, 2% Tween-20, and 0.004% merthiolate. Extracts (10%-20%) of stool specimens from infants and young children with acute gastroenteritis were tested by culture, IEM, EIA and/or RIA (Halonen et al. 1980; Hierholzer and Gary, 1979; Hierholzer et al. 1988b; Meurman et al. 1983; Sarkkinen et al. 1980; Vesikari et al. 1981). Conjunctival specimens were obtained from numerous outbreaks of acute hemorrhagic conjunctivitis in the western hemisphere (Hierholzer and Pallansch, 1989).

Enzyme Immunoassay

Details of the EIA procedures used are given elsewhere (Hierholzer et al. 1987, 1989). EIA Format 1 was the all-monoclonal test designed to parallel that used for the TR-FIA as closely as possible; the steps are outlined in Table 1. Biotinylation of IgG, which had been purified as for TR-FIA, was done with Enzotin reagent (N-biotinyl-ω-aminocaproic-acid-N-hydroxysuccinimide ester) at pH 8.5. EIA Format 2 for each virus was a polyclonal test adapted from one previously evaluated (Anderson et al. 1983, 1984, 1985; Hierholzer et al. 1989); the steps are outlined in Table 2. A second option in Format 2 was to avoid the biotin-avidin reaction, and use a MAb mouse ascitic fluid for the detector antibody followed by goat anti-mouse IgG-peroxidase as the conjugate.

Time-resolved Fluoroimmunoassay. Purification and labeling of MAbs

A 5 ml sample of mouse ascitic fluid was clarified, dialyzed against 0.05 M Tris-HCl buffer, pH 8.0, and loaded onto a DEAE-Sephacel column followed by

TABLE 1. The All-Monoclonal Capture EIA Test (Format 1)

1. Solid phase is a 96-well flat-bottom polystyrene microtiter plate (e.g., Immulon-2, Dynatech Laboratories, Alexandria, VA).

2. Add 75 μl/well of specific monoclonal antibody (MAb) (as capture antibody) as purified IgG diluted in carbonate buffer, pH 9.6; incubate overnight at 4ºC; wash 3x with 0.01 M PBS, pH 7.2, containing 0.15% Tween-20.

3. Add 75 μl/well of specimen (NPA, etc.) at a low dilution (e.g., 1:5, 1:10, 1:20) in wash buffer containing 0.5% gelatin; incubate 1.5 hr at 37ºC; wash 3x as in step 2.

4. Add 75 μl/well of same or different MAb-biotin (as biotinylated detector antibody), diluted in 0.01 M PBS, pH 7.2, with 0.5% gelatin and 0.15% Tween-20 (2% normal goat serum can be added if necessary to reduce background); incubate 1 hr at 37ºC; wash 3x as in step 2.

5. Add 75 μl/well of streptavidin/peroxidase (as conjugate), diluted in the diluent used for step 4; incubate 10 min at room temperature; wash 6x as in step 2.

6. Add 125 μl/well of substrate system: 0.1 mg/ml of 3,3',5,5'-tetramethylbenzidine (TMB) in 2% DMSO in 0.1 M acetate/citrate buffer, pH 5.5 with 0.005% hydrogen peroxide added when used (Hancock and Tsang, 1986); incubate 20 min at ambient temperature; stop color reaction with 2 M sulfuric acid; read at 450 nm in an automated EIA reader.

a salt gradient from 0-0.225 M NaC1 in the Tris buffer (Hierholzer et al. 1987). Eluted proteins were tested by EIA for IgG activity at the 10^{-2} and 10^{-4} dilutions. IgG positive fractions were pooled, concentrated to 10 ml by ultrafiltration, and precipitated by 50% ammonium sulfate. The protein pellet after centrifugation was resuspended in 2 ml of PBS and dialyzed overnight against PBS. Protein concentration was estimated by absorbance measurements at 280 nm. Alternatively, IgG was purified by affinity chromatography using GammaBind G-Agarose (Genex Corp. Gaithersburg, MD), in which the IgG was bound at neutral pH, eluted with 8 M urea or 0.5 M ammonium acetate, pH 3.0, and then, neutralized as appropriate. The purified antibody was divided into 3 parts: some was kept plain to use as capture antibody in both EIA and TR-FIA tests; some was conjugated with biotin to use in biotin-avidin EIAs; and some was conjugated with Eu^{3+} to use as detector antibody in TR-FIA.

For europium labeling, the chelate [N'-diethylene triaminopentaacetic acid (DTPA)-Europium] was prepared as a 1 mg/150 μl solution in distilled water. It was mixed with the IgG at a molar excess of 50 in 1 M carbonate buffer, pH 9.2. The reaction was incubated overnight at 4ºC, brought to ambient temperature for 2 hr, and then stopped by purifying the labeled complex by exclusion chromatography (Hierholzer et al. 1987). Fractions of 1 ml were diluted 1:5000 in enhancement solution (DELFIA/TR-FIA #1244-105, Wallac Oy, Turku, Finland), and counted in a single-photon-counting Model 1230 Arcus fluorometer (LKB/Wallac Oy, Turku). The peak fractions containing IgG (by absorbance at 280 nm) were pooled and BSA was sometimes added to 1% concentration for storage at -70ºC.

TABLE 2. The Polyclonal Capture EIA Test (Format 2)

1. Solid phase is a 96-well microtiter plate (see Table 1, step 1).
2. Add 75 µl/well of polyclonal antiserum as purified IgG (see Table 1, step 2).
3. Add 75 µl/well of specimen (NPA, etc.) at a low dilution (see Table 1, step 3).
4. Add 75 µl/well of a different species antiserum (as detector antibody), diluted in 0.01 M PBS, pH 7.2, with 0.5% gelatin, 0.15% Tween-20, and 2% normal species serum if needed; incubate 1 hr at 37°C; wash 3x as in step 2.
5. Add 75 µl/well of anti-species IgG/peroxidase (as conjugate) in the diluent used for step 4; incubate 1 hr at 37°C; wash 6x as in step 2.
6. Add 125 µl/well of substrate system: 0.1 mg/ml of TMB in 2% DMSO in 0.1 M acetate/citrate buffer, pH 5.5, with 0.005% peroxide added when used; incubate 20 min at ambient temperature; stop color reaction with 2 M sulfuric acid; read at 450 nm.

TABLE 3. Outline of the All-Monoclonal, One-Incubation TR-FIA Test

1. Solid phase is a 12-well flat microtiter strip (e.g., Flow Titertek).
2. Add 250 µl/well of specific monoclonal (as capture antibody) as purified IgG diluted in carbonate buffer, pH 9.6; adsorb overnight at room temperature; wash 3x with aqueous 0.9% NaC1/0.05% Tween-20.
3. Add 250 µl/well of 0.1% gelatin in 0.05% M Tris/0.9% NaC1/0.05% NaN$_3$ buffer, pH 7.75; incubate overnight at room temperature to saturate wells with gelatin; wash 3x as in step 2. Strips can be stored at this point for up to 1-1/2 years at 4°C in slightly moist bag with no loss of activity.
4. Add 100 µl/well of specimen (antigen) at low dilution (e.g., 1:5, 1:10, 1:20) in TR-FIA antigen buffer, + 100 µl/well of purified Eu^{3+} labeled monoclonal IgG (as detector antibody) also diluted in antigen buffer; incubate 1 hr at 37°C; wash 10x (see step 2).
5. Add 200 µl/well of enhancement solution; agitate on a shaker for 10 min at room temperature; count in fluorometer.

One-Incubation TR-FIA Procedure

The test is outlined in Table 3. Purified monoclonal IgG, diluted to optimal concentration in pH 9.6 carbonate buffer, was added to the wells of 12-well strips (Titertek #78-591-99, Flow Laboratories/Eflab Oy, Finland) in 250 µl volumes and adsorbed overnight at ambient temperature in a moist chamber. The wells were washed 3x with aqueous 0.9% NaCl/0.05% Tween-20 and saturated with 250 µl of 0.1% gelatin (Difco, Detroit, MI) in 0.05 M Tris/0.9% NaCl/0.05% NaN$_3$ buffer, pH 7.75, again with overnight incubation at ambient temperature. As in EIA, the wash steps were critical and were performed with care.

The wells were washed 3x, and 100 µl each of antigen and Eu-detector antibody were added to appropriate wells. The antigen (NPA or cell culture harvest) was diluted 1:10 or 1:20 in specimen diluent, consisting of 50 mM Tris, pH 7.75, 0.9% NaC1, 0.01% NaN$_3$, 0.5% gelatin, 0.01% Tween-40, 20 μM DTPA, and 2% BSA. The Eu-detector antibody was diluted to the appropriate concentration in the same diluent. After the antigen and antibody were added, the strips were incubated for 1 hr at 37°C in a moist chamber.

The strips were carefully washed 10x, and 200 μl per well of enhancement solution was added. The plates were gently agitated on a shaker for 10 min at ambient temperature and then placed in the Arcus fluorometer for counting. The fluorometer was programmed to take the mean and coefficient of variation (CV) of 12 reagent blanks and to take the mean and CV of the duplicates or triplicates of each specimen minus the reagent blanks. We then further analyzed the printed data by computing the mean and standard deviation (SD) of the negative specimens run in the same test, and used this mean + 3 SDs as the cut-off value for positive specimens (Hierholzer et al. 1987, 1989).

RESULTS

Formatting the EIAs

The monoclonal EIA (Format 1) for each virus using the same antibodies as TR-FIA was optimized by checkerboard titration in usual fashion. The optimal conditions are given in Table 4. As a verification of the TR-FIA, we also biotinylated the other purified antibodies (listed in Table 5) and tested all possible combinations in EIA. For adenovirus, absorbance values up to 1.39 were obtained for Ad2 HEp-2 cell culture supernatants (at a 1:1000 dilution), compared to negative specimen values of ~0.05. Good signal-to-background ratios were obtained with clinical samples as well. For RSV and the PI viruses, several capture/detector combinations were equally sensitive in detecting multiple strains of virus, but none were more sensitive than the combination chosen for the best TR-FIA for each virus. All PI-2 combinations had relatively high backgrounds, but were still type specific.

The polyclonal EIAs (Format 2) also were optimized as shown in Table 4. The tests selected all had low backgrounds; but, like the monoclonal EIA, the RSV test had the lowest background and the PI-2 test had the highest. For adenovirus, the biotin-avidin EIA had values up to 1.14 for positive cultures and 0.03 for HEp-2 cell controls. The option in Format 2 of using MAb 2/6 in PBS-GT/2% NRS diluent as detector antibody, followed by goat anti-mouse IgG peroxidase in PBS-GT/2% NGS/5% NRS diluent as developer, gave somewhat poorer results (positive = 1.26, negative = 0.06) and was excluded from further study. MAb 20/11 was somewhat less sensitive than MAb 2/6 in all formats.

Coronavirus monoclonal tests gave mean EIA absorbance values of 1.683 for 229E grown in human embryonic lung diploid fibroblast (HLF) cells (0.011 absorbance for HLF negative cell controls) and 1.165 for mouse-brain-adapted OC43 grown in human rhabdomyosarcoma (RD) cells (0.045 for RD controls), and polyclonal tests gave maximum absorbance values of 1.099 (0.091 for HLF controls) and 1.402 (0.034 for RD controls) for the two viruses, respectively. Coronaviruses were more susceptible to freeze-thaw cycles and storage at 4°C than were the other virus groups tested, such that EIA titers dropped significantly with repeated use of an NPA or cell culture specimen.

Enterovirus EIA tests were not satisfactory, despite efforts to increase their sensitivity. Mean absorbance values for the best-formatted monoclonal tests

TABLE 4. Reagents and Parameters for EIA Formats[a]

Capture antibody		Antigen	Detector antibody		Developing system	
Antiserum	Dilution	Dilution	Antiserum	Dilution	Conjugate	Dilution

Format 1: Monoclonal sera

Adeno: 20/11	1:3,000	1:20	2/6-biotin	1:1,000	strep/av.perox.	1:3,000
RSV: 130-8F	1:10,000	1:20	131-2A/biotin	1:300	ibid	
PI1: 253-9D	1:30,000	1:10	251-12A/biotin	1:1,000	ibid	
PI2 233-4D	1:10,000	1:10	233-4D/biotin	1:2,000	ibid	
PI3: 341-7H	1:30,000	1:10	343-5G/biotin	1:1,000	ibid	
229E: 401-3C	1:3,000	1:10	401-4A/biotin	1:10,000	ibid	
OC43: 541-8F	1:3,000	1:10	542-7D/biotin	1:3,000	ibid	
EV70: 74-5G	1:1,000	1:3	74-5G/biotin	1:300	ibid	
CA24: 550-3A	1:10,000	1:3	550-3A/biotin	1:3,000	ibid	

Format 2: Polyclonal sera

Ad(hex.)rab.	1:3,000	1:20	MAb 2/6-biotin	1:1,000	strep/av.perox.	1:3,000
RSV: horse	1:1,000	1:20	bovine	1:1,000	anti-bov.perox.	1:3,000
PI1: horse	1:3,000	1:10	rabbit	1:30,000	anti-rab.perox.	1:3,000
PI2: horse	1:10,000	1:10	rabbit	1:3,000	ibid	
PI3: horse	1:3,000	1:10	rabbit	1:3,000	ibid	
229E: g. pig	1:10,000	1:10	rabbit	1:3,000	ibid	
OC43: mouse	1:3,000	1:10	rabbit	1:10,000	ibid	
EV70: rabbit	1:10,000	1:3	MAb 74-5G/biot	1:300	strep/av.perox.	1:3,000
CA24: rabbit	1:30,000	1:3	Mab 550-3A/biot	1:3,000	strep/av.perox.	1:3,000

[a] Diluent for capture antibodies was pH 9.6 carbonate buffer, with incubation overnight at 4°C. Antigen was diluted in TR-FIA antigen diluent
(see "Techniques"); incubation was 1-1/2 hr, 37°C (except EV70 and CA24
were incubated overnight at 23°C). Detector antibody in Format 1 was
diluted in PBS-GT (with 2% NGS if needed), and in Format 2 was diluted
in PBS-GT/2% normal rabbit or horse, as appropriate (N.B., 229E needed
2% goat serum in detector diluent; OC43 needed no serum in diluent);
incubation was 1 hr, 37°C. Developing system for Format 1 was diluted
in PBS-GT, incubated 10 min at room temperature, and for Format 2 was
diluted in PBS-GT/2% normal rabbit or horse serum, incubated 1 hr at
37°C (see Tables 1 and 2 for details).

TABLE 5. Characteristics of Monoclonals and Labeled Antibodies[a]

Virus	Monoclone number	MAb properties			Purified aby		Eu^{3+} labeled aby	
		Immun. strain	Protein specif.	IgG subcl.	Protein (mg/ml)	EIA titer	Eu/IgG ratio	Protein mg/ml
Adeno:	2/6	Ad3 (GB)	hexon	1κ	11.27	7.8	8.6	0.213
	20/11	"	hexon	1κ	5.54	6.7	6.5	0.320
RSV:	100-7G	18537	N	1κ	3.52	5.6	8.0	0.164
	130-8F	A2	F	1κ	4.06	6.0	6.6	0.204
	130-12H	"	N	2Aκ	2.75	6.0	8.3	0.298
	131-2A	"	F	2Aκ	2.14	6.0	7.3	0.247
	131-4G	"	N	2Aκ	4.14	6.2	9.6	0.222
	133-1H	"	F	2Aκ	4.59	6.2	6.8	0.236
	170-1E	V670	N	2Aκ	2.79	6.5	6.8	0.192
PI-1:	251-12A	C35	F	2Aκ	14.31	6.2	7.6	0.418
	253-9D	"	F	2Aκ	10.07	6.0	7.5	0.269
	353-12A	"	F	2Aκ	13.66	6.2	6.8	0.313
	354-8B	"	F	1κ	16.09	6.4	12.5	0.283
PI-2:	230-9D	Greer	HN	2Aκ	0.47	3.8	9.0	0.412
	230-11H	"	HN	2Aκ	8.23	6.4	12.3	0.296
	231-9F	"	HN	1κ	2.75	5.4	11.6	0.186
	233-4D	"	HN	2Aκ	9.93	5.8	13.1	0.302
	233-9A	"	HN	1κ	7.72	5.8	13.2	0.258
PI-3:	240-12D	C243	HN	2Bκ	4.67	4.8	12.2	0.290
	241-1H	"	HN	2Bκ	3.00	4.0	11.3	0.297
	242-7A	"	HN	2Bκ	7.36	5.6	14.4	0.288
	260-10B	"	HN	1κ	11.41	5.8	12.3	0.373
	261-7H	"	HN	2Aκ	9.06	6.0	12.1	0.406
	341-7H	"	HN	2Aκ	6.09	5.6	8.2	0.228
	342-11A	"	HN	1κ	9.57	6.0	9.4	0.359
	343-5G	"	HN	2Aκ	7.39	5.6	9.1	0.250
C.299E:	400-10H	proto.	N	1κ	13.88	3.0	7.4	0.310
	401-3C	"	N	2Aκ	8.15	5.2	7.3	0.305
	401-4A	"	N	1κ	18.27	7.2	6.8	0.297
	402-6F	"	N	1κ	13.43	6.4	7.9	0.286
	402-8H	"	N	2Aκ	16.90	5.2	5.6	0.365
C.OC43:	540-4D	proto.	N	2Aκ	4.60	5.7	7.5	0.290
	541-8F	"	N	1κ	3.26	5.5	10.0	0.332
	541-11H	"	N	2Aκ	3.41	4.0	8.2	0.166
	542-7D	"	N	1κ	10.18	6.2	10.1	0.354
	543-10E	"	N	2Aκ	3.04	3.3	8.4	0.184
	543-11F	"	N	2Aκ	3.55	5.2	8.6	0.260
EV70:	72-5E	KW-43	VP1	3κ	1.92	3.0	5.7	0.161
	73-7H	"	VP1	1κ	7.03	4.0	6.3	0.263
	74-5G	"	VP1	1κ	9.46	4.8	7.2	0.225
CA24var:	550-3A	EH24/70	VP1	2Aκ	16.88	5.7	9.1	0.164
	552-12F	"	VP1	(IgM)	2.14	3.8	8.4	0.110

[a] MAbs chosen for high reactivity with multiple strains by IFA and EIA, except for the enterovirus monoclonals chosen by EIA alone.

were 1.114 for HLF-grown EV70 (0.042 for HLF cell controls) and 0.648 for HLF-grown CA24var (0.030 for cell controls), and for the optimized polyclonal tests were 1.322 for EV70 (0.018 for controls) and 0.445 for CA24var (0.046 for controls). These tests included overnight 23°C incubation of the capture antibody/antigen step instead of the usual 1-1/2 hr 37°C incubation step, because we have found repeatedly higher absorbance values for picornavirus EIA tests after the prolonged attachment period.

Formatting the TR-FIA

The one-incubation TR-FIA was formatted for each virus by testing various combinations of MAbs at serial dilutions. The optimization utilized checkerboard titrations of each MAb at both capture and detector positions in the test against NPA specimens and against tissue culture isolates, the same as for optimizing the EIA tests. Of 42 MAbs evaluated in this study, 24 were found to be usable in either capture or detector position in the TR-FIA. Characterization and labeling data for these MAbs are listed in Table 5. Most had type-specific EIA titers of 10^{-6} after purification; their Eu:IgG molar ratios varied from 5.6 to 14.4. In the TR-FIA, system background was the mean of 12 reagent (system) controls and was automatically subtracted out when specimens were run. Test background was measured as the fluorescence values in known negative specimens that were tested in parallel with positive specimens. Thus, a P/N ratio could be calculated for each specimen. The P/N value was the mean of the replicate tests for a positive specimen divided by the mean of all the negative specimens tested (in replicates) in the run.

The P/N values were used to ascertain the best format in TR-FIA (Table 6). For adenovirus, MAb 20/11 in capture position gave a mean P/N value of 11,580 with MAb 2/6 in detector position, which was 34% greater than the next closest combination (MAb 2/6 in both positions). For RSV, 3 anti-fusion (F) protein MAbs and 4 anti-nucleoprotein (N) MAbs were selected for TR-FIA trials on the basis of broad reactivity with diverse RSV strains by EIA and IFA (Anderson et al. 1985, 1986). All were evaluated in capture and detector antibody positions in checkerboard titrations. Although all of the F MAbs were reasonably active in either position in the test, the best combination was MAb 130-8F as capture with MAb 131-2A as detector. This combination gave positive values as high as 2.5×10^6 counts per second (cps) and a mean P/N value of 4408 cps with NPA specimens (Table 6). This was 5% greater than the next closest combination (MAb 133-1H vs. 131-2A). The mean P/N is also shown for cell culture isolates in this table, although the tests were optimized for NPA specimens because the primary emphasis was on rapid diagnosis. The anti-N monoclonals were remarkably less sensitive in this assay, yielding mean P/N values of 25-67 with the different combinations; they were not tested further.

Similar results were obtained for the parainfluenza viruses, except that the P/N values were much lower. The 4 MAbs for PI-1 were equally reactive as capture antibody, but 251-12A was clearly the most sensitive detector antibody. Thus, the best combination was 253-9D capture with 251-12A detector, giving a mean P/N of 264 with NPA specimens. The highest P/N for PI-1 cell cultures was 263, obtained with 354-8B as capture and 251-12A as detector. Of 5 PI-2 MAbs, the best combination was 233-4D in both positions, giving a mean P/N of 339 with NPA specimens and 285 with cell culture isolates. A different combination was the most sensitive for the isolates: 233-9A as capture with 231-9F as detector, giving a mean P/N of 905. Of 8 PI-3 MAbs, only 3 gave good results in the TR-FIA. MAb 341-7H as capture with 343-5G as detector was the best combination, giving a mean P/N with NPA specimens of 573. This was 30% greater than the next closest combination (342-11A vs. 343-5G). The best combination for PI-3 cell culture harvests was 342-11A capture and 343-5G detector, with a P/N of 460. Maximum activity for all the parainfluenza viruses in 1:10 diluted NPA specimens ranged from 60,000 to 600,000 cps, and for 1:10 diluted cell culture isolates ranged from 170,000 to 940,000 cps. All MAbs tested were type-specific by TR-FIA.

TABLE 6. Comparison of the Most Sensitive Combinations of Monoclonals for TR-FIA Tests[a]

Virus	Capture antibody	Conc. (µg/well)	Detector antibody	Conc. (ng/100µl)	Mean P/N of specimens clinical	cultures
Adeno:	*20/11*	*0.5*	*2/6*	*25*	11580	6743
	2/6	0.5	2/6	50	7675	5881
RSV:	*130-8F*	*0.5*	*131-2A*	*25*	4408	2342
	133-1H	0.5	131-2A	50	4198	1765
	130-8F	0.5	130-8F	50	2622	317
PI-1:	*253-9D*	*0.25*	*251-12A*	*25*	264	207
	353-12A	0.25	251-12A	50	217	217
	354-8B	0.25	251-12A	25	195	263
PI-2:	*233-4D*	*0.5*	*233-4D*	*25*	339	285
	230-11H	0.5	233-4D	25	184	53
	231-9F	0.5	233-4D	50	178	284
	233-9A	1.0	231-9F	50	101	905
PI-3:	*341-7H*	*0.25*	*343-5G*	*25*	573	340
	342-11A	0.5	343-5G	25	404	460
	240-12D	1.0	240-12D	6.25	41	162
C.299E:	*401-3C*	*0.5*	*401-4A*	*25*	10	32
	401-4A	0.5	401-4A	50	6	47
C.OC43:	*541-8F*	*0.5*	*542-7D*	*50*	174	790
	541-8F	0.5	542-7D	25	142	792
	541-8F	0.5	541-8F	50	131	470
EV70:	*74-5G*	*0.5*	*74-5G*	*25*	5	59
	73-7H	0.5	74-5G	50	2	18
CA24var:	*550-3A*	*0.5*	*550-3A*	*25*	30	96
	550-3A	0.5	550-3A	50	24	31

[a] Italics denotes optimal test system for direct detection TR-FIA

During the checkerboard titrations in the above tests, the capture antibodies were tested at 0.25, 0.5 and 1.0 µg/well, and the detectors at 6.25, 12.5, 25, 50, and 100 ng/100 µl. For each concentration of capture antibody, the optimal concentration of detector antibody was determined by the P/N calculation for representative positive and negative NPA specimens to obtain the greatest sensitivity. In the examples shown (Figure 1), the optimal combination for adenovirus was MAb 20/11 as capture antibody at 0.5 µg/well, and MAb 2/6 as Eu-labeled detector at 25 ng/well (P/N = 25,035); and for parainfluenzavirus 1 was MAb 253-9D as capture at 0.25 µg/well and MAb 251-12A as detector at 25 ng/well (P/N = 308).

The log-log relationship of signal-to-antigen concentration is shown for 3 viruses in Figure 2. These figures are typical of the dose-response curves included periodically to verify the sensitivity of the test. The curves thus also

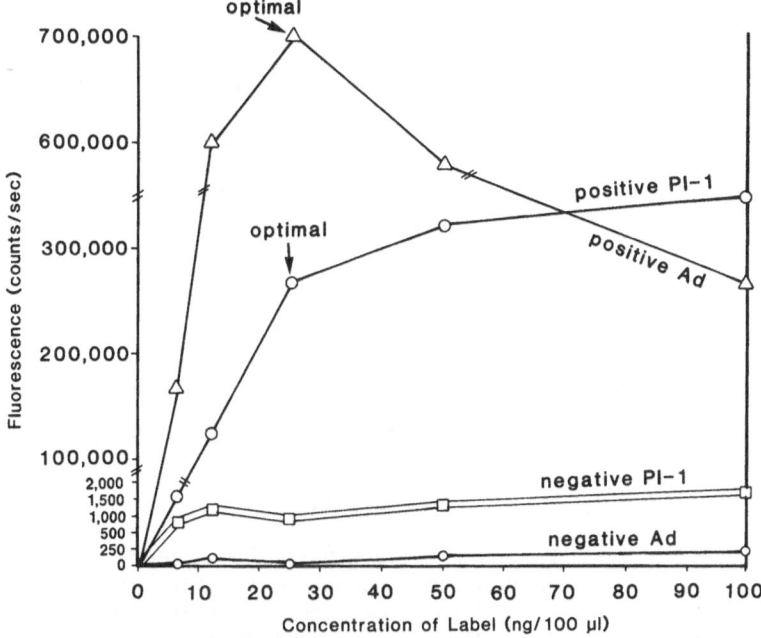

Figure 1. Optimization of capture and detector antibodies. With capture antibody at 0.5 μg/well, peak fluorescence (in counts per second) in adenovirus-positive specimen and low fluorescence in adenovirus-negative specimen indicate optimal concentration of detector (label). For PI-1, the capture antibody was at 0.25 μg/well. Both detectors optimized at 25 ng/well (see Table 8 for identification of antibodies). Adenovirus specimens were diluted 1:20 and PI specimens 1:5.

serve as standard curves for the hexon antigen (adenovirus) and for whole virus (all others). Defining positive as ≥background mean + 3 SDs, the all-monoclonal TR-FIA detected a minimum of 0.05 ng/ml of hexon antigen, 0.8 ng/ml of RSV, 5 ng/ml of PI-1, 1 ng/ml of PI-2, and 1 ng/ml of PI-3. Similar studies have not yet been completed for the coronaviruses and enteroviruses.

The last parameter studied was the choice of protein stabilizer in the specimen diluent. We compared 2% normal rabbit serum, used in previous studies, with 2% BSA and with no stabilizer. In all tests, the 2% BSA clearly gave the highest P/N values and reduction in nonspecific signal, particularly in stool specimens. BSA was then adopted routinely for all future tests.

Comparison of TR-FIA with EIAs Using Clinical Specimens

Using the optimum reagents and dilutions just described, the TR-FIA was compared with EIA Formats 1 and 2 in testing original clinical specimens (Table 7). For adenovirus, the full comparative evaluation included 33 Ad-positive NPA or other non-stool specimens, 39 Ad-negative NPA specimens, 36 Ad-positive stool specimens, and 26 Ad-negative stool specimens. Specimens were known to be Ad-positive or -negative by culture and subsequent typing or by prior EIA or RIA tests. Some Ad-negative specimens were positive for other viruses and thus were important controls in the study. All specimens were tested at 1:20 dilution in the specimen diluent to conserve specimen for the many tests required.

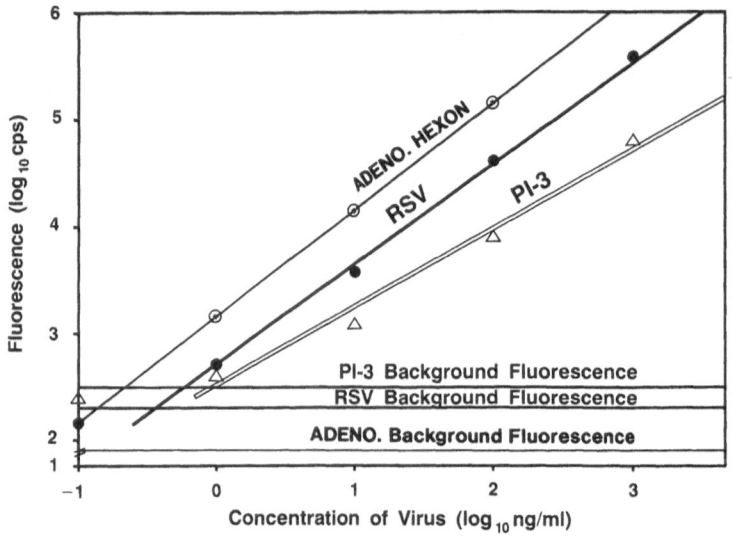

Figure 2. Typical dose-response curves. Purified adenovirus hexon antigen and purified whole RSV and PI-3 virions were tested in tenfold dilution series with the optimized TR-FIA formats (see Table 8). The adenovirus TR-FIA can detect a minimum of 0.05 ng of hexon antigen per ml; the RSV test can detect 0.8 ng of virus per ml; and the PI-3 test can detect 1.0 ng of virus per ml.

In the NPA group, 28/33 known Ad-positive specimens were positive in TR-FIA at the cutoff value of ≥ 3 SDs above the mean of the known negative specimens. The 28 TR-FIA-positive specimens included 2 lung and 1 pancreas suspension and included Ad types 1, 2, 3, 5, 7, 31, 34 and 35 of the 13 specimens that were serotyped. The 5 Ad-positive, TR-FIA-negative specimens included 3 NPA (types 2, 5), 1 urine (type 34), and 1 kidney (type 35) specimen. This gave a false-negative rate of 15%. One culture-negative NPA specimen was TR-FIA-positive, even upon repeat testing, to give a false-positive rate of 2%. The remaining 38 Ad-negative specimens were also negative by TR-FIA, and included 7 specimens positive for PI-1, 2, 3, RSV, and cytomegalovirus.

These data compared favorably with the EIA data. In Format 1, 26/33 Ad-positive specimens were also positive ($\geq$mean of negatives + 3 SDs) by EIA, and all Ad-negative specimens were negative, for an overall correlation of 90%. By Format 2, 29/33 Ad-positive specimens were also positive by EIA, and all negative specimens negative, for an overall correlation of 94%. There was some correlation by specimen among the known Ad-positive specimens that were not identified by these tests. Two culture-positive specimens were identified by TR-FIA but not by either EIA test; 2 were identified by EIA Format 2 but not by TR-FIA or Format 1; and 1 specimen was not identified by any test. The other misidentified specimens were scattered among the 3 tests. TR-FIA used in conjunction with EIA Format 2 identified 71/72 positive and negative specimens.

In the stool specimen group, TR-FIA identified all 36 known Ad-positive specimens. These included Ad types 2, 15, 31, 40, and 41 of the 15 specimens that were serotyped. The 26 Ad-negative specimens were all correctly identified as negative; these included 8 specimens positive for echovirus, rotavirus, and reovirus. The TR-FIA data with stool specimens (100% correlation) far exceeded the results by either EIA format. Only 28/36 known Ad-positive specimens

TABLE 7. Comparison of TR-FIA with EIA for Detection of Antigen in Clinical Specimens[a]

Specimen	TR-FIA[a]	EIA-1 (mono.)	EIA-2 (poly.)
Ad (resp.) (1:20):			
Negative (n=39) range	-62 to + 439	0.006 to 0.068	0.003 to 0.053
Mean & SD	114 ± 106	0.027 ± 0.014	0.017 ± 0.010
Mean + 3 SDs	432	0.069	0.046
Positive (n=33) range	372 to 1454000	0.041 to 1.463	0.036 to 1.369
Number (%) positive	28 (85)	26 (79)	29 (88)
Ad (stool) (1:20):			
Negative (n=26) range	-72 to +181	0.027 to 0.091	0.019 to 0.087
Mean & SD	12 ± 51	0.039 ± 0.012	0.047 ± 0.010
Mean + 3 SDs	165	0.074	0.078
Positive (n=36) range	190 to 942749	0.049 to 1.245	0.056 to 1.354
Number (%) positive	36 (100)	28 (78)	27 (75)
RSV (1:20):			
Negative (n=40) range	-15 to + 527	-0.002 to +0.056	0.000 to 0.065
Mean & SD	188 ± 158	.008 ± .011	0.024 ± 0.014
Mean + 3 SDs	662	0.040	0.067
Positive (n=50) range	636 to 2638094	0.009 to 1.340	0.016 to 1.500
Number (%) positive	46 (92)	31 (62)	38 (76)
PI-1 (1:10):			
Negative (n=31) range	104 to 1702	0.025 to 0.135	0.045 to 0.118
Mean & SD	582 ± 475	0 .071 ± 0.022	0.072 ± 0.016
Mean + 3 SDs	2008	0.136	0.121
Positive (n=35) range	1444 to 57372	0.068 to 1.105	0.076 to 1.008
Number (%) positive	33 (94)	29 (83)	23 (66)
PI-2 (1:10):			
Negative (n=26) range	300 to 797	0.179 to 0.383	0.038 to 0.174
Mean & SD	471 ± 129	0.311 ± 0.041	0.098 ± 0.042
Mean + 3 SDs	857	0.434	0.223
Positive (n=28) range	1109 to 71992	0.324 to 1.362	0.171 to 1.500
Number (%) positive	28 (100)	21 (75)	25 (89)
PI-3 (1:10):			
Negative (n=32) range	151 to 1085	0.055 to 0.258	0.025 to 0.134
Mean & SD	332 ±194	0.115 ± 0.042	0.065 ± 0.027
Mean + 3 SDs	915	0.241	0.145
Positive (n=38) range	429 to 589529	0.175 to 1.500	0.065 to 1.500
Number (%) positive	36 (95)	34 (89)	36 (95)
229E (1:10):			
Negative (n=20) range	84 to 380	0.007 to 0.021	0.089 to 0.146
Mean & SD	242 ± 81	0.011 ± 0.003	0.132 ± 0.012
Mean + 3 SDs	485	0.020	0.168
Positive (n=13) range	596 to 3623	0.011 to 0.157	0.093 to 0.281
Number (%) positive	13 (100)	9 (69)	7 (54)
OC43 (1:10):			
Negative (n=10) range	739 to 2885	0.011 to 0.022	0.018 to 0.079
Mean & SD	1568 ± 542	0.017 ± 0.002	0.035 ± 0.012
Mean + 3 SDs	3194	0.025	0.071
Positive (n=10) range	3784 to 692850	0.024 to 0.750	0.048 to 0.330
Number (%) positive	10 (100)	9 (90)	8 (80)
EV70 (1:3):			
Negative (n=18) range	507 to 796	0.034 to 0.065	0.011 to 0.040
Mean & SD	626 ± 76	0.043 ± 0.006	0.019 ± 0.009
Mean + 3 SDs	854	0.062	0.046
Positive (n=20) range	756 to 3738	0.042 to 0.354	0.020 to 0.731
Number (%) positive	15 (75)	11 (55)	5 (25)
CA24 (1:3):			
Negative (n=63) range	1585 to 3534	0.015 to 0.034	0.024 to 0.057
Mean & SD	2286 ± 506	0.022 ± 0.007	0.036 ± 0.012
Mean + 3 SDs	3804	0.043	0.073
Positive (n=38) range	1901 to 86982	0.020 to 0.244	0.033 to 0.460
Number (%) positive	26 (68)	12 (32)	12 (32)

[a] TR-FIA data are mean counts/second; EIA data are mean absorbance at 450 nm.

were found positive in Format 1 and 27 were positive in Format 2. The same specimens were misidentified in both formats, except that one additional specimen was missed by Format 2. All of the Ad-negative specimens tested negative by both EIAs.

In most instances, the specimens that tested false-negative in any of the 3 tests were those that had very low titers of virus, as suggested by the absorbance readings in a positive test and the number of passages required to recover the virus (Hierholzer et al. 1987). Thus, all 3 tests should still be considered capable of detecting small quantities of Ad antigen in clinical specimens.

For RSV, the comparative evaluation included 40 negative and 50 positive NPA specimens; there were somewhat fewer than that for PI-1, 2 and 3 (Table 7). Ten "negative" specimens were positive for other viruses (viz., adenovirus, cytomegalovirus, influenza A, mumps, and parainfluenza 4A). Specimens were tested at a 1:20 or 1:10 dilution in the specimen diluent, again to conserve specimen. In the RSV group, 92% of known RSV-positive specimens were positive in TR-FIA at the cutoff value of ≥ 3 SDs above the mean of the known RSV negatives. This was clearly better than 62% positive for the monoclonal EIA or 76% for the polyclonal EIA. There were no false-positives in any of the assays.

For PI-1, 94% of known PI-1 specimens were positive by TR-FIA, 83% by monoclonal EIA, and 66% by polyclonal EIA. For PI-2, 100% of known positive specimens were positive by TR-FIA, compared with 75% for monoclonal EIA and 89% for polyclonal EIA. The PI-3 TR-FIA test was closer to EIA, correctly identifying PI-3 in 95% of positive specimens, the same as in the polyclonal EIA, and compared with 89% in the monoclonal EIA. As in the RSV tests, there were no false-positives in any of the assays; the lower percentage rates in the EIA tests appeared to be due to low-titered specimens which the less-sensitive tests could not detect. This was also suggested by the fact that all of the *positive* specimens that were negative by TR-FIA were also negative by both EIA formats.

For the respiratory coronaviruses, finding NPA specimens that were adequately tested yet still "fresh" enough to contain intact virus was a problem. Nevertheless, the optimized TR-FIA detected type 229E in all 13 positive specimens, compared to a 69% detection rate for the monoclonal EIA and a 54% detection rate for the polyclonal EIA. Among the 20 "negative" NPA specimens tested were 10 specimens positive for type OC43, adenovirus, RSV, and parainfluenza 1, which confirmed the type-specificity of the 229E monoclonals. Similar results were obtained for type OC43: TR-FIA correctly identified OC43 in all 10 positive specimens and in none of the 10 OC43-negative NPA specimens (of which 5 were positive for 229E, RSV, and parainfluenza 3). The EIA formats also were type-specific, and detected OC43 in 90% and 80% of OC43-positive specimens in the monoclonal and polyclonal tests, respectively. No false positive results were obtained in any of the coronavirus tests. Best results were obtained with specimens which had not been frozen/thawed multiple times or stored at 4°C, which is similar to findings in other tests for these highly labile viruses (Hierholzer and Tannock, 1989).

For the agents of acute hemorrhagic conjunctivitis (AHC), tests for enterovirus type 70 and coxsackievirus A24 variant were less sensitive than comparable tests for adenovirus or paramyxoviruses, but were type-specific and were shown to be particularly useful for strains that were not recoverable in cell culture. The EV70 TR-FIA detected virus in 75% of known positive eye swab specimens and 91% of cell culture supernatants. The test also identified EV 70 in 26 of 100 eye swabs from two outbreaks which were proven serologically as EV70-related but for which culture was unsuccessful (Hierholzer and Pallansch, 1989). EIA formats 1 and 2 were similar to TR-FIA in detecting EV70 in cell culture supernatants (94% and 97%, respectively), but were much less sensitive for detecting EV70 in eye swab specimens (55% and 25%, respectively). The comparative tests for CA24var gave values similar to those for EV70 (Table 7). The best-formatted TR-FIA identified CA24var in 68% of known positive eye swab specimens, compared to 32% positivity rates for both monoclonal and polyclonal EIAs. As in the EV70 tests, negative specimens included eye swab and NPA specimens that were positive for adenoviruses, paramyxoviruses, and the opposite enterovirus, which served as important specificity controls. No false positive values were obtained in any of the picornavirus tests, although some cell culture supernatants and cell controls gave borderline values that appeared to be related to the amount of serum proteins present in the maintenance medium. Interestingly, eye swab specimens stored (-20°C) for shorter time periods (1-3 years) were nearly all identified as CA24var, whereas specimens stored for >3 years were not identified in half the cases; this confirms anecdotal information that the CA24var is not as stable as most other enteroviruses.

TABLE 8.   Direct Antigen Detection in Routine Use of TR-FIA Batteries[a]

| Virus | Capture aby (strip) | | | Detector aby (label) | | | No. of specimens identified |
	MAb	IgG isotype	Concentrn. (µg/well)	MAb	IgG isotype	Concentrn. (ng/well)	
Adenovirus	20/11	1κ	0.5	2/6	1κ	25	137
Respir. syncytial	130/8F	1κ	0.5	131-2A	2Aκ	25	73
Parainfluenza-1	253-9D	2Aκ	0.25	251-12A	2Aκ	25	29
Parainfluenza-2	233-4D	2Aκ	0.5	233-4D	2Aκ	25	26
Parainfluenza-3	341-7H	2Aκ	0.25	343-5G	2Aκ	25	47
Coronavirus 229E	401-3C	2Aκ	0.5	401-4A	1κ	25	5
Coronavirus OC43	541-8F	1κ	0.5	542-7D	1κ	50	4
Enterovirus-70	74-5G	1κ	0.5	74-5G	1κ	25	68
Coxsackie A24var.	550-3A	2Aκ	0.5	550-3A	2Aκ	25	12

[a] Respiratory specimens were routinely diluted 1:5 in the TR-FIA antigen diluent, and conjunctival specimens were diluted 1:3. Specimens were not pre-treated by centrifugation, concentration, or sonication.

Application of TR-FIA to Routine Diagnosis

The final phase of evaluation of the TR-FIA was the identification of viral antigens in respiratory and conjunctival specimens obtained during routine outbreak investigations over the past 4 years (Table 8). The TR-FIA battery consisted of tests for adenovirus, RSV, parainfluenza types 1, 2, and 3, and coronaviruses 229E and OC43 for respiratory specimens, and enterovirus type 70, coxsackievirus A24 variant, and adenovirus for conjunctival specimens. Strips containing capture antibody for these tests were kept prepared in advance and stored in moist chambers at 4°C until needed. We have found no drop in sensitivity, as determined with standard positive controls, in strips stored up to 1-1/2 years in this manner.

Using this battery, adenovirus was detected in all specimens that eventually yielded an isolate, and hexon antigen was identified in many cell culture harvests before the virus could be grown enough for serotyping. The other agents likewise were detected at different times in respiratory or conjunctival secretions from various etiologic studies. As for adenovirus, the battery was also useful in identifying cell culture harvests (Table 6) because TR-FIA testing was readily available and results could be obtained within 2 hrs. Thus, the TR-FIA proved to be a reliable and practical test for large numbers of specimens on a routine basis.

We recently had the opportunity to directly compare culture, IFA, TR-FIA, monoclonal EIA, and polyclonal EIA on two groups of specimens obtained during outbreak investigations (Table 9). The adenoviruses were from eye swab specimens from an outbreak of epidemic keratoconjunctivitis in an ophthalmologist's office, and the parainfluenzaviruses were from NPA specimens from children during an outbreak of URI and croup in a grammar school. Both sets of specimens were less than ideal, as indicated by the length of time required for the viruses to grow out in cell culture. The adenoviruses took from 3 to 7 weeks for all 12 to grow, and were all typed by HI as Ad8, and the parainfluenza 1 strains took from 2 to 5 weeks to grow. Although this is a small comparison in terms of diagnostic specimens, the data are consistent with our general observations about the utility of the various procedures.

CONCLUSION

Rapid diagnosis of respiratory virus infections has made great strides in recent years. Fluorescent antibody tests, the first for rapid diagnosis of respiratory viral infections (Gardner and McQuillan, 1968), have been followed by current RIA and EIA tests and by various labeled probe tests. With the availability of improved reagents, enzyme immunoassays in general have proven more sensitive, reliable, and adaptable to large numbers of specimens than have IFA, immunoelectron microscopy, and other specialized tests for direct antigen detection (Ahluwalia et al. 1987; Anestad et al. 1983; Cranage et al. 1981; Freke et al. 1986; Grandien et al. 1985; Halonen et al. 1983, 1985; Harmon et al. 1979; Harmon and Pawlik, 1982; Hornsleth et al. 1981, 1986, 1988; Hughes et al. 1988; Jalowayski et al. 1987; Johansson et al. 1985; Lehtomaki et al. 1986; Leite et al. 1985; Macnaughton et al. 1983; Parkinson et al. 1982; Pereira et al. 1985; Popow-Kraupp et al. 1986; Richman et al. 1984; Wong et al. 1982; Zrein et al. 1986). EIA has also

TABLE 9. Parallel Comparison of TR-FIA with Other Tests for Virus Identification[a]

| Etiologic agent | No. spec. | Number Identified By | | | | |
		Culture	TR-FIA	mono. EIA	poly.EIA	IFA
Adenovirus	16	12	11	9	10	5
Parainfluenza-1	14	10	9	6	4	4

[a] ES or NPA specimens were first clarified by light centrifugation to obtain cells for IFA; the cells were washed once and resuspended in PBS, and added to slides for reaction with pooled Ad antihexon monoclonals 2/6 and 20/11 or pooled parainfluenza 1 monoclonals 251-12A and 253-9D, followed by rabbit anti-mouse IgG fluorescein conjugate. The supernatants were divided, with one part diluted 1:5 in TR-FIA antigen buffer for TR-FIA and both EIA tests, and the other part treated with antibiotics and inoculated onto primary human embryonic kidney (HEK) cells for adenovirus or human lung mucoepidermoid carcinoma (NCI-H292) cells for parainfluenzavirus.

been shown to be equal in sensitivity for antigen detection with RIA, DNA sandwich hybridization, and other labeled-DNA probe tests (Gomes et al. 1985; Lehtomaki et al. 1986; Sarkkinen et al. 1980, 1981a, 1981b; Stalhandske et al. 1985; Takiff et al. 1985; van Dyke and Murphy-Corb, 1989; Virtanen et al. 1983). The advent of MAbs increased the potential for highly-specific EIA tests capable of detecting nanogram quantities of antigen (Anderson et al. 1983; Freke et al. 1986; Halonen et al. 1985; Hendry et al. 1985a, 1985b; Kadi et al. 1986; Kao et al. 1984; Kim et al. 1983; Orvell et al. 1986; Pothier et al. 1985; Routledge et al. 1985; Rydbeck et al. 1986; Singh-Naz and Naz, 1986; Waner et al. 1985).

For example, several EIA formats using a capture antibody have been described for adenoviruses. One early test was a direct EIA in which anti-Ad serum was adsorbed to plates, nasal washings were added, and phosphatase-conjugated anti-Ad IgG was added as the direct detector; the test was developed with p-nitrophenyl phosphate as substrate and absorbance was read at 400 nm (Harmon et al. 1979). Most tests were of the indirect EIA format, in which a polyclonal capture antibody was adsorbed to 96-well plates as the solid phase, antigens or specimens were added, a second polyclonal antibody (detector) from a different animal species was added, and enzyme-conjugated anti-species antibody was added, followed by the substrate and color development. The enzyme system was either alkaline phosphatase with p-nitrophenyl phosphate substrate, with absorbance read at 403-405 nm (Harmon and Pawlik, 1982; Johansson et al. 1980, 1985; Uhnoo et al. 1983, 1984, 1986), or was horseradish peroxidase with 2,2'-azino-di-[3-ethylbenzthiazoline sulfonate] (Harmon and Pawlik, 1982) or o-phenylenediamine substrate, read at 490-492 nm (Anderson et al. 1983; Halonen et al. 1980, 1985; Lehtomaki et al. 1986; Leite et al. 1985; Pereira et al. 1985; Sarkkinen et al. 1980, 1981a). We presently use streptavidin/peroxidase as the conjugate with TMB as the substrate because it is a more sensitive and considerably less toxic dye (Hancock and Tsang, 1986; Hierholzer et al. 1989).

The capture indirect EIA has worked well with polyclonal antisera in both capture and detector positions for nasopharyngeal specimens (Anderson et al. 1983; Halonen et al. 1985; Harmon and Pawlik, 1982; Lehtomaki et al. 1986; Sarkkinen et al. 1981a) and for stool specimens (Halonen et al. 1980; Johansson et al. 1980, 1985; Leite et al. 1985; Pereira et al. 1985; Sarkkinen et al. 1980; Uhnoo et al. 1983, 1984, 1986; Yolken and Franklin, 1985). Thus, the test has been particularly useful for diagnosing enteric adenoviruses in infant gastroenteritis cases.

Sensitivity in EIA was recently increased by using monoclonal antibodies in the test at the detector antibody position (Anderson et al. 1983; Halonen et al. 1985; Singh-Naz and Naz, 1986). Theoretically at least, using MAbs in both positions in the assay would improve specificity and lower the background signal; but an all-monoclonal test necessitates labeling the detector, such as with ^{125}I, biotin, or europium, to distinguish it from the capture antibody. We recently found that biotinylation provided a simple and quick method to label MAbs and that the biotin-avidin reaction in place of the anti-species antibody did in fact increase the sensitivity of some assays (Hierholzer et al. 1987; Hornsleth et al. 1986).

The ultimate goal of rapid diagnostic virology is to maximize sensitivity, speed and simplicity. After its development in 1981, the TR-FIA appeared to provide a significant improvement in rapid diagnosis by capitalizing on these goals. Further, it had an inherent advantage by comprising a very sensitive detection system based on metal chelate chemistry and time-lapse fluorometry (Halonen et al. 1983; Lovgren et al. 1985; Soini, 1985; Soini and Hemmila, 1979; Soini and Kojola, 1983). It has elements of the IFA test in that a fluorophore is excited at one wavelength and detected at another, plus elements of the EIA test in that a specific antigen is immunologically bound to a labeled antibody which is then quantitated. In the monoclonal, one-incubation TR-FIA, the fluorescent label is constructed as a EU^{3+} chelate conjugated to purified IgG from a monoclonal antibody. The purification of the monoclonal IgG from mouse ascitic fluid is done by standard chromatography. Then the IgG at a protein concentration of 1 mg/ml is conjugated with a molar excess of EU^{3+} chelate (N'-diethylene triaminopentaacetic acid-europium) at pH 9.2 overnight at 4°C, followed by purifying the labeled complex by exclusion chromatography. The label has a fluorescent decay time of ~1 millisecond (or 1000 μs).

Once these reagents are prepared and optimized in serial dilutions, the test is set up as outlined in Table 3. Basically, the purified IgG is adsorbed overnight to the wells of microtiter strips; then the wells are saturated overnight with gelatin; the specimen (antigen) and the europium-labeled detector antibody are added at the same time (thus the term "one-incubation"); and finally the enhancement solution is added. To start the counting cycle in the fluorometer, the europium is excited with a 1-μs light pulse from a xenon lamp at a wavelength of 340 nm. The fluorometer then measures a 400-μs delay time, during which the autofluorescence of the reagents and the background fluorescence in the plastic wells and specimens have time to disappear. Then, the longer-lived specific fluorescence is measured for 400 μs at an emission wavelength of 613 nm. After a 200-μs pause, the xenon lamp again excites the probe, and the cycle is repeated. There are ~1000 cycles during the total counting time of 1 second; these are averaged by the fluorometer, and the mean counts per second is printed out. Thus, the test is called "time-resolved" fluoroimmunoassay because it is an immunoassay in which the specific fluorescence is measured *after* all background fluorescence has disappeared.

The TR-FIA has been applied to the detection of viral antibodies (Meurman et al. 1982) and antigens (Halonen et al. 1983; Siitari et al. 1983), serum gammaglobulins (Kuo et al. 1985), peptide hormones in serum (Kaihola et al. 1985; Lawson et al. 1986; Lovgren et al. 1985; Toivonen et al. 1986), and viral DNA in clinical material (Dahlen et al. 1988). We showed previously that the test, using polyclonal antisera in both antibody positions, was more sensitive and specific for respiratory virus antigens than IFA, EIA, or RIA (Halonen et al. 1983, 1985). We then found the TR-FIA to be even more sensitive for antigen detection when monoclonal antibodies were used in both positions, as applied to influenza virus (Walls et al. 1986), adenovirus (Hierholzer et al. 1987), RSV and parainfluenza virus (Hierholzer et al. 1989). The all-MAb TR-FIA even out-performed the comparably constructed all-MAb biotin/avidin EIAs for all of these viruses.

For adenoviruses, the TR-FIA identified 85% of Ad-positive nasopharyngeal aspirate and other non-stool specimens, compared with 79% for the monoclonal EIA and 88% for the polyclonal EIA; TR-FIA identified adenovirus in 100% of positive stool samples, compared with 78% and 75% for the two EIAs. All serotypes were equally well identified because of the group-reactive properties of the hexon component. For RSV and parainfluenza virus types 2 and 3, the most sensitive EIA was a polyclonal assay using horse capture antibodies and bovine or rabbit detector antibodies with anti-species peroxidase; for parainfluenza type 1, the most sensitive EIA was the monoclonal assay with biotin-labeled detector antibody and streptavidin-peroxidase conjugate. Yet, the TR-FIA detected RSV antigen in 92% of the specimens positive by culture, which was a decidedly higher sensitivity than either the monoclonal or polyclonal EIA format (62% and 76%, respectively). For the parainfluenza viruses, the TR-FIA detected type-specific antigen in 94-100% of culture-positive specimens, which again was more sensitive than the all-monoclonal EIAs (75-89%) or all-polyclonal EIAs (66-95%). All tests were evaluated with nasopharyngeal aspirate specimens from respiratory illnesses and with cell culture harvests of multiple strains of each virus isolated over many years. The levels of sensitivity for the TR-FIA tests described here are particularly important because of the role of adenoviruses in gastroenteritis and because adenoviruses, RSV, parainfluenza viruses, and coronaviruses are major causative agents of upper and lower respiratory illness in infants and young children (Albert, 1986; Bennett et al. 1987; Brown et al. 1984; de Jong et al. 1983; Hierholzer and Tannock, 1986a, 1986b, Macnaughton et al. 1983; Madeley, 1986; McIntosh et al. 1978; Schmitz et al. 1983; Stott and Taylor, 1985; Wadell, 1984; Welliver et al. 1986; Wigand and Adrian, 1986).

The F-protein MAbs of RSV were dramatically more sensitive in TR-FIA than the nucleocapsid MAbs. This was surprising on the one hand because they had similar sensitivities in EIA and IFA tests with multiple strains isolated in cell cultures. On the other hand, the predominant antigen present in nasal secretions during natural, active infection is the F glycoprotein (Hendry et al. 1985b). Taken together, the F and G glycoproteins and their precursors and aggregates, all of which are surface antigens of the virus, far outnumber the nucleocapsid protein, which would be in greater concentration inside the cells (Anderson et al. 1985, 1986; Hendry et al. 1985a; Stott and Taylor, 1985). As expected, the F and HN surface glycoproteins of the parainfluenza viruses were active in the TR-FIA as in EIA and IFA, owing to their well-known biological properties and role in virus entry and induction of protective antibody (Coelingh and Tierney, 1989; Cowley and Barry, 1983; Ito et al. 1987; Orvell et al. 1986; Ray and Compans, 1986; Rydbeck et al. 1986).

The all-MAb format is attractive because it utilizes an unlimited supply of reagents of unchanging characteristics for all components of the assay. The specificity, and to some extent sensitivity, can be modulated by the choice of the MAb. By eliminating the enzyme-conjugated anti-species antibody used in many EIA tests, the TR-FIA eliminates one source of nonspecific reactions.

35

The TR-FIA proved to be the most sensitive of the assays we evaluated, and the one-incubation TR-FIA, an indirect test in that capture and detector antibodies were used, was chosen for its relative simplicity. However, the initial investment in laboratory time to prepare reagents and set up the test is extensive. The test requires highly purified antibody, some of which is labeled and repurified, and the use of an expensive enhancement solution. The initial evaluations of the antibody in terms of protein content, specific antibody activity, successful labeling, and test format require the expertise of a multidisciplinary laboratory. Additionally, the photon-counting fluorometer and data handling system constitute a major purchase for the laboratory.

However, these investments in time and instruments are worthwhile because a) all reagents, buffers, and supplies are stable under ordinary storage conditions; b) the Eu^{3+} chelate is inexpensive; c) high dilutions and small volumes of reagents are used; d) a large number of specimens can be tested at one time in a convenient manner; and e) the start-up costs are spread over a long time. Further, the ability to store large numbers of strips with different capture antibodies (moist, in the refrigerator) for at least 1-1/2 years allows diagnostic tests to be set up on short notice and concluded within 2 hr, all handling and diluting of specimens included. The short test time is due also to the objective measurement of fluorescence in an automated strip-feeding photometer and to the instant analysis by a program that subtracts the background, averages the replicates, and then prints the data. The use of expendable supplies for TR-FIA, such as buffers, filters, microtiter strips, and enhancement solution, adds to the cost of the test, but no more so than for EIA. Time and cost considered, the TR-FIA edges closer to the overall objective of rapid diagnosis, that is, to be able to accurately detect viral antigen in close to 100% of specimens that would be positive by culture.

REFERENCES

Ahluwalia G, Embree, J. McNicol P, Law B, Hammond GW (1987) Comparison of nasopharyngeal aspirate and nasopharyngeal swab specimens for respiratory syncytial virus diagnosis by cell culture, indirect immunofluorescence assay, and enzyme-linked immunosorbent assay. J Clin Microbiol 25:763-767.

Albert MJ (1986) Enteric adenoviruses: Brief review. Arch Virol 88:1-17.

Anderson LJ, Godfrey E, McIntosh K, Hierholzer JC (1983) Comparison of a monoclonal antibody with a polyclonal serum in an enzyme-linked immunosorbent assay for detecting adenovirus. J Clin Microbiol 18:463-468.

Anderson LJ, Hatch MH, Flemister MR, Marchetti GE (1984) Detection of enterovirus 70 with monoclonal antibodies. J Clin Microbiol 20:405-408.

Anderson LJ, Hierholzer JC, Tsou C, Hendry RM, Fernie BF, Stone Y, McIntosh K (1985) Antigenic characterization of respiratory syncytial virus strains with monoclonal antibodies. J Infect Dis 151:626-633.

Anderson LJ, Hierholzer JC, Stone YO, Tsou C, Fernie BF (1986) Identification of epitopes on respiratory syncytial virus proteins by competitive binding immunoassay. J Clin Microbiol 23:475-480.

Anestad G, Breivik N, Thoresen T (1983) Rapid diagnosis of respiratory syncytial virus and influenza A virus infections by immunofluorescence: Experience with a simplified procedure for the preparation of cell smears from nasopharyngeal secretions. Acta Path Microb Immunol Scand (B) 91:267-271.

Bennett JV, Holmberg SD, Rogers MF, Solomon SL (1987) Infectious and parasitic diseases. In: Closing the Gap: The Burden of Unncessary Illness - A Study of the Carter Center, Emory University. R.W. Amler and H.B. Dull (eds) pp. 102-114, Oxford University Press, New York.

Brown M, Petric M, Middleton PJ (1984) Diagnosis of fastidious enteric adenoviruses 40 and 41 in stool specimens. J Clin Microbiol 20:334-338.

Coelingh KLV, Tierney EL (1989) Antigenic and functional organization of human parainfluenza virus type 3 fusion glycoprotein. J Virol 63:375-382.

Cowley JA, Barry RD (1983) Characterization of human parainfluenza viruses. I. The structural proteins of parainfluenza virus 2 and their synthesis in infected cells. J Gen Virol 64:2117-2125.

Cranage MP, Stott EJ, Nagington J, Coombs RR (1981) A reverse passive haemagglutination test for the detection of respiratory syncytial virus in nasal secretions from infants. J Med Virol 8:153-160.

Dahlen P, Hurskainen P, Lovgren T, Hyypia T (1988) Time-resolved fluorometry for the identification of viral DNA in clinical specimens. J Clin Microbiol 26:2434-2436.

de Jong JC, Wigand R, Kidd AH, Wadell G, Kapsenberg JG, Muzerie CJ, Wermenbol AG, Firtzlaff RG (1983) Candidate adenoviruses 40 and 41: Fastidious adenoviruses from human infant stool. J Med Virol 11:215-231.

Freke A, Stott EJ, Roome AP, Caul EO (1986) The detection of respiratory syncytial virus in nasopharyngeal aspirates: Assessment, formulation, and evaluation of monoclonal antibodies as a diagnostic reagent. J. Med Virol 18:181-191.

Gardner PS, McQuillin J (1968) Application of immunofluorescent antibody technique in rapid diagnosis of respiratory syncytial virus infection. Brit Med J 3:340-343.

Gomes SA, Pereira HG, Russell WC (1985) In situ hybridization with biotinylated DNA probes: A rapid diagnostic test for adenovirus. J Virol Meth 12:105-110.

Grandien M, Pettersson CA, Gardner PS, Linde A, Stanton A (1985) Rapid viral diagnosis of acute respiratory infections: Comparison of enzyme-linked immunosorbent assay and the immunofluorescence technique for detection of viral antigens in nasopharyngeal secretions. J Clin Microbiol 22:757-760.

Gray KG, McFarlane DE, Sommerville RG (1968) Direct immunofluorescent identification of respiratory syncytial virus in throat swabs from children with respiratory illness. Lancet 1:446-448.

Halonen P, Meurman O, Lovgren T, Hemmila I, Soini E (1983) Detection of viral antigens by time-resolved fluoroimmunoassay. Curr Top Microbiol Immunol 104:133-146.

Halonen P, Obert G, Hierholzer JC (1985) Direct detection of viral antigens in respiratory infections by immunoassays: A four year experience and new developments. In: Medical Virology IV, pp. 65-83, L.M. de la Maza, E.M. Peterson (eds) Lawrence Earlbaum, Hillsdale, New Jersey.

Halonen P, Sarkkinen H, Arstila P, Hjertsson E, Torfason E (1980) Four-layer radioimmunoassay for detection of adenovirus in stool. J Clin Microbiol 11:614-617.

Hancock K, Tsang VC (1986) Development and optimization of the FAST-ELISA for detecting antibodies to Schistosoma mansoni. J Immunol Meth 92:167-176.

Harmon MW, Pawlik KM (1982) Enzyme immunoassay for direct detection of influenza type A and adenovirus antigens in clinical specimens. J Clin Microbiol 15:5-11.

Harmon MW, Drake S, Kasel JA (1979) Detection of adenovirus by enzyme-linked immunosorbent assay. J Clin Microbiol 9:342-346.

Hendry RM, Fernie BF, Anderson LJ, Godfrey E, McIntosh K (1985a) Monoclonal capture antibody ELISA for respiratory syncytial virus: Detection of individual viral antigens and determination of monoclonal antibody specificities. J Immunol Meth 77:247-258.

Hendry RM, Godfrey E, Anderson LJ, Fernie BF, McIntosh K (1985b) Quantification of respiratory syncytial virus polypeptides in nasal secretions by monoclonal antibodies. J Gen Virol 66:1705-1714.

Hierholzer JC (1973) Further subgrouping of the human adenoviruses by differential hemagglutination. J Infect Dis 128:541-550.

Hierholzer JC, Gary GW (1979) Properties of the "noncultivable" adenoviruses associated with human infantile diarrhea. INSERM 90:103-108.

Hierholzer JC, Pallansch MA (1989) Acute hemorrhagic conjunctivitis in the western hemisphere (1980-87) In: Acute Hemorrhagic Conjunctivitis: Etiology, Epidemiology and Clinical Manifestations. Ishii K, Uchida Y, Miyamura K, Yamazaki S (eds) pp. 49-56, University of Tokyo Press, Tokyo.

Hierholzer JC, Tannock GA (1986a) Adenoviruses. In: Manual of Clinical Laboratory Immunology, 3rd edition, NR Rose, H Friedman, JL Fahey (eds). pp. 527-531. American Society for Microbiology, Washington, D.C.

Hierholzer JC, Tannock GA (1986b) Respiratory syncytial virus: A review of the virus, its epidemiology, immune response and laboratory diagnosis. Austr Paediatr J 22:77-82.

Hierholzer JC, Tannock GA (1988) Coronaviridae: The coronaviruses. In: Laboratory Diagnosis of Infectious Diseases. Principles and Practice. Vol. II. EH Lennette, P. Halonen, FA Murphy (eds). pp. 451-483; Springer-Verlag, New York.

Hierholzer JC, Suggs MT, Hall EC (1969) Standardized viral hemagglutination and hemagglutination-inhibition tests. II. Description and statistical evaluation. Appl Microbiol 18:824-833.

Hierholzer JC, Gamble WC, Dowdle WR (1975) Reference equine antisera to 33 human adenovirus types: Homologous and heterologous titers. J Clin Microbiol 1:65-74.

Hierholzer JC, Phillips DJ, Humphrey DD, Coombs RA, Reimer CB (1984) Application of a solid-phase immunofluorometric assay to the selection of monoclonal antibody specific for the adenovirus group-reactive hexon antigen. Arch Virol 80:1-10.

Hierholzer JC, Johansson KH, Anderson LJ, Tsou CJ, Halonen PE (1987) Comparison of monoclonal time-resolved fluoroimmunoassay with monoclonal capture-biotinylated detector enzyme immunoassay for adenovirus antigen detection. J Clin Microbiol 25:1662-1667.

Hierholzer JC, Wigand R, Anderson LJ, Adrian T, Gold JW (1988a) Adenoviruses from patients with AIDS: A plethora of serotypes and a description of five new serotypes of subgenus D (types 43-47). J. Infect Dis 158:804-813.

Hierholzer JC, Wigand R, de Jong JC (1988b) Evaluation of human adenoviruses 38, 39, 40 and 41 as new serotypes. Intervirology 29:1-10.

Hierholzer JC, Bingham PG, Coombs, RA, Johansson KH, Anderson LJ, Halonen PE (1989) Comparison of monoclonal antibody time-resolved fluoroimmunoassay with monoclonal antibody capture-biotinylated detector enzyme immunoassay for repiratory syncytial virus and parainfluenza virus antigen detection. J Clin Microbiol 27:1243-1249.

Hornsleth A, Aaen K, Gundestrup M (1988) Detection of respiratory syncytial virus and rotavirus by enhanced chemiluminescence enzyme-linked immunosorbent assay. J Clin Microbiol 26:630-635.

Hornsleth A, Brenoe E, Friis B, Knudsen FU, Uldall P (1981) Detection of respiratory syncytial virus in nasopharyngeal secretions by inhibition of enzyme-linked immunosorbent assay. J Clin Microbiol 14:510-515.

Hornsleth A, Friis B, Krasilnikof PA (1986) Detection of respiratory syncytial virus in nasopharyngeal secretions by a biotin-avidin ELISA more sensitive than the fluorescent antibody technique. J Med Virol 18:113-117.

Hughes JH, Mann DR, Hamparian VV (1988) Detection of respiratory syncytial virus in clinical specimens by viral culture, direct and indirect immunofluorescence, and enzyme immunoassay. J Clin Microbiol 26:588-591.

Ito Y, Tsurudome M, Hishiyama M (1987) The polypeptides of human parainfluenza type 2 virus and their synthesis in infected cells. Arch Virol 95:211-224.

Jalowayski AA, England BL, Temm CJ, Nunemacher TJ, Bastian JF, MacPherson GA, Dankner WM, Straube RC, Connor JD (1987) Peroxidase-antiperoxidase assay for rapid detection of respiratory syncytial virus in nasal epithelial specimens from infants and children. J Clin Microbiol 25:722-725.

Johansson ME, Uhnoo I, Kidd AH, Madeley CR, Wadell G (1980) Direct identification of enteric adenovirus, a candidate new serotype, associated with infantile gastroenteritis. J Clin Microbiol 12:95-100.

Johansson ME, Uhnoo I, Svensson L, Pettersson CA, Wadell G (1985) Enzyme-linked immunosorbent assay for detection of enteric adenovirus 41. J Med Virol 17:19-27.

Kadi Z, Dali S, Bakouri S, Bouguermouh A (1986) Rapid diagnosis of RSV infection by antigen immunofluorescence detection with monoclonal antibodies and immunoglobulin M immunofluorescence test. J Clin Microbiol 24:1038-1040.

Kaihola HL, Irjala K, Viikari J, Nanto V (1985) Determination of thyrotropin in serum by time-resolved fluoroimmunoassay evaluated. Clin Chem 31:1706-1709.

Kao CL, McIntosh K, Fernie B, Talis A, Pierik L, Anderson L (1984) Monoclonal antibodies for the rapid diagnosis of respiratory syncytial virus infection by immunofluorescence. Diag Microbiol Infect Dis 2:199-206.

Kim HW, Wyatt RG, Fernie BF, Brandt CD, Arrobio JO, Jeffries BC, Parrott RH (1983) Respiratory syncytial virus detection by immunofluorescence in nasal secretions with monoclonal antibodies against selected surface and internal proteins. J Clin Microbiol 18:1399-1404.

Kuo JE, Milby KH, Hinsberg WD, Poole PR, McGuffin VL, Zare RN (1985) Direct measurement of antigens in serum by time-resolved fluoroimmunoassay. Clin Chem 31:50-53.

Lawson N, Mike N, Wilson R, Pandov H (1986) Assessment of a time-resolved fluoroimmunoassay for thyrotropin in routine clinical practice. Clin Chem 32:684-686.

Lehtomaki K, Julkunen I, Sandelin K, Salonen J, Virtanen M, Ranki M, Hovi T (1986) Rapid diagnosis of respiratory adenovirus infections in young adult men. J Clin Microbiol 24:108-111.

Leite JP, Pereira HG, Azeredo RS, Schatzmayr HG (1985) Adenoviruses in faeces of children with acute gastroenteritis in Rio de Janeiro, Brazil. J Med Virol 15:203-209.

Lovgren T, Hemmila I, Pettersson K, Halonen P (1985) Time-resolved fluorometry in immunoassay. In: Alternative Immunoassays. Collins WP (Ed) pp. 203-217. John Wiley & Sons, Ltd, New York, NY.

Macnaughton MR, Flowers D, Isaacs D (1983) Diagnosis of human coronavirus infections in children using enzyme-linked immunosorbent assay. J Med Virol 11:319-326.

Madeley CR (1986) The emerging role of adenoviruses as inducers of gastroenteritis. Ped Infect Dis 5:S63-S74.

McIntosh K, McQuillin J, Reed SE, Gardner, PS (1978) Diagnosis of human coronavirus infection by immunofluorescence: Method and application to respiratory disease in hospitalized children. J Med Virol 2:341-346.

Meurman OH, Hemmila IA, Lovgren TN, Halonen PE (1982) Time-resolved fluoroimmunoassay: A new test for rubella antibodies. J Clin Microbiol 16:920-925.

Meurman O, Ruuskanen O, Sarkkinen H (1983) Immunoassay diagnosis of adenovirus infections in children. J Clin Microbiol 18:1190-1195.

Meurman O, Sarkkinen H, Ruuskanen O, Hanninen P, Halonen P (1984) Diagnosis of respiratory syncytial virus infection in children: Comparison of viral antigen detection and serology. J Med Virol 14:61-65.

Orvell C, Rydbeck R, Love A (1986) Immunological relationships between mumps virus and parainfluenzaviruses studied with monoclonal antibodies. J Gen Virol 67:1929-1939.

Parkinson AJ, Scott EN, Muchmore HG (1982) Identification of parainfluenza virus serotypes by indirect solid-phase enzyme immunoassay. J Clin Microbiol 15:538-541.

Pereira HG, Azeredo RS, Leite JP, Andrade ZP, deCastro L (1985) A combined enzyme immunoassay for rotavirus and adenovirus (EIARA). J Virol Meth 10:21-28.

Popow-Kraupp T, Kern G, Binder C, Tuma W, Kundi M, Kunz C (1986) Detection of respiratory syncytial virus in nasopharyngeal secretions by enzyme-linked immunosorbent assay, indirect immunofluorescence, and virus isolation: A comparative study. J Med Virol 19:123-134.

Pothier P, Nicholas JC, de Saint-Maur GP, Ghim S, Kazmierczak A, Bricout F (1985) Monoclonal antibodies against respiratory syncytial virus and their use for rapid detection of virus in nasopharyngeal secretions. J Clin Microbiol 21:286-287.

Ray R, Compans RW (1986) Monoclonal antibodies reveal extensive antigenic differences between the hemagglutinin-neuraminidase glycoproteins of human and bovine parainfluenza 3 viruses. Virology 148:232-236.

Richman DD, Cleveland PH, Redfield DC, Oxman MN, Wahl GM (1984) Rapid viral diagnosis. J Infect Dis 149:298-310.

Routledge EG, McQuillin J, Samson AC, Toms GL (1985) The development of monoclonal antibodies to respiratory syncytial virus and their use in diagnosis by indirect immunofluorescence. J Med Virol 15:305-320.

Ruuskanen O, Sarkkinen H, Meurman O, Hurme P, Rossi T, Halonen P, Hanninen P (1984) Rapid diagnosis of adenoviral tonsillitis: A prospective clinical study. J Pediatr 104:725-728.

Rydbeck R, Orvell C, Love A, Norrby E (1986) Characterization of four parainfluenza virus type 3 proteins by use of monoclonal antibodies. J Gen Virol 67:1531-1542.

Sarkkinen HK, Tuokko H, Halonen PE (1980) Comparison of enzyme-immunoassay and radioimmunoassay for detection of human rotaviruses and adenoviruses from stool specimens. J Virol Meth 1:331-341.

Sarkkinen HK, Halonen PE, Arstila PP, Salmi AA (1981a) Detection of respiratory syncytial, parainfluenza type 2, and adenovirus antigens by radioimmunoassay and enzyme immunoassay on nasopharyngeal specimens from children with acute respiratory disease. J Clin Microbiol 13:258-265.

Sarkkinen HK, Halonen PE, Salmi AA (1981b) Type-specific detection of parainfluenza viruses by enzyme immunoassay and radioimmunoassay in nasopharyngeal specimens of patients with acute respiratory disease. J Gen Virol 56:49-57.

Schmitz H, Wigand R, Heinrich W (1983) Worldwide epidemiology of human adenovirus infections. Am J Epidemiol 117:455-466.

Siitari H, Hemmila I, Soini E, Lovgren T, Koistinen V (1983) Detection of hepatitis-B surface antigen using time-resolved fluoroimmunoassay. Nature 301:258-260.

Singh-Naz N, Naz RK (1986) Development and application of monoclonal antibodies for specific detection of human enteric adenoviruses. J Clin Microbiol 23:840-842.

Soini E (1985) Instrumentation: Photometric and photon emission immunoassays. In: Alternative Immunoassays. Collins WP (Ed) pp. 87-102. John Wiley & Sons, Ltd, New York, NY.

Soini E, Hemmila I (1979) Fluoroimmunoassay: Present status and key problems. Clin Chem 25:353-361.

Soini E, Kojola H (1983) Time-resolved fluorometer for lanthanide chelates - A new generation of nonisotopic immunoassays. Clin Chem 29:65-68.

Stalhandske P, Hyypia T, Allard A, Halonen P, Pettersson U (1985) Detection of adenoviruses in stool specimens by nucleic acid spot hybridization. J Med Virol 16:213-218.

Stott EJ, Taylor G (1985) Respiratory syncytial virus: Brief review. Arch Virol 84:1-52.

Takiff HE, Yolken R, Straus SE (1985) Detection of enteric adenoviruses by dot-blot hybridization using a molecularly-cloned DNA probe. J Med Virol 16:107-118.

Toivonen E, Hemmila I, Marniemi J, Jorgensen PN, Zeuthen J, Lovgren T (1986) Two-site time-resolved immunofluorometric assay of human insulin. Clin Chem 32:637-640.

Uhnoo I, Wadell G, Svensson L, Johansson M (1983) Two new serotypes of enteric adenovirus causing infantile diarrhoea. Develop Biol Standard 53:311-318.

Uhnoo I, Wadell G, Svensson L, Johansson M (1984) Importance of enteric adenoviruses 40 and 41 in acute gastroenteritis in infants and young children. J Clin Microbiol 20:365-372.

Uhnoo I, Wadell G, Svensson L, Olding-Stenkvist E, Ekwall E, Molby R (1986) Aetiology and epidemiology of acute gastroenteritis in Swedish children. J Infect 13:73-89.

Van Dyke RB, Murphy-Corb M (1989) Detection of repiratory syncytial virus in nasopharyngeal secretions by DNA-RNA hybridization. J Clin Microbiol 27:1739-1743.

Vesikari T, Maki M, Sarkkinen HK, Arstila PP, Halonen PE (1981) Rotavirus, adenovirus, and nonviral enteropathogens in diarrhea. Arch Dis Child 56:264-270.

Virtanen M, Palva A, Laaksonen M, Halonen P, Soderlund H, Ranki M (1983) Novel test for rapid viral diagnosis: Detection of adenovirus in nasopharyngeal mucus aspirates by means of nucleic-acid sandwich hybridization. Lancet 1:381-383.

Wadell G (1984) Molecular epidemiology of human adenoviruses. Current Topics Microb Immun 110:191-220.

Walls HH, Johansson KH, Harmon MW, Halonen PE, Kendal AP (1986) Time-resolved fluoroimmunoassay with monoclonal antibodies for rapid diagnosis of influenza infections. J Clin Microbiol 24:907-912.

Waner JL, Whitehurst NJ, Downs T, Graves DG (1985) Production of monoclonal antibodies against parainfluenza 3 virus and their use in diagnosis by immunofluorescence. J Clin Microbiol 22:535-538.

Welliver RC, Wong DT, Sun M, McCarthy N (1986) Parainfluenza virus bronchiolitis: Epidemiology and pathogenesis. Am J Dis Child 140:34-40.

Wigand R, Adrian T (1986) Classification and epidemiology of adenoviruses. In: Adenovirus DNA. pp. 409-441. Doerfler W (ed.) Martinus Nijhoff Publ. Co., Boston, MA.

Wong DT, Welliver RC, Riddlesberger KR, Sun MS, Ogra PL (1982) Rapid diagnosis of parainfluenza virus infection in children. J Clin Microbiol 16:164-167.

Yolken RH, Franklin CC (1985) Gastrointestinal adenovirus: an important cause of morbidity in patients with necrotizing enterocolitis and gastrointestinal surgery. Ped Infect Dis 4:42-47.

Zrein M, Obert G, van Regenmortel MH (1986) Use of egg-yolk antibody for detection of respiratory syncytial virus in nasal secretions by ELISA. Arch Virol 90:197-206.

DISCUSSION

Wright J, (Gull Foundation, Salt Lake City, UT):

What volumes were your initial specimens being collected in?

Hierholzer J (Centers for Disease Control, Atlanta, GA):

The specimens ranged between 0.5 and 1 ml. They were nasopharyngeal aspirates from children, and because of this, we had variable dilutions. If we had a lot of specimen and we were looking for just a few agents, we might only dilute the specimens 1:5. On the other hand, if we had a little bit of specimen, and we had to look for many viral antigens, then we would dilute it 1:10 or 1:20.

Wright J:

Have you looked at monoclonal - polyclonal combinations in the TR-FIA?

Hierholzer J:

We did in a study with Pekka Halonen several years ago, and the test simply didn't compare to the all-monoclonal test for sensitivity.

Wright J:

How much technology here is wrapped up in patents? In other words, what is going to keep it from getting into the U.S. market?

Hierholzer J:

I really don't know. The company who makes our fluorometer is LKB/Wallac which is a solid firm. Another fluorometer is made by Cyber Fluor, Inc. of Toronto, Canada. Those instruments are readily available. Many hospitals have bought them for serum proteins like thyroglobulin. I don't really see any problem with someone marketing the strips that have the capture antibody already on them, because all of these monoclonals were developed in our laboratory.

Wright J:

How expensive are those machines?

Hierholzer J:

The price has been coming down. I think they're running between $20,000-$40,000 now.

Wright J:

So, do you anticipate that this technology could make it into the average clinical virology lab?

Hierholzer J:

It would be practical for large labs. Small labs would probably be better off buying kits that are being marketed now for ELISA and FA.

Riepenhoff-Talty M (Children's Hospital, Buffalo, NY):

What about influenza?

Hierholzer J:

Influenza was formatted for the TR-FIA by another group at CDC. That's why I didn't present their data. They have published that and and the reference is Walls, H.H. et al. 1986 (see References).

Lee I (Burroughs-Wellcome Co, Research Triangle Park, NC):

Are there any false-negatives at very high antigen concentrations?

Hierholzer J:

No, I have not found any false-negatives at high antigen concentrations. It may be, in the last slide that some of those specimens of insufficient volume to test by culture or ELISA, were false-negatives, and I would not know it. But, in all of these samples that we clearly knew from the start were positive at high antigen concentration, we had no false-negatives.

Al-Nakib W (Kuwait University, Kuwait, Kuwait):

How did you treat your specimens for the coronavirus detection system?

Hierholzer J:

I didn't treat any of these specimens. We started with a dilution, say of 1:5, which is in an antigen diluent that contains 0.01% Tween-40 which helps to keep the protein solubilized, but it's not enough to break the viruses apart. The coronaviruses are more labile than all of the other viruses I've shown, even RSV. I don't yet have enough experience with field specimens to know if I am going to have a problem with the coronaviruses breaking down to the point I cannot detect them. I know that the 229E and DC43 monoclonals are directed to the N ribonucleoprotein, and it still appears necessary to have intact virus in the samples.

Al-Nakib W:

Are these specimens from adults or children?

Hierholzer J:

These have all been NPA samples from children.

Al-Nakib W:

One thing you have not mentioned regarding coronaviruses, there are some very good serological tests available, especially ELISA. For example, it is used in some laboratories in terms of establishing a diagnosis, at least, serologically.

Hierholzer J:

These are the EIA tests developed by M. Macnaughton's group in England? (Infec Immun 38:419-423, 1982)

Al-Nakib W:

No, the antibody tests developed by K. Callow (J Hyg 95:173-189, 1985).

Hierholzer J:

Are those tests commercially available?

Al-Nakib W:

No.

Hierholzer J:

That is why I don't regard the coronavirus tests to be generally available. I only know of a few labs who are using the tests, but you are right, they are good serologic tests for coronavirus if a lab gets them going.

de la Maza (University of California Irvine Medical Center, Orange, CA):

I am not sure that you have any data there discussing the specificity of the test. Are you assuming that everything that is positive is true-positive?

Hierholzer J:

The tests were all first developed with NPA samples from which a particular virus had been grown out. We had to start with culture-positive being the gold standard. In other words, if a sample had adenovirus in it and it grew out, than that was called an adenovirus-positive specimen, and of course, that means that no other virus grew out. The negative specimens, likewise, were cultured extensively, and were also tested by whatever EIA tests and FA tests were being used in the lab at the time, and they were specimens that were called negative. I used them in this study as negative specimens. We felt that was as good as we could do in having truly positive and negative specimens.

de la Maza L:

Based on that data, your specificity has been 100%?

Hierholzer J:

Yes; all of the monoclonal antibodies used in this study had specificities of 100%. Sensitivities in detecting viral antigens ranged from 100% for parainfluenza 2 to 68% for CA24 variant. That meant, for instance, that all of the specimens from which we had isolated parainfluenza. 2 were positive in the TR-FIA, but only 75-89% of them were positive by the EIA tests.

THE USE OF THE POLYMERASE CHAIN REACTION IN THE DETECTION, QUANTIFICATION AND CHARACTERIZATION OF HUMAN RETROVIRUSES

Bernard J. Poiesz[1], Garth D. Ehrlich[1], Bruce C. Byrne[1], Keith Wells[1], Shirley Kwok[2] and John Sninsky[2]

[1] State University of New York
Division of Hematology/Oncology
Health Science Center
Syracuse, New York 13210 USA and
[2] Cetus Corporation
1400 Fifty Third Street
Emeryville, California 94608, USA

THE RETROVIRUSES

Retroviruses are the etiologic agents of a host of diseases found in verte-brates. These include malignancies such as lymphomas, leukemias, sarcomas and carcinomas; autoimmune diseases such as arthritis and lupus; and cyto-pathic diseases leading to anemias and immune deficiency states. There are four well characterized human retroviruses (Figure 1). Human T-cell lym-phoma/leukemia virus type I and II (HTLV-I and II) are oncornaviruses. HTLV-I is believed to be the etiologic agent of adult T-cell lympho-ma/leukemia (ATLL) and a progressive neurological disorder formerly called tropical spastic paraparesis and now termed HTLV-I associated myelopathy (HAM) (Bhagavati et al. 1988; Ehrlich and Poiesz 1988; Poiesz et al. 1980). HTLV-I is also associated with immune deficiency. HTLV-I infection is endem-ic in Southern Japan, the Caribbean and Central Africa. In the United States, HTLV-I infection is most prevalent in the Southeastern section of the country among rural blacks, but is has also been identified in patients throughout America and appears to be increasing in prevalence among intravenous drug abusers (IVDA) (Ehrlich and Poiesz, 1988; Ratner and Poiesz, 1988). HTLV-II is approximately 65% homologous to HTLV-I (Kalyanaraman et al. 1982). It has been associated with rare cases of T-cell and hairy cell leukemias, but its exact disease association and endemic area of infection are unclear at this time. HTLV-II infection is also being diagnosed with greater frequency in American IVDA (Ehrlich et al. 1989). HTLV-V is a recently described retrovirus with lim-ited homology to HTLV-I (Manzari et al. 1987). It has been identified in a few patients with CD4+ cutaneous T-cell lymphoma, but the virus has not been completely characterized and its disease association is poorly understood.

HTLV-I: Adult T-cell leukemia, myelopathy, immunodeficiency

HTLV-II: Hairy T-cell leukemia; CD8 leukemia

HTLV-V: ?; Cutaneous T-cell lymphoma

HIV-1: Acquired Immunodeficiency Syndrome

HIV-2: Acquired Immunodeficiency Syndrome

Figure 1. The known or suspected human retroviruses and their associated diseases.

Human immunodeficiency virus types 1 and 2 (HIV-1 and 2) are the etiologic agents of the acquired immune deficiency syndrome (AIDS) (Barre-Sinoussi et al. 1983; Clavel et al. 1986). HIV-2 infection is primarily confined to Western Africa, while HIV-1 is responsible for the current pandemic of AIDS (Wells et al. 1990).

PROBLEMS ASSOCIATED WITH DIAGNOSIS

Most humans infected with a retrovirus manifest no overt symptoms for many years following infection. Some, but not all, eventually evolve through protracted prodromal stages of their disease prior to the development of a terminal illness. The therapeutic options for patients with retrovirally induced terminal illnesses are few, and none to date are curative (Wells and Poiesz, 1990). Hence, considerable effort has been made to identify infected asymptomatic patients in order to prevent spread of the infection to others and to specify the exact nature of their illnesses (Duggan et al. 1988), or to initiate earlier antiviral treatment strategies. Considerable use has been made of serological assays to detect infected individuals. Such techniques have served society well as screening assays to protect recipients of donated blood products. However, because of the latency associated with retroviral infections, the possibility of a seronegative infected state exists. Further, for a number of reasons (eg. pronostics, evaluating treatment efficacy and/or analyzing genetic variation of different virus isolates), it is important to have the capability to actually test for the presence of the retroviruses themselves. Accordingly, we have adapted the polymerase chain reaction (PCR) technique for the detection, quantification and characterization of human retroviruses (Karpatkin et al. 1988; Kwok, et al. 1987, 1988a, 1988b, 1988c).

SENSITIVE AND SPECIFIC DETECTION BY PCR

Primer pairs have been developed which allow for the sensitive and specific detection of HIV-1, HIV-2, HTLV-I and HTLV-II proviral DNAs using either the Klenow fragment of *E. coli* DNA polymerase I or more recently, *Taq* DNA polymerase to support enzymatic amplification (Figures 2, 3, 4). Similarly, these techniques have been employed for the detection of retroviral RNA after first producing cDNA, using cloned Moloney murine leukemia virus (MuLV)

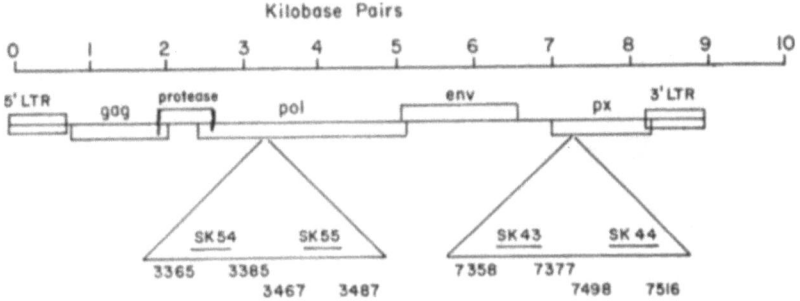

Figure 2. Typical primer pairs which will amplify the *pol* and *tax* genes respectively, of HTLV-I. Detector sequences for hybridization to the amplified DNA products of PCR would be located between the two primers.

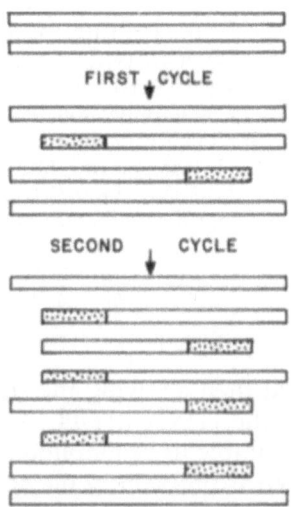

Figure 3. Schematic of the PCR through two amplification cycles. Double-stranded human DNA, which presumably contains retroviral DNA is denatured to yield two single strands of DNA. Primer pairs (stripped) consist of small oligonucleotides to 5' regions of opposite strands of a portion of the retroviral genome. Typically, these primer pairs define a region of 100 to 300 bp in the viral genome. As the reaction is cooled, the primers will hybridize to homologous DNA sequences before the larger single strands of human DNA can reanneal; this is due to the small size and relative abundance of the primers. At the optimal temperature and in the presence of a DNA-dependent DNA polymerase and all of the nucleotide triphosphates, the single-stranded DNA is copied 3' to the annealed primer to complete the cycle. At the end of the first cycle there are now four copies of the DNA defined by the primer pair. At the end of the two cycles there are eight copies, and so forth. As can be seen, over many cycles there will be a geometric increase in the amount of the specific amplified viral DNA.

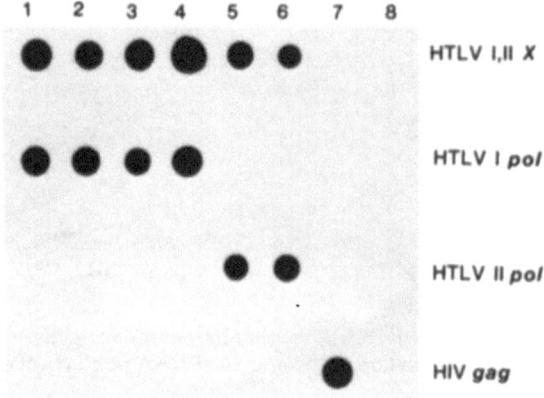

Figure 4. Analysis of PCR-amplified DNA from cells infected with various human retroviruses. PCR amplification was performed using 30 cycles and *Taq* polymerase. Columns 1 through 8 represent DNA extracted from cell lines containing integrated HTLV-I (columns 1 through 4), HTLV-II (columns 5,6), HIV-1 (column 7), and an uninfected T-cell line, Molt-4 (column 8). The first row represents PCR amplification that utilized primers SK43/44 and probe SK45 that contain sequences conserved between HTLV-I and HTLV-II; second row, SK54/55 and SK56 that are HTLV-I specific; third row, SK58/59 and SK60 that are HTLV-II specific; and fourth row SK38/39 and SK 19 that are HIV-1 specific.

reverse transcriptase (RT), and then allowing the standard *Taq* DNA polymerase based system to amplify the cDNA (RT-PCR) (Figures 5, 6) (Byrne et al. 1988). Prior treatment of the sample to be amplified with DNAse allows for the determination of levels of RNA expression in the specimen. Detection systems utilized have included: direct incorporation of radioactive isotope into the amplified product (Figure 7); or hybridization techniques employing spot or slot blot (Figures 4, 5 and 6), liquid hybridization (Figure 8) or oligostriction detection formats (Figure 9) (Abbott et al. 1988; Ehrlich et al. 1990). Direct incorporation of radioisotope works for certain retroviral primer pairs; however, for most primer pairs we have chosen to ultimately perform a hybridization detec-

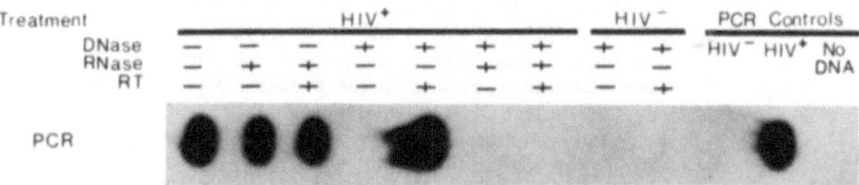

Figure 5. Slot blot hybridization of nucleic acids extracted from HIV-1 infected (HIV⁺) and uninfected (HIV⁻) HUT 78 cells which contain both DNA and RNA. DNAse treated HIV⁺ samples cannot be amplified by conventional PCR. However, if first incubated with purified Maloney MuLV RT they can subsequently be amplified. The fact that this phenomenon can be negated by prior treatment of the original sample with RNAse confirms the source of the product as RNA.

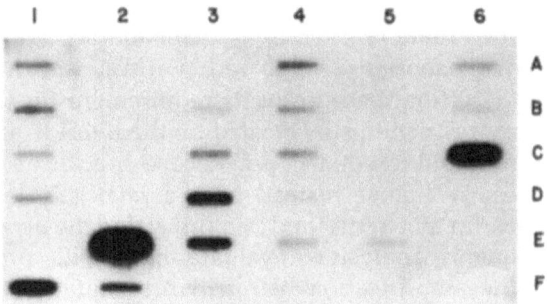

Figure 6. Slot blot hybridizations of DNA products amplified via PCR and RT-PCR on nucleic acids extracted from the peripheral blood mononuclear cells of eight HIV-1 seropositive and culture positive individuals (A1-3, B1-3, C1-3, D1-3, A4-6, B4-6, C4-6, D4-6). Lanes 1 and 4 A-D represent conventional DNA driven PCR. Lanes 2 and 5 A-D are the same samples treated with DNAse and subjected to conventional PCR. Lanes 3 and 6 A-D are the same samples treated with DNAse and subjected to RNA driven RT/PCR. Lanes E represent conventional PCR; Lane F RT PCR. Lane E1 is uninfected HUT 78 DNA; Lane E2-6 is a serial dilution of HIV-1 infected HUT 78 DNA, while F1-6 is a serial dilution of HIV-1 infected HUT 78 cellular RNA.

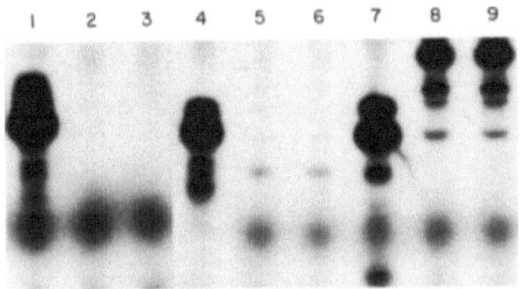

Figure 7. Autoradiograph of polyacrylamide gel electrophoresis of products derived from PCR of clinical samples. Instead of performing a liquid hybridization step to detect the products, PCR was performed in the presence of $\alpha[^{32}P]$ dCTP thereby allowing for the direct incorporation of radioactive nucleoside monophosphate into the amplified DNA. Lanes 1-3 were amplified with the HTLV-1 *tax* primer pair, SK43-44 and corresponding to HTLV-I positive, HTLV-I and HIV-1 negative, and HIV-1 infected DNA respectively. Lanes 4-6 were amplified with the HIV-1 specific *gag* primer pair, SK38-39 and correspond to HIV-1 positive, HTLV-I and HIV-1 negative, and HTLV-1 positive DNA, respectively. Lanes 7-9 were amplified with the HTLV-I *pol* specific primers SK54-55 and corresponding to HTLV-I positive, HTLV-I and HIV-1 negative and HIV-1 positive DNA, respectively. The relatively high molecular weight products seen in Lanes 8 and 9 are present in all human beings tested to date. They do not hybridize with the HTLV-I detector oligonucleotide under stringent conditions. Also, as seen in Lane 7, they are not produced in the presence of high copy prototype HTLV-I sequences. We speculate that these products detected in Lanes 8 and 9 are derived from endogenous retroviral sequences which share homology to HTLV-I *pol*.

tion step because of the often observed nonspecific amplification of presumably endogenous retroviral sequences located within the human genome. Oligomer restriction detection, because of its requirement for the generation of a specific endonuclease restriction site, created by the hybridization of the detector sequence to the amplified target, has proven too specific as a screening technique for retroviruses. We have instead used a relatively less stringent slot blot to screen samples for positivity and a more stringent liquid hybridization technique and/or sequence analysis to evaluate the specific product. Our routine conditions for the amplification of retroviral sequences and the primer pairs and probes used to support and detect the amplified products are outlined in the Appendix.

Quantification of the amplified product can be achieved by comparing a laser generated densitometric scan of the autoradiographic signal of a patient's sample versus a serially diluted standard control (Figure 10). Similar analysis can be performed for human genes (eg. β-globin) and results can be expressed as copies of retroviral DNA per unit of human β-globin amplified.

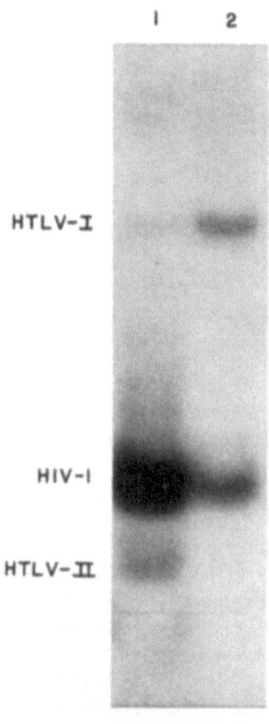

Figure 8. Liquid hybridization of DNA extracted from the peripheral blood mononuclear cells of two IVDA. The samples were simultaneously amplified for HTLV-I, HTLV-II and HIV-1 sequences. The amplification products for each virus are a distinct and different size. They were subjected to liquid hybridization separated on a gel as outlined in the appendix. As can be seen, patient 1 is infected with all three viruses while patient 2 is infected with HTLV-I and HIV-1.

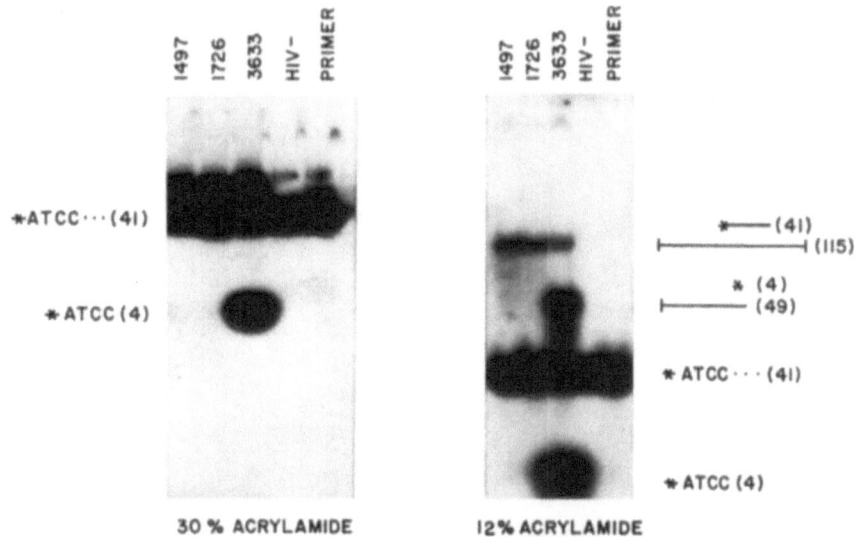

Figure 9. Oligomer restriction hybridization is similar in format to liquid hybridization (see appendix). Illustrated above is oligomer restriction hybridization of five samples amplified with the HIV-1 *gag* gene specific primer pairs SK38,39. Samples 1494, 1726, and 3633 are DNAs from the peripheral blood mononuclear cells of three seropositive HIV-1 culture positive AIDS patients. The next two samples are DNA from a normal HIV-1 negative blood donor and the primer control. When the 41 base long [^{32}P] end labeled detector sequences hybridizes to the 115 base long prototype *gag* gene product, it produces a restriction endonuclease site. When the resulting hybrid is digested with the appropriate enzyme, it generates a labeled four base fragment which is easily distinguished in a 30% acrylamide gel (sample 3633). Samples 1497 and 1726 as well as the controls would appear to be negative in this gel. However, a 12% acrylamide gel resolves the larger DNA species into their various components. In the 12% acrylamide gel it can be seen that samples 1497 and 1726 do yield a positive liquid hybridization species of the predicted size, but the samples are not digested by the restriction endonuclease, suggesting sequence variability at the restriction site. It is apparent that the 12% polyacrylamide gel oligomer restriction format could be used as a quick screen for both positivity and sequence variability.

Ultimate characterization of amplified product has been conducted via cloning of the amplified products into M-13 and subsequently performing sequence analysis (Kwok et al. 1988b).

Results, to date, indicate that PCR detection is a highly sensitive format. Using 1 μg of fresh peripheral blood mononuclear cell DNA (≈150,000 cells) and 30 cycles of amplification and at least 2 viral specific primer pairs, our detection rates for HIV-1 and HTLV-I in seropositive individuals have approximated 80% and 100% respectively. We believe that the relatively lower rates of HIV-1 detection are secondary to lower *in vivo* copy numbers, and sequence variability between the various HIV-1 isolates versus the HTLV-I isolates. Indeed, when the DNA from 10^6 cells of HIV-1 infected individuals is analyzed, the sensitivity of the PCR system increases significantly.

One important use of PCR has proven to be its ability to distinguish between HTLV-I and HTLV-II infections (Kwok et al. 1988a, 1988b). In a recent survey of 169 IVDA in New York City, 17 (10%) persons tested positive for antibodies to both HTLV-I and HTLV-II. PCR analysis of this cohort indicate that 15 (9%) and 19(11%) of these subjects were positive for HTLV-I and HTLV-II se-

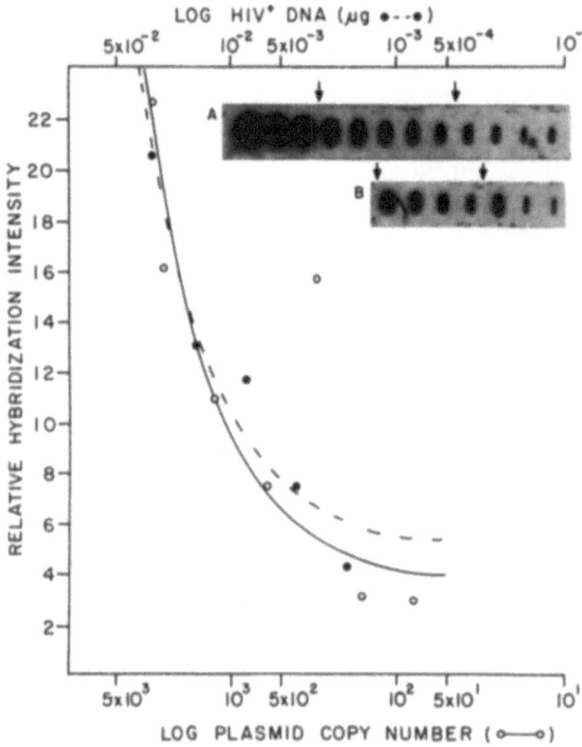

Figure 10. Quantitative analysis of the reaction products of amplified HIV-1 positive DNA diluted into negative DNA. Decreasing concentrations of HIV-1-positive DNA samples were amplified in a total of 1 mg of DNA per reaction mixture. Amplification was accomplished using the HIV-1 *gag*-specific primer-detector system (SK38/39, SK19). The amplified DNA was slot blotted and hybridized with the ^{32}P labeled detector. Curves were fitted by linear regression using only the densitometric scans resulting from those dilutions within the *arrows* on the inserts; coefficient of coincidence $r^2 = 0.94$ for both plasmid and virus sets. *Insert A*, and ----, dilutions of DNA isolated from the chronically infected cell line HUT 78/HIV$_{AAV}$. The film demonstrates a quantitative decrease in hybridization intensity as target HIV-positive DNA is diluted from 200 ng per reaction mixture to 100 pg. *Insert B*, O and ——, dilutions of DNA from a plasmid carrying a 2.2 kilobase insert encompassing the amplified region. The film depicts hybridization resulting from amplification of dilutions from 3,000 to 80 copies of HIV DNA per reaction mixture.

quences, respectively (Ehrlich et al. 1989). Perhaps of even greater interest is the fact that a total of 11 (7%) subjects were found to be seronegative but PCR positive for either HTLV-I or HTLV-II.

The use of PCR theoretically, could facilitate the detection of variant retroviruses that are closely or distantly related to the known human retroviruses. This could be accomplished by demonstrating that certain primer pairs from one retrovirus will support amplification from a novel target DNA, but other primers from the same prototype virus would not (Figure 4). A more

```
                    Group
               Specific Primer                                        Generic Primer
               ⌐‾‾‾‾‾‾‾‾‾¬                                          ⌐‾‾‾‾‾¬
HTLV-I    L P Q G F L N S P T L F E M Q L A H I L Q P I R Q A F P Q C T I L Q Y M D D I
HTLV-II   L P Q G F K N S P T L F E Q Q L A A V L N P M R K M F P T S T I V Q Y M D D I
BLV       L P Q G F I N S P A L F E R A L Q E P L R Q V S A A F S Q S L L V S Y M D D I
MULV      L P Q G F K N S P T L F D E A L H R D L A D F R I Q H P A L I L L G Y V D D L
AKV       L P Q G F K N S P T L F D E A L H R D L A D F R I Q H P D L I L L Q Y V D D I
FeLV      L P Q G R K N S P T L F D E A L H S D L A D F R B R Y P A L V L L Q Y V D D I
          + + + + +a + + +   + +       +       +                       +     + +

HuRRS-P   L P W G F - D I L H L F G Q A L S K D L T E F - - S H L Q V K I L Q Y V G D I

MMTV      L P Q G M K N S P T L C Q K F V D F A I L T V R D K Y Q D S Y I V H Y M D D I
SRV-1     L P Q R M A N S P T L C Q K Y V A T A I H K V R H A W K Q M Y I I H Y M D D I
MPMV      L P Q G M A N S P T L C Q K Y V A T A I H K V R H A W K Q M Y I I H Y M D D I
IAP       L P Q G M A N S P T I C Q L Y V Q E A L E P I R K Q F T S L I V I H Y M D D I
RSV       L P Q G M T C S P T I C Q L V V G Q V L E P L R L K H P S L C M L H Y M D D L
          + + +   +     + + +   + +       +               +             + + + +

HERV-K    L P Q G M L N S P T I C Q T F V G R A L Q P V R E K F S D C Y I I H Y I D D I
HM16      L P Q G M L N S P I I C Q T F V A Q V L Q P V R D K F S D C Y V I H Y V D - I

HIV-I     L P Q G N K G S P A I F Q S S M T K I L E P F K K Q N P D I V I Y Q Y M D D L
HIV-II    L P Q G W K G S P A I F Q H T M R Q V L E P F R K A N K D V I I I Q Y M D D I
SIV       L P Q G W K G S P A I F Q Y T M R H V L E P F R K A N P D V T L V Q Y M D D I
EIAV      L P Q G F V L S P Y I Y Q K T L Q E I L Q P F R E R Y P E V Q L Y Q Y M D D L
CAEV      L P Q G W K L S P S V Y Q F T M Q E I L G E W I Q E H P E I Q F R I Y M D D I
VISNA     L P Q G W K L S P A V Y Q F T M Q K I L R G W I E E H P M I Q F G I Y M D D I
          + + + +   + +     +               +                       + + + +
          * * *b       * *                                           * *c* *

HBV       I P M G V G L S P F L L A Q F T S A I C S V V R R A F P H C L A F S Y M D D V
WHV       L P M G V G L S P F L L A Q F T S A L A S M V R R N F P H C V V F A Y M D D L
GSHV      L P M G V G L S P F L L A Q F T S A L T S M V R R N F P H C L A F A Y M D D L
DHBV      L P M G V G L S P F L L H L F T T A L G S E I S R R F N V - W T F T Y M D D F
          + + + + + + + + + + +   + +     +           +     +             + + +
          §   §b     § §                                                 § §c§
```

Figure 11.   Amino acid sequences of relatively conserved retroviral *pol* gene products including the hepatitis viruses (bottom group) of man and various vertebrates (bottom group). Degenerate primer pairs which could be homologous to all observed and conceivable sequence permutations within 5' group specific region and a generic region could be used to amplify intervening sequences of potentially all retroviruses. As shown, however, the sequences between the primers are more variable and detection would require subgroup specific detectors. Alternatively, all amplified fragments could be cloned and ultimately sequenced. The obvious drawback of this system is its high sensitivity in that it will most likely identify innumerable endogenous retroviral sequences. To that end, it would be the ultimate method to identify many such sequences in man.

generic screen for retroviruses would make use of degenerate primer pairs to conserved regions within retroviral *pol* sequences which are theoretically capable of amplifying all *pol* containing retroviral sequences (Mack and Sninsky, 1988) (Figure 11). Further characterization of sequences flanking a fragment of homologous DNA detection by routine PCR or PCR using degenerate primers could be achieved by techniques such as inverse PCR (Ochman et al. 1988) (Figure 12) or ligation mediated PCR (Mueller and Wold, 1989). Theoretically, one could literally "march" in both directions away from a detected sequence using successive and overlapping rounds of inverse PCR, cloning and sequencing to define a new retroviral sequence.

PCR has also been utilized to evaluate the *in vitro* effect of anti-retroviral agents. The analysis of the synthesis of proviral DNA and or RNA over time allows for the dissection of the viral life cycle and dissection of mechanisms of action of antiretroviral agents (Figures 13, 14 and 15).

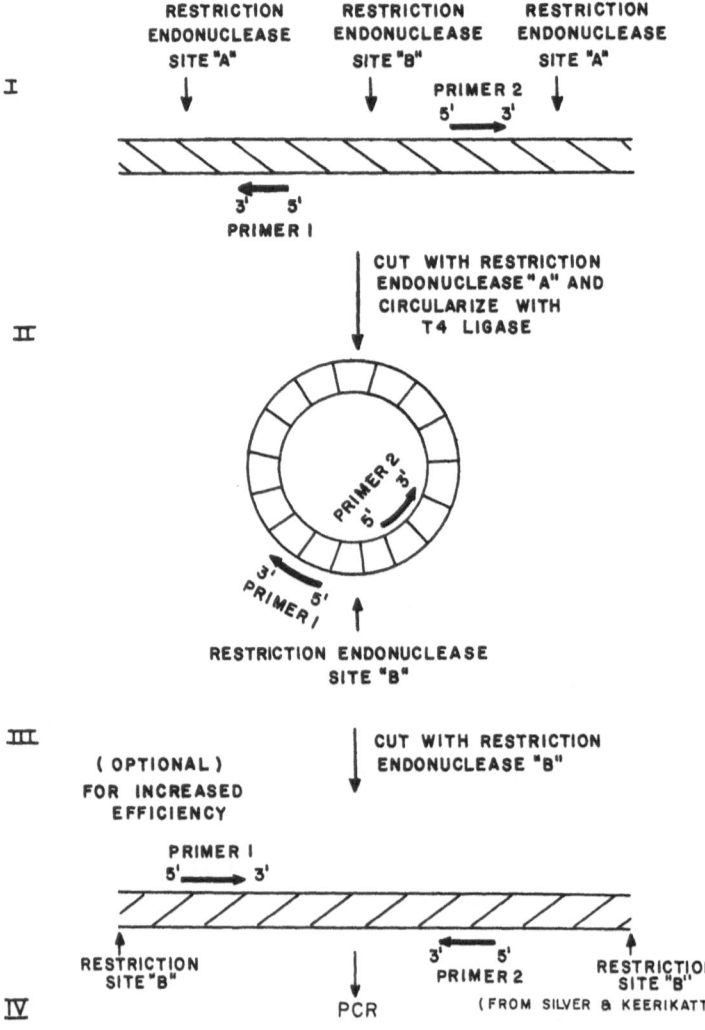

Figure 12. Schematic for inverse PCR. Given that one has detected and sequenced a particular piece of DNA, primers could be designed which would be oriented in the wrong direction for conventional PCR but would prime DNA synthesis into the region flanking the known sequence (I). Upon digestion of the target DNA with restriction endonuclease A at sites located outside the known sequence, subsequent circularization with T4 ligase, and digestion with restriction endonuclease B at a site located within the region of known sequence and between the two primers, a linear fragment of DNA would be produced which upon denaturation and annealing of the primers would now have them facing in the appropriate direction and would have the unknown flanking sequences now located between them. The ultimate product of conventional PCR could be cloned and sequenced to define the original flanking DNA.

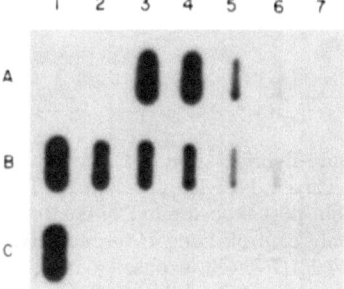

Figure 13. Slot blot hybridization of DNAs amplified for the HIV-1 gag gene. Lane A1 HIV-1 negative HUT 78 cells; A2 primer only control; A3-7 serial dilution of HUT 78 cells stably infected with HIV-1 (10ng to 1pg of cellular DNA input); B1-7 HUT 78 cells 6 hour post exposure to DNAse-treated cell-free HIV-1 virions (each slot represents a successive 1:2 dilution of input virus); C1 untreated cell free HIV-1 virions; C2 DNAse treated, cell-free HIV-1 virions.

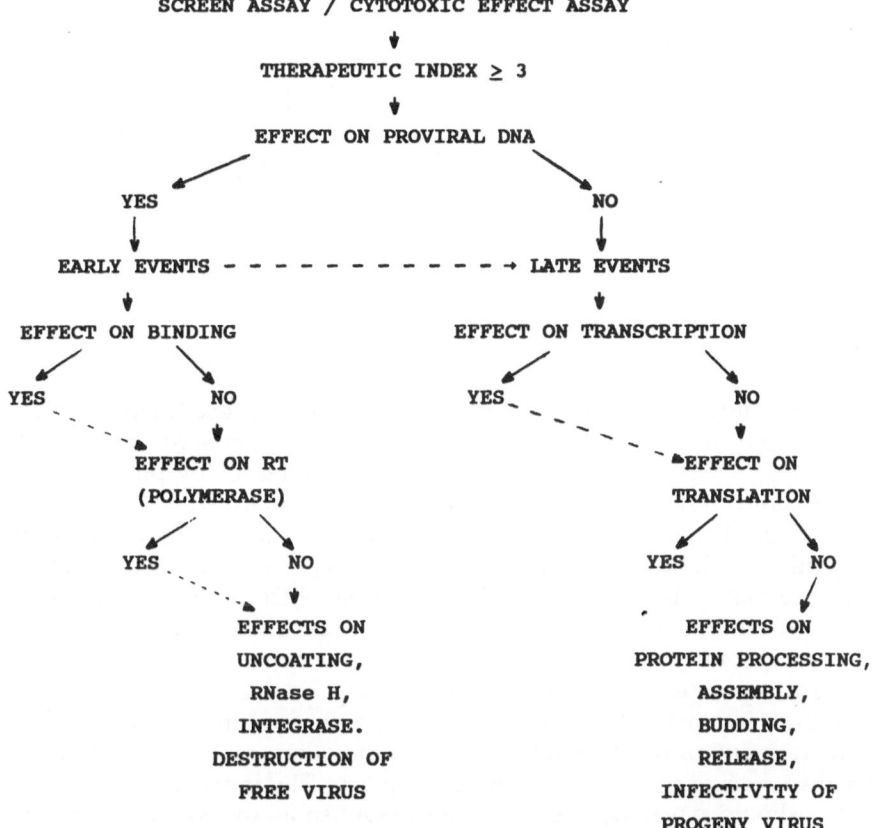

Figure 14. Algorithm used in our laboratory to dissect the mechanism of action of anti-HIV-1 agents. All compounds are screened in a microtiter assay wherein uninfected HUT 78 cells are treated for 24 hours with varying concentrations of the agent in question. The cells are then exposed for 4 hours to a known inoculation of HIV-1 virions. The cells are then washed and placed in microculture. On day 4 of the culture, cell free condition media is analyzed for HIV-1 p24 antigen content. A similar thymidine incorporation assay is also performed using the same concentration of drug and HUT 78 cells minus the virus; the 50% inhibitory dose for the viral and cytotoxicity assays are then calculated and a therapeutic index determined. Interesting compounds demonstrating a greater antiviral than cytopathic effect are evaluated further. Since the screen assay would be affected by inhibitors during any part of the virus life cycle, we use the compounds effect on proviral DNA formation at 6 hour post infection as described in Figure 13 to ascertain whether the agent is affecting early or late events in the virus life cycle.

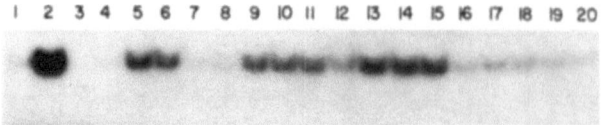

Figure 15. Liquid hybridization of DNAs amplified for HIV-1 gag gene DNA. Lane 1, HIV-1 uninfected HUT 78 cells; Lane 2, HUT 78 cells 6 hour post exposure to DNAse treated HIV-1 virus; Lane 3 primer only control; Lane 4, seronegative normal blood donor. Lanes 5-20, HUT 78 cells exposed to 0.01, 0.1, 1 and 10 mM, respectively, of putative RT inhibitors (Lanes 5-8, azidothymidine; Lanes 9-12, 2-phosphoryl-methoxyethyladenine; Lanes 13-16, 2'3'-dideoxy-2'3'didehydro-thymidine; Lanes 17-20, 3' fluoro-dideoxythymidine.

In summary, PCR has been adapted for the facile, quantitative and specific amplification of retroviral nucleic acids. The clinical use of this technology should prove invaluable in the detection, counseling and treatment of infected individuals. PCR should also allow for relatively rapid characterization of retroviral variants and offers theoretical promise for the discovery of new retroviruses and/or retroviral disease associations in man.

APPENDIX 1

Technically, the most difficult aspect of PCR is to prevent cross contamination of samples or carryover. Carryover occurs either by aerosolization of amplified samples when tightly capped tubes are opened or by pipetting devices which do not physically separate the barrel of the device from the material being pipetted. If the amplified DNA is viewed as analogous to an infectious agent with the ability to replicate under the proper environmental conditions (i.e. the thermal cycler), it is relatively easy to establish certain physical and procedural precautions which will greatly reduce the chance of molecular carryover. Any laboratory using PCR for diagnostic or epidemiologic studies should separate its operations into three physically separate rooms; two of which are clean rooms, ie. no amplified product, and one dirty room. There should be no airflow from any of these rooms into any of the others. One clean room should be used to receive and process clinical samples. The individual(s) involved in this step should never handle amplified DNAs so as to reduce to almost zero the chance of up front contamination. Further, all samples should be aliquoted prior to freezing. Most samples can be stored indefinitely at -70ºC while awaiting extraction or lysis. A second clean room is used for DNA extraction and purification. This room is also used for setting up the reaction cocktails prior to PCR; it is never used for the actual amplification, or for the assessment of the products of amplification. The third room is the amplification room where the PCR is performed and the determination of the results are made. The tightly capped tubes to be amplified are never opened in this room until all amplification is complete. These tubes should never be opened and reclosed once amplification has begun. Likewise, they should never be taken back into the clean room after amplification has been allowed to proceed. Nothing from this room should ever go back into the first two rooms.

Each of the rooms should be equipped with its own set of pipetting and storage devices. There should be no trafficking of pipettors or other reagents and equipment between labs. To further reduce the chance of carryover posed by aerosolized target DNA, we recommend the use of positive displacement pipetting devices which have disposable tips and plungers, such as Rainin micromen or Digitron pipettors when pipetting the target DNA. Contamination with target DNA is much less likely since it is, as opposed to amplified DNA, generally of high molecular weight. Standard pipettors are appropriate for dispensing non-DNA containing solutions and buffers.

In addition to physical precautions, a number of procedural safeguards should be employed. A diagnosis of infection should never be made on the result of a single PCR reaction performed with a single primer pair in the absence of some corroborating evidence such as serological or antigenic data. It is recommended in all diagnostic cases that at least two sets of primer pairs produce a positive result, particularly if the patient is seronegative. Further, it is suggested that all such results be confirmed from a second DNA sample prepared at another time for a separate frozen aliquot. When DNA is to be extracted from clinical material only a single aliquot is used for the initial screen. If a given DNA is positive, then go back and extract DNA from a second aliquot for confirmational analysis.

The following methods developed over the last three years in our laboratories have proved useful in a wide range of applications. These procedures can be used to amplify DNA in simple cellular lysates or organically extracted DNA preparations. We have found that boiling and quenching on ice of the high molecular weight target DNA prior to PCR greatly enhances the amplification process (Abbott et al. 1988). Full denaturation of the chromosomal DNA does not take place at $94^{\circ}C$; therefore, the binding of primers is limited in the first critical rounds of amplification which greatly reduces the quantity of the final product. This is of particular importance when the target sequence is very rare as is the case with many asymptomatic carriers. Boiling of samples prior to PCR also reduces the amount of primer oligonucleotide, and therefore, the cost, which must be used to obtain a given signal.

PCR Amplification Protocol

A) Stock solutions and reagents for PCR and hybridization
     Solution #1 - 1 M Tris pH 8.3, autoclaved
     Solution #2 - 2% Sigma Gelatin (swine skin type), autoclaved
     Solution #3 - 3.73 gms KCl and 0.51 gms $MgCl_2$ (hydrous) in 80 mls
                   water (0.625 mM KCl; 31 mM $MgCl_2$), autoclaved

10X Taq Buffer -   Using aseptic technique, add 10 mls each of solutions #1 and #2 to solution #3, bringing the total volume to 100 ml. Filter through gene screen plus (Dupont, Boston, MA) aliquot and store frozen at $-20^{\circ}C$. Use each aliquot for a single set of reactions only, then discard. The concentrations of this final solution are: 500 mM KCl; 25 mM MgCl$_2$; 100 mM Tris pH 8.3; and 0.2% Gelatin.

20X SSPE - 174 gm NaCl
        24 gm NaH$_2$PO$_4$ (monobasic)
        7.4 gm Na$_2$EDTA
        pH to 7.4 with NaOH,
        QS to 1 liter

100X TE - Make a 100x stock with 60.5 gm Tris
        18.6 gm EDTA
        pH to 7.5 and qs. to 500 ml with distilled water

1X TE - 10 mM Tris pH 7.5
        1 mM Na$_2$EDTA

Nucleotides (dNTPs) - Nucleotides are purchased from Pharmacia as 100 mM stocks. Add 38 µl of each of the four dNTPs to 848 µl of water. The final concentration is 3.75 mM of each nucleotide in 1 ml.

Amplification primers- Dilute to a final concentration of 10 pmoles/µl, aliquot and store frozen in dH$_2$0 at -20°C.

Denaturation Solution - 0.5 M NaOH   50 ml 10 M stock and
                      1.5 M NaCl   300 ml 5 M stock up to 1 1.

Neutralization Solution - 0.5 M Tris pH 7.5   167 ml 3 M stock
                      1.5 M NaCl   300 ml 5 M stock up to 1 1.

Prehybridization
Solution -   150 ml 20x SSPE   To make 1x TE dilute 100x
                0.1 gm ficol   with dH$_2$0, filter through
                0.1 gm PVP   gene screen plus, aliquot
                0.1 gm BSA   and store frozen at -20°C
                1.0 ml 0.5 M EDTA pH 8
                12.5 ml 20% SDS
                qs to 500 ml

Hybridization -   Put 200 ml of prehybridization into a separate beaker and stir in 20 gm dextran sulfate while beaker is on a hot plate (medium-low heat). After it has gone into solution, transfer to a bottle and store at 4°C. Store prehybridization likewise.

10x TBE -   108 gm Tris
            55 gm Boric Acid
            9.3 gm EDTA
            qs to 1 liter

Gel Loading Dye -   25 ml water
                     50 ml glycerol
                     20 ml 0.5 M EDTA pH 8
                     0.1 gm bromophenol blue
                     0.1 gm xylene cyanol
                            pass through 3MM filter paper before use!

B)   PCR Reactions

1. Use 1 µg of DNA for experimental samples and negative controls; for positive controls use a range of dilutions which will give strong to weak signals (i.e. 4×10^{-4} to 8×10^{-6} of single copy DNA for HTLV-I). First determine how many µl of DNA are to be used. Next determine quantity of TE to be used. Do this by subtracting µl of DNA from 50 µl total volume, this value will be amount in µl of TE to be use.

        i.e. 216 µg/ml = 1/.216 = 4.6 µl = 1 µl

           .. use 5 µl DNA

        50 µl (total volume) - 5 µl DNA = 45 µl sterile TE

2. Use sterile, genescreen plus filtered TE for PCR runs. (When preparing new stocks aliquot TE first with clean room pipette and store at -20°C in clearly marked box). TE can be stored aliquoted in 2 ml sterile tubes in a -20°C freezer. Discard aliquot of TE when done! Do not put back in -20°C freezer!

3. Aliquot DNA and TE into 500µl Eppendorf tubes using positive displacement position pipetts (Rainin; Microman). Always pipet positive controls last. Mix DNA/TE dilution.

4. Boil DNA for 5 mins. Place on ice ~ 1 min. Spin down condensation in microfuge.

5. While DNA is boiling: prepare cocktail as below

- 10x Taq Buffer: stored aliquoted, like TE, in -20°C freezer 10 µl/sample (not reusable i.e. use once and throw away).

- dNTP's: 6 µl/sample (each individual should have their own aliquot stored at -20°C).

- Primers: use 10 pmoles of each primer to be used/reaction. Therefore, if amplifying with 1 primer pair you will use a total of 2 primers or 2 µl of primers per reaction, if you are amplifying with 3 primer pairs you would use a total of six primers or 6 µl of primers/reaction. Everyone should have their *own* primers! Do not use someone else's primers.

- Taq Enzyme: 0.4 µl/sample (2 units) put name on aliquot of enzyme, everyone has their own. Store at -20°C.

- 1xTE: calculate TE to allow for cocktail volume to equal 50 µl per sample. Also, it may be a good idea to allow for 1 to 2 extra samples, depending on total number of your samples to allow for pipetting errors.
i.e. to PCR 20 samples calculate for 22 samples.
. . 220 µl 10x Taq buffer
   132 µl dNTPS
    22 µl of each primer
    11 µl TAQ EDnz. 1x
   <u>693</u> µl TE
1,100   Total cocktail volume ÷ 50 µl = 22 samples

6. Add 50 µl of cocktail to each boiled DNA sample using clean room Pipetman. Pipet cocktail slowly onto side of sample tube. Do not immerse tip in DNA solution. Do not release plunger, ie. aspirate, until tip is completely withdrawn from tube and then release slowly and dispose of tip. Keep caps closed as much as possible.

7. Add approximately 75 µl of light mineral oil slowly down side of tube using plastic disposable pasteur pipets to overlay the reaction cocktail.

8. Take tightly capped tubes into PCR lab and place in thermal cycler (do not open tubes once in PCR lab to add any reagents; do not take back into setup lab after starting PCR). For all primer pairs listed in Table 1

the following ramping profile will produce a positive signal with 2-3 input molecules of target DNA following 30 cycles of amplification.

File X - ambient temp to 68°C - 1 sec   - 1 cycle

File Y - 68°C - 94°C 1 sec; hold 15 sec 94°C ⌐

          94°C - 53°C 1 sec; hold 30 sec 53°C | - 30 cycles

          53°C - 68°C 1 sec; hold 30 sec 68°C ⌐

File Z - 68°C - 10 minutes

C) **Liquid Hybridization**
1. Follow directions for PCR-DNA amplification.

2. After amplification, add 150 μl chloroform: isoamyl alcohol (24:1). Shake tubes vigorously. Let tubes set until 2 layers are visible. The amplified DNA will be the top layer.

3. Take out 30 μl of the PCR-DNA, place in another 0.5 ml Eppendorf tube.

4. Add 20 μl of liquid hybridization cocktail.

Hybridization Cocktail:
Probe - use enough probe to allow 250,000 cpm/sample
salt, 1.5M NaCl; 25mM Na_2 EDTA 1/10 total volume, i.e. 5 μl
TE - varies depending on volume of probe. Should use enough TE to allow 20 μl aliquots per sample.
Example for 20 samples: Probe = 118,769 cpm for 2 μl sample

                        ÷   2
                  59,384.5 cpm for 1μl sample

            need 250,000 cpm/sample
               250,000 ÷ 59,384.5 = 4.2 ~4 μl probe

            So set up cocktail for 20 samples:
            use 4 μl/sample probe - 80 μl
            use 5 μl sample salt -   100 μl
            Bring vol up with TE - <u>140 μl</u>
                            400μl total volume needed to
                                  allow for 20 μl
                                  aliquots/sample

5. Boil probe/amplified DNA samples for 5 minutes. Place in 55°C water bath approximately for 30 minutes.

6. Add 1/10 volume (5 μl) gel loading dye/sample

7. Run samples on a 8% polyacrylamide gel.
    Gel
    16 ml of 30% acrylamide gel
    6 ml of 10xTBE
    37.5 ml of dH$_2$0
    420 μl of 10% amonium persulfate
    50 μl of TEMED

8. Load samples on gel. Run at 200 volts for approximately 60 min. NOTE: with shorter probes it may be necessary to cool the gel or run at lower voltage to prevent denaturation of hybrid.

9. Autoradiography: cover gel with Saran wrap, place individually wrapped film (Kodak XAR-5 film) over gel and weight with glass plate. Expose at room temperature ≈ 2 hour - O/N.

D) **Spot Blot Method of Hybridization**

1. Take a piece of genescreen plus - cut out only what you'll need, do all cutting between the green interleafing sheets. Do not touch gene-screen plus with hands or powdered gloves. Using blue ball point pen only, draw circles where each sample will run.

2. Treat genescreen plus by soaking in water until wet, then pretreat in 2x SSPE for 10-15 min. Air dry genescreen plus until paper is damp only, not wet.

3. Dot each circle with 10 μl of PCR's DNA sample. Be careful when adding sample, genescreen must be fairly dry so sample will stay within outlined circle. If genescreen is saturated, sample will run outside circle.

4. Denature DNA sample on genescreen for 5 min. Do this by placing a piece of 3MM Whatmann paper on a clean glass plate. Saturate the Whatmann paper with premade denaturation solution. Place your

genescreen paper on top of wet Whatmann paper, making sure there are no air bubbles. Let set for 5 min.

5. Neutralize genescreen for 5 min. Do this by placing a piece of 3M Whatmann paper on a clean glass plate. Saturate Whatmann paper with premade neutralizing solution. Place your genescreen paper on top of wet Whatmann paper, making sure there are no air bubbles.

6. Dry genscreen by placing under white lamps for 15 minutes. Once genescreen is completely dry proceed with prehybridization.

    Prehybridization:
    150 ml - 20xSSPE
    0.1 gm - ficol
    0.1 gm - PVP
    0.1 gm - BSA
    1.0 ml - 0.5M EDTA
    12.5 ml - 20% SDS

    Bring up to 500 ml with ddH$_2$0
    Put 300 ml into a bottle labeled Pre-hybridization solution.
    Put 200 ml into a separate beaker and add 20 gm of dextran sulfate (keep in refrigerator). Stir into solution and pour in bottle labeled hybridization solution.

To Prepare:
1. Cut a heat sealable pouch to fit your genescreen paper. Put gene-screen in pouch, making sure three sides of the pouch are sealed. Put prehybridization solution (amount will vary on size of blot) in pouch. Seal 4th side. Prehybridize at 55°C for a minimum of 60 min; can go O/N.

2. Hybridization: Prepare hybridization solution by determining the number of ml of solution you will require, typically 4-5, and measure out in 15 ml orange cap tube. Add end labeled probe to the hybridization solution at the rate of 10^6 cpm/ml, i.e. for 4 ml hybridization add 4 x 10^6 cpm. Cut one corner or side of pouch and squeeze out prehybridization solution. Add hybridization solution with probe already added (amount will vary depending on size of blot). Seal pouch. Hybridize at 55°C for a minimum of three hours - can hybridize O/N.

Washes:
1. Wash at 55ºC, 30 min. with 2xSSPE, 0.1% SDS.
2. Wash a second time at 55ºC, 10 min. with (new) 2xSSPE, 0.1% SDS. After 10 min. check genescreen with hand held GM-counter. If gene-screen appears to be hot, place back in wash and continue washing until negatives appear clean.

APPENDIX 2

Each grouping represents a set of primers and probes which can be used in various combinations to analyze a given region of a given human retrovirus.

HTLV-I LTR

SG160 +PRIMER: CCCGGGGGCTTAGAGCCTCCAGT (51-74)

SG161 -PRIMER: GAATTCTCTCCTGAGAGTGCTATA (768-745)

SG162 -DETECTOR:
        TCAGGTAGGGCGGCGGGCGCGTGAAGGAGAGATGCGAGCC (415-376)

SG163 +DETECTOR:
        GGCTCGCATCTCTCCTTCACGCGCCCGCCGCCCTACCTGA (376-415)

HTLV-I GAG

SG166 +PRIMER: CTGCAGTACCTTTGCTCCTCCCTC (1388-1411)

SG295 +PRIMER: CTTACCACGCCTTCGTAGAA (1641-1660)

SG296 -PRIMER: TTCTACGAAGGCGTGGTAAG (1660-1641)

SG167 -PRIMER: CCCGGGGGGGGGACGAGGCTGAGT (1957-1934)

SG168 +DETECTOR:
        GACCCTTCCTGGGCCTCTATCCTCCAAGGCCTGGAGGAGC (1601-1639)

SG169 -DETECTOR:
        TGCCTTCTGGCAGCCCATTGTCAAGAGCTATGTTGAGGCG (1700-1661)

SG242 -DETECTOR:
    ATATAAGGCTATCTAGCTGCTGGTGATGGAGGGAAGCCAC(1451-1412)

HTLV-I POL

SG231 +PRIMER: CCCGGGCCCCCTGACTTGTC (2801-2820)

SG238 -PRIMER: CTGCAGGATATGGGCCAGCT (3037-3018)

SG232 +DETECTOR:
    CAGCCTGCCAACCACACTAGCCCACTTGCAAACTATAGAC (2821-2860)

SG237 -DETECTOR:
    GAACAGGGTGGGACTAGTTTTAAACCCTTGGGGTAGTACT (3010-2971)

SK54 +PRIMER: CTTCACAGTCTCTACTGTGC (3365-3384)

SK55 - PRIMER: CGGCAGTTCTGTGACAGGG (3483-3465)

SK56 -DETECTOR:
    CCGCAGCTGCACTAATGATTGAACTTGAGAAGGAT (3460-3426)
HTLV-I/II POL

SK110 +PRIMER: CCCTACAATCCAACCAGCTCAG (4757-4778) (HTLV-I)
            CCATACAACCCCACCAGCTCAG (4735-4756) (HTLV-II)

SK112 +DETECTOR:
    GTACTTTACTGACAAACCCGACCTAC (4825-4850) (HTLV-1)

SK188 +PRIMER: TCATGAACCCCAGTGGTAAA (4880)-4898) (HTLV-II)

SK115 +DETECTOR: CATAGCCCTATGGACAATCAACCACCTGAATGT (HTLV-I)
            CAAAGCCCTTTGGACTCTCAATCAGCTAAATGT (HTLV-II)

SK111 -PRIMER: GTGGTGAAGCTGCCATCGGGTTTT (4942-4919) (HTLV-I)
            GTGGTGGATTTGCCATCGGGTTTT (4920-4897) (HTLV-II)

HTLV-I ENV

SG219 +PRIMER: CCCCAGCTGCTATACTCTCACAA (5270-5292)

67

SG293 +PRIMER: CTTGTTCCTTAAAGTGCCCA (5521-5540)

SG294 -PRIMER: TGGGCACTTTAAGGAACAAG (5540-5521)

SG220 -PRIMER: CTCGAGGATGTGGTCTAGGT (5823-5804)

SG224 +DETECTOR:
   TCCTCATACCACTCTAAACCCTGCAATCCTGCCCAGCCAG (5301-5340)

SG225 -DETECTOR:
   ATATTGAGGCGTGAAACTTCTTGAGTAAAATTGACATCGT (5650-5611)

SG534 +PRIMER: ATCCTCGAGCCCTCTATACCATG (5796-5818)

SG535 -PRIMER: GCGGGATCCTAGGGTGGGAACAG (6127-6106)

SG228 +DETECTOR:
   GTCCAGTTAACCCTACAAAGCACTAATTATACTTGCATTG (5841-5880)

SG229 -DETECTOR:
   AGCTTGAATCTGGGGGTCAAAGCAGTGGGTCCAGTTAAAT (6050-6011)

HTLV-I gp21

SG608 +PRIMER: GTCACCTGGGCCCACCCTAGGATCCCGCTC (6101-6130)

SG277 +PRIMER: AAATTGCGCAGTATGCTGCC (6310-6329)

SG645 +PRIMER:
   CCACAAAGTCGACTCAAAATTGCGCAGTATG (6293-6324)

SG278 -PRIMER: TGGGAATTGGTAATATTCGG (6427-6408)

SG609 -PRIMER: GGCCTCTCGAGCTCACTGTGAGAGGCCAAG (6527-6498)

SG279 +DETECTOR:
   CAGAACAGACGAGGCCTTGATCTCCTGTTCTGGGAGCAA (6330-6368)

SB280 1 -DETECTOR:
   CTTGCTCCCAGAACAGGAGATCAAGGCCTCGTCTGTTCT (66369-6331)

HTLV-1 TAX/P40

SK43 +PRIMER: CGGATACCCAGTCTACGTGT (7358-7377)

SK44 -PRIMER: GAGCCGATAACGCGTCCATCG (7516-7496)

SK45 +DETECTOR:
    ACGCCCTACTGGCCACCTGTCCAGAGCATCAGATCACCTG (7447-7468)

HTLV-II - GAG

GE61 +PRIMER: GGGATTTGAATTCCTCCATTC (779-799)

GE62 -PRIMER: GCTGCTGGAAGTCGAAATCGGAGGGCC (942-916)

GE +DETECTOR:
    GCTATCAACCCACCACTGGCTTAACTTTCTCCAGGCTGC (860-898)

HTLV-II POL

SK58 +PRIMER: ATCTACCTCCACCATGTCCG (4198-4217)

SK59 -PRIMER: TCAGGGGAACAAGGGGAGCT (4300-4281)

SK60 -DETECTOR:
    CTGGGTTAAAGGTGGAAGTTACTTATGTGTCTGAGGGAAT (4276-4237)

HTLV-II ENV

SG638 +PRIMER: TCAAATTTATCCAGCTGACC (5799-5818)

SG639 -PRIMER: GCCACACTGCTAAAGGAACG (6125-6106)

SG640 +DETECTOR:
    GCATGGTTTGCGTGGATAGATCCAGCCTCTCATCCTGGGA (5841-5880)

HIV-1 LTR

SK29 +PRIMER: ACTAGGGAACCCACTGCT (501-518)

SK30 -PRIMER: GGTCTGAGGGATCTCTA (605-589)

SK31 +DETECTOR: ACCAGAGTCACACAACAGACGGGCACACACTACT

HIV-1 GAG

SK38 +PRIMER: ATAATCCACCTATCCCAGTAGGAGAAAT (1551-1578)

SK39 -PRIMER: TTTGGTCCTTGTCTTATGTCCAGAATGC (1665-1638)

SK19 +DETECTOR:
ATCCTGGGATTAAATAAAATAGTAAGAATGTATAGCCCTAC (1595-1635)

HIV-1 POL

SK32 +PRIMER: ACCTGCCACCTGTAGTAG (4316-4333)

SK33 -PRIMER: GCCATATTCCTGGACTACAG (4420-4401)

SK34 +DETECTOR:
TAGTAGCCAGCTGTGATAAATGTCAGCTAAAAGGAGAAGCC (4343-4383)

HIV-1 ENV

SK68 +PRIMER: AGCAGCAGGAAGCACTATGG (7801-7820)

SK69 -PRIMER: CCAGACTGTGAGTTGCAACAG (7942-7921)

SK70 +DETECTOR:
ACGGTACAGGCCAGACAATTATTGTCTGGTATAGT (7841-7875)

SK47 +PRIMER: GGATTTGGGGTTGCTCTGGA (8009-8028)

SK48 -PRIMER: CCAGAGATTTATTACTCCA (8087-8069)

SK49 +DETECTOR: GCATTCCAAGGCACAGCAGTGGTGCAAATGAGTT

ACKNOWLEDGEMENTS

The authors would like to thank Lynn Zaumetzer, Virginia Bryz-Gornia, Therese Dean, Barbara Jones, Mary Rubert and Mark Abbott for technical support, and Lori Raven for manuscript preparation. Work in this manuscript was supported by Contract No. NO1HB67021.

REFERENCES

Abbott M, Poiesz B, Sninsky J, Kwok S, Byrne B and Ehrlich G (1988) A comparison of methods for the detection and quantification of the polymerase chain reaction. J Infect Dis 1158-1169.

Barre-Sinoussi F, Chermann JC, Rey F, Nugeyne MT, Chamaret S, Greust J, Douget C, Axler-Blin C, Vezinet-Brun F, Rouzioux C, Rozenbaum W and Montagnier L (1983) Isolation of a T-lymphotropic retrovirus from a patient at risk for acquired immune deficiency syndrome (AIDS). Science 220: 868-870.

Bhagavati S, Ehrlich G, Kula R, Kwok S, Sninsky J, Udani V, and Poiesz B (1988) Detection of human T-cell lymphoma/leukemia virus-type I (HTLV-I) in the spinal fluid and blood of cases of chronic progressive meylopathy and a clinical, radiological and electrophysiological profile of HTLV-I associated myelopathy. N Engl J Med 318:1141-1147.

Byrne BC, Li JJ, Sninsky JJ, Poiez BJ (1988) Detection of HIV-1 RNA sequences by *in vitro* DNA amplification. Nucl Acid Res 16:4165.

Clavel F, Geutard D, Vezinet-Brun F, Chamaret S, Rey MA, Santos-Ferreira MO, Laurent AG, Douget C, Katlama C, Rouzioux C, Klatzmann D, Champalimaud JL and Montagnier L (1986) Isolation of a new human retrovirus from West African patients with AIDS. Science 233:343-346.

Duggan D, Ehrlich G, Davey F, Kwok S, Sninsky J, Goldberg J, Baltrucki L, and Poiesz B (1988) HTLV-I induced lymphoma mimicking Hodgkins's disease: Diagnosis by polymerase chain reaction amplification of specific HTLV-I sequences in tumor DNA. Blood 71:1027-1032.

Ehrlich G and Poiesz B (1988) Clinical and molecular parameters of HTLV-I infection. Clin Lab Med 8:65-84.

Ehrlich GD, Glaser JB, LaVigne K, Quan D, Mildvan D, Sninsky JJ, Kwok S, Papsidero L, Poiesz BJ (1989) Prevalence of human T-cell leukemia/lymphoma virus (HTLV) type II infection among high-risk individuals with type specific identification of HTLVs by polymerase chain reaction. Blood 74:1658-1664.

Ehrlich GD, Greenberg S, Abbott MA (1990) Detection of human T-cell lymphoma/leukemia viruses in PCR protocols. In: Innis MA, Gelfand DH, Sninsky JJ, White TJ (eds), A Guide to Methods and Applications, Academic Press, p. 325-336.

Kalyanaraman VS, Sarngadharan MG, Robert-Guroff M, Miyoshi I, Golde D, and Gallo RC (1982) A new subtype of human T-cell leukemia virus (HTLV-II) associated with a T-cell variant of hairy cell leukemia. Science 218:571-573.

Karpatkin S, Nardi M, Lennette E, Byrne B, and Poiesz B (1988) Anti-HIV-1 antibody complexes on platelets of seropositive thrombocytopenic homosexuals and narcotic addicts. Proc Natl Acad Sci USA 85:9763-9767.

Kwok S, Mack D, Mullis K, Poiesz B, Ehrlich G, Blair D, Friedman-Kien A, and Sninsky J (1987) Identification of HIV viral sequences using *in vitro* enzymatic amplification and oligomer cleavage detection. J Virol 1690-1694.

Kwok S, Mack D, Ehrlich G, Poiesz B, Dock N, Alter H, Mildvan D, Grieco M, and Sninsky J (1988a) Diagnosis of human immunodeficiency virus in seropositive individuals: Enzymatic amplification of HIV viral sequences in peripheral blood mononuclear cells. In: Luciw PA, Steimer KS (eds)

Genetic Engineering Approaches to AIDS Diagnosis, Decker Inc, p. 2410-253.

Kwok S, Ehrlich G, Poiesz B, Kalish R, and Sninsky J (1988b) Enzymatic amplification of HTLV-I viral sequences from peripheral blood mononuclear cells and infected tissues. Blood 72:1117-1123.

Kwok S, Ehrlich G, Poiesz B, Bhagavati S, and Sninsky J (1988c) Characterization of HTLV-I sequence from a patient with chronic progressive myelopathy. J Infect Dis 158:1193-1197.

Mack DH and Sninsky JJ (1988) A novel approach to the identification of new members of known virus groups: Hepadnavirus model system. Proc Natl Acad Sci USA 85:6977-6981.

Manzari V, Gismondi A, Barillani G, Morrone S, Modesti A, Albonicil L, De-Marchis L, Fuzio V, Gradilone A, Zari M, Frati L, Santoni A (1987) HTLV-V: A new human retrovirus isolated in a tac-negative T-cell lymphoma/leukemia. Science 238:1581.

Mueller PR and Wold B (1989) In vivo footprinting of a muscle specific enhancer by ligation mediated PCR. Science 246:780-786.

Ochman H, Gerber AS and Hartl DL (1988) Genetic applications of an inverse polymerase chain reaction. Genetics 120:621-623.

Poiesz BJ, Ruscetti FW, Gazdar AF, Bunn PA, Minna JD and Gallo RC (1980) Detection and isolation of type C retrovirus particles from fresh and cultured lymphocytes of a patient with cutaneous T-cell lymphoma. Proc Nat Acad Sci USA 77:7514-7419.

Poiesz B, Ehrlich G, Papsidero L, and Sninsky J (1988) Detection of human retroviruses. In DeVita V, Hellman S, Rosenberg S (eds), AIDS: Etiology, Diagnosis, Treatment and Prevention, Lippincott, Philadelphia, P. 137-154.

Ratner L and Poiesz B (1988) Human T-cell lymphotropic virus type I associated leukemias in a non-endemic region. Medicine 67:401-422.

Wells KH and Poiesz BJ (1990) Detection, molecular biology and treatment of retroviral infection. Amer J Dermatology, in press.

Wells KH, Byrne BC and Poiesz BJ (1990) Detection, prevention and treatment of retroviral infections. Seminars in Oncology, in press.

DISCUSSION

Bone D (Dupont Company, Wilmington, DE):

You talked about finding HTLV-II in a higher proportion of domestic individuals than people thought. You talked about IV drug abusers. Is there anything else that you can add about the frequency with which HTLV-II is found in this country?

Poiesz B:

We're involved in several large clinical trials of IV drug abusers from around the country, and endemic populations from around the world. Also, with Al Williams at the Red Cross, we are already involved in evaluating the DNA from those persons who have scored positive in the HTLV-I serology screen, done in different blood banks across the country. The overall finding in the blood banks and the IV drug abusers, particularly in New York City, is that about half the patients are HTLV-I positive and half of them are HTLV-II positive. That varies from what part of the country you go to. In New York City, most of the blood donors who score positive are actually HTLV-I positive. Here in Alameda County, most of them are HTLV-II positive. We've done that with Jim Lipka here at Stanford. So it seems to vary. We still don't know what are the endemic groups for HTLV-II and it's a little difficult to study it in the IV drug abusers. We find it in black, hispanic and caucasian IV drug abusers. In one meeting we tried to sort out where they were all abusing their drugs. We had all the shooting galleries enumerated, etc. Our hope was that we could find the endemic group for HTLV-II. That is clearly difficult in that group. I have a feeling that when the demographics are looked at in the blood donor population, we'll get a clue as to what groups should have HTLV-II in them or not, and then study them further and confirm this. The other plan is to bring in their family members and to start looking at these people clinically such that we can see what diseases they might have. We ourselves, upon screening large populations have identified two more HTLV-II positive people that have a disease that was first described by the group at UCLA, Irwin Chen and his colleagues. They had a CD-8 leukemia, are neutropenic, CD4-penic and they are anemic and died from pancytopenia. Their CD-8 cells type out as NK cells and they are LGL's. The virus copy number does not predict, nor to the Southern blots, that their leukemic cells are positive for the virus. So, it does not seem to be the leukemia population that's infected. We suppose that it is the CD4-penic population, but we don't know this for a fact. That prompted us to do a study on LGL leukemia. We thought that it was a lead. We looked at 20 from Seattle and those two were the only positive patients. So, there is a fair amount of HTLV-II in the American population. But, what it's doing and who all has it, I think will take the next year or two to shake out.

de la Maza L (University of California Irvine Medical Center, Orange, CA):

Bernie, if you have a patient, let's say a high risk patient, that comes to see you and you test for antibodies, the antibodies come out negative for HIV-1, you test for a p24, it comes out negative, then you go ahead and do a PCR and it's positive. How will you handle that patient?

Poiesz B:

Number one, we have observed that phenomenon. I would say it seems to occur more frequently when you look for HTLV-I than when you look for HIV-1. We looked at 4,000 AIDS and ARC patients, and could only find four that were seronegative and PCR positive. We looked at 500 ATL patients in Japan and the United States and found seven who were seronegative and PCR positive. We also are finding seronegative asymptomatic patients PCR positive, particularly children born to already defined infected people and in the IV drug abuser population. Again, this phenomenon is more common for HTLV-I and HTLV-II than for HIV-1.

Of course one has to prove that such a PCR result is not a false positive. What we do clinically, first to sort that out is, again, we have multiple aliquots of that sample. We pull out the aliquot from that person that hasn't been used yet, and run it again to make sure it's positive. Ultimately, we can sequence the amplified product now and compare it to known viral sequences. We now know that there are base changes in all the human retroviruses that occur with a relative frequency. If you go through the envelope region, every isolate is quite distinct from the other, not just for HIV-I but on HTLV-I and II as well. If you look at their envelope genes, you will find between any one person and another, over a 200 base stretch, maybe 3 base pair changes. If our PCR-positive seronegative sample has a unique sequence then it cannot be attributed to carryover. The best method is to get another sample from the patient and run it again.

We're finding that, as expected, because of the epidemiology in Japan, many of the seronegative children of HTLV-I infected people are PCR positive. And that had to be. If they are seronegative up to puberty in Japan, the culture is such that they are not picking up virus at that time through sexual or drug promiscuity. Whatever is in them they acquire prenatally, and we presume that it's just that the virus is that latent and they have not made antibodies. We have also found and have published in Blood this past month, that many of these HTLV-I seronegative patients actually make antibody to tax antigen. Tax antigen, the nuclear binding protein that is an up-regulator of HTLV-I is not packaged in the virion, hence, it is not present in the ELISA assay formats and the routine Western blot formats. It's present if you do RIAs on cell lysates, and that's how we picked this up. With Triton here in the Bay Area, using recombinant tax, we proved that people have antibody to tax, and they were repeatedly PCR positive. You can obviate the false negative ELISA assay by simply adding the tax to the antigen prep and then getting them positive.

So the phenomenon of seronegativity occurs. In people with disease, it's a relatively low frequency event. We're now looking at mother-baby pairs for both HIV-I and HTLV-I in New York City and in Central New York to track how long it takes for them to seroconvert versus how long it takes for them to be DNA positive and RNA positive. Because we can do the RNA PCR, we want to prove that the reason that they are regulatory antibody positive first, is because the RNA transcripts of those regulatory genes are made first. The hypothesis is that maybe the down regulators are synthesized first and suppress viral replication, and only when the up-regulators come on do you get the rest of the viral proteins. That is an hypothesis that is testable using the PCR.

de la Maza L:

So, at this point, you would basically advise the patient to come back in x number of weeks and repeat the test, or how are you handling those patients?

Poiesz B:

We repeat the test. We also calculate a copy number too. You have to pay attention to your copy number and not get fooled by an insensitive assay yielding a false negative result. In each PCR run we analyze a standard dilution series to determine that the assay was sensitive enough to detect a certain minimal DNA copy number. Again, because we in effect are using some of this data clinically.

Because all these events are happening. I see the patients myself. They go to the Red Cross, they get tested and then they come back, and they are using the information clinically. You are advising them what to do. You may tell a person not to have unprotected sex with his wife because he harbors the retrovirus. It's a big discussion. We already do it a lot with HIV infected patients. We have to do it with HTLV-I and HTLV-II infected patients as well. So, we had better be right about it. So, I segregate at the lab. Upon doing that, I feel relatively comfortable that the signal was positive. For HTLV-I we require that the patient be positive for three individual primer pairs run on a separate sample, done on a different day. The probability of getting carry-over into that is very small, and quite frankly, we haven't found anybody positive now, using that routine, that doesn't make sense. It all makes sense that they have it. Once finding them positive in doing it a few times, we advise them about what we know about their disease, association with the particular retrovirus, and we advise them about safe sex, not to donate blood and that their body fluids might be biohazardous.

ROUND TABLE: POLYMERASE CHAIN REACTION (PCR)

de la Maza, L:

How are you handling the carry-over problem?

Poiesz B:

If you separate the room and the personnel where you're doing the actual PCR and where the people are handling the amplified product from the room and technical staff that are setting up the PCR reaction, the probability of the amplified product getting into the reaction is virtually nil. Our clinical specimens which are usually either plasma or whole heparinized blood, come into a separate room. We are a large operation, and handle thousands of samples a week. We segregated putative HTLV-I positive, putative HIV positive, and normal donors into separate biohazard hoods, pipette aids, etc. Then the sample plasmas go to one room to be stored and handled later, cause we are doing reverse transcriptase (RT) directed PCR for RNA in the plasma, and to my surprise, there actually is a fair amount of detectable retroviral RNA in the plasma of these patients which allows for retrospective studies on plasma samples. The cells that get frozen down for DNA go into a different room. They're handled by a group of technicians that either freeze them or count the cells and put them directly into lysis buffer or they do chloroform phenol extraction. Our primer pairs are all made in a separate room upstairs by a different group of technicians, and our retrovirus cultures that are maintained over time are in a completely different building. We've segregated everything. We've run in these large clinical studies primer pair controls, and we have pedigreed normal blood donors from the Syracuse Red Cross that we've studied now consistently over time. We know their demographics, their behavior, etc. They get placed as alternate samples throughout the process. If they were to start coming up positive for primer pairs that they were previously negative for, then something has gone wrong with the system. It took a while, emotionally, to convince everybody. If you have a smaller lab, it's hard. If you have to set up everything and do it, and you don't have an army of technicians, you do indeed, have to do something different. We have the technicians gown up, glove up, they wear surgical caps and surgical gowns and take them off at the end of the day. When the same technicians were doing the complete process, we didn't have them amplify in the morning and then go back and set of the same PCR in the afternoon. In my opinion, the more you do it and the closer you have everything, the more the likelihood of it occurring. There are techniques to minimize and get rid of stuff from reagents, passing your reagents through "Gene screen" to get DNA is an attempt to make them relatively amplified

Medical Virology 9
Edited by L.M. de la Maza and E.M. Peterson
Plenum Press, New York, 1990

product-free. We are working on techniques to allow you to identify carry-over when you have it, but a lot of it is in the experimental stage.

Manos M:

Bernie has focused on carry-over from PCR product, and that is the major problem, as I mentioned in my talk. I also want to bring your attention to sample preparation and cross-contamination between samples. Because if you have a very hot sample, in terms of your infectious agent, that can act as a vehicle for a different type of carry-over. So, it's important to strategize in terms of exactly how you are going to physically set up every technique you do. We even talk about where each rack will be and where your arm goes and where you throw things away. Really think about aerosols, think about when you're going to open and close tubes, think about moving things to separate racks, think about spacing tubes within racks. There are a lot of little things like that that sound incredibly boring, but they're very important. That's important at every step of this whole process. From logging the samples into your lab, to preparing them for PCR, to doing the PCR, that type of care has to be taken.

Chang R (University of California, Davis, CA):

I want to ask Dr. Manos. This morning you showed us a specimen that had three types of papillomas. How do you know if there is no carry-over?

Manos M:

That's a very good question. I must say, I did sneak a look at the questionnaire of that particular patient, and I wasn't surprised. The only thing we can do in those situations, as was described earlier, is to include many negative controls. If there is no carry-over in any of those, then we assume the experiment was clean. Also, we took two samples from each patient, a cervical and a vulvar swab. That particular person had the same pattern of the three different size PCR products on both of those swabs. We processed those swabs on different days in different areas. They were amplified on different days, etc. That's about all we can say about that.

de la Maza L:

I know that one of the approaches that some people are taking to minimize specimen handling is to have all your reagents aliquoted. My question is, what is your experience if you aliquot Taq, the oligos and everything and freeze it at -70ºC, and then just pulling a vial and adding it directly to the patient sample. Do you have any experience with that and what do you recommend?

Manos M:

I can tell you that for several of the systems being used at Cetus, I know for the DQ-α system, and I think now for the HIV system, SK-38 and 39, those PCR reaction mixes have been stored for one to two months successfully. With our particular system, that doesn't work as well, although we are able to store things for a few weeks, and we do that routinely, for the larger studies. I think it just depends on how stable your primers are, what kinds of concentrations of magnesium you might have in there, and also the concentration of your primers. I think that is the difference between those systems. But, I think you should try it.

de la Maza L:

Are you storing at -70°C?

Manos M:

Actually, I think the work with the DQ-α, etc. was done only at -20°C, and that was successful.

de la Maza L:

Bernie, are you using that approach?

Poiesz B:

For some primer pairs, as was mentioned. If all you were going to screen for was say, SK-38, 39, which is the conserved gag primer pair that most of us now use for HIV-I, that stores fairly well. It's when you get into all these multiplex primer pairs that storing them for all of them, and the exact combination starts to fall down. They don't all, for some reason, store as well. I'm not sure I totally understand that. We didn't get into the concept of primer dimers and all the funny things. If there is homology, the primers themselves to each other, sometimes they'll come back on each other. I think the only suggestion you could make, if you want to do that in any system that you set up for this, is set it up, take out known positive and negative controls and dilution series, store the reagents over time and prove to yourself that it actually behaves the way you want it to behave. You almost have to, after a while, set this up like you are really running a clinical pathology lab. I think it has to be given that rigor. It's so complex. If you set up the routine, as was mentioned, it helps. Molecular biologists and researchers rebel a tad against that concept, I would say, even in our own little merry group. They're not used to doing it that way. But, it works out better in the long run.

Porter-Jordan K (The George Washington Medical Center, Washington, D.C:

We found that our principle contamination problem was not amplified fragments or larger fragments, but actually smaller degrading fragments. Most of which, I suspect, were produced by autoclaving which just breaks down DNA.

Manos M:

I haven't tried nesting in our system. As I think I have mentioned to you, it's very difficult to do that in the area of the genome that we're working in because we need consensus primers. So, that's not an option. I think you're right though, in the situation that you're describing. Nesting could be useful. I think different laboratories have different problems to tackle in terms of the major form of contamination. There are a lot of labs who want to do PCR who have been working with HPV plasmids for years, and they don't have the option of changing labs. So they then have to worry about that contamination as well as all the PCR products once they get started. Again, the nested primers isn't necessarily going to address that. I think it may be useful for the thing that you're describing, especially with what I believe is from the autoclaving?

Poiesz B:

I agree with what's said. You're right, the autoclave doesn't get rid of it. They're all different size pieces in the autoclave at the end of it all, and it comes from all your different sources. Again, the nesting primer pairs located within the original primer pairs tends to eliminate that type of false positive. It works for a while when everything is very conserved. If you're using the process to look for variance, as we're using it to look for variant retroviruses, etc., the sequences may be off in the nest. You've gotten them positive with the original primer pairs, and using the nested primer pairs and they don't work. Then you get confused. Depending upon your purpose, nested primer pairs are a very efficient way to eliminate some carry-over.

de la Maza L:

Bernie, what method of detection are you using? Are you using acrylamide gels and ^{32}P, or what are your doing?

Poiesz B:

We have philosophical debates in our laboratory on different projects. Certain folks like the slot blot screen. They just screen through a series of specimen and then they confirm on liquid hybridization and run the product down the gel and take a radiograph of that, and confirm that the piece was actually of a certain fragment, or do a Southern blot on that specimen. Routinely, we're either doing liquid hybridization, taking an autoradiograph of the size pro-duct, or we're doing slot blot to screen. The slot blot we've rigged up just for convenience to allow us to quantitate the product, but you can actually take the audioradiographs of the the liquid hybridization analysis and quantitate the positive bands on a densitometer. I am of the mind that the original analysis where we showed that some of the primer pairs allowed us to look at amplified product that is not highly homologous with our original intended sequences is important in the search for new human retroviruses or distinct variant. In some of our studies where we're looking for novel retroviruses, we screen by direct incorporating and doing either liquid scintillation counts, essentially a DNA polymerase assay if you will, or doing the radiographs of the amplified product. If we see an interesting piece, we pull that out, then we will amplify that, clone it, sequence it, and the plan is, if we have a novel sequence, use the inverse PCR to further identify it and take that novel sequence and go back in and try to determine the true copy number in the sample. If it's an endogenous retrovirus, it should be there at least one copy per cell. If it's less than that, when you actually have the true sequence, then maybe it's exogenous. That's a different use than what other people might use the system for. But that to me, would be casting a very broad net in a search for new infectious agents. You can turn up or down the gain of that stringency just by what ever hybridization technique you use.

Participant:

We work in a relatively small laboratory and would like to start doing PCR. We just started doing a few runs. What kind of recommendations would you have for a small laboratory that doesn't have a lot of people to segregate everything into different areas? Is there anything we can do to minimize carry-over?

Poiesz B:

Everything that we said about separation of at least the aliquoted material is important. This probably does work better if you set up a laminar air-flow hood that's vented, say to the outside via fume hood, and there's a barrier between most of your body and the samples being tested.

Participant:

We have access to several of those, but we just don't have the personnel or the space.

Poiesz B:

Then, you're really going to have to make an effort, I think, to make sure that their hands are free of amplified product. There's a lot of work going on that would be able to prove for sure whether you have amplified product in there or not. I showed you one little potential control, if you got a positive result you can at least size the DNA where it originally came from. You can always sequence it to prove it. It's a fair amount of work. There's a lot of work going on to make that type of phenomenon, proving whether you have carry-over or not much easier. But right now, if you set up your controls and your controls start coming up positive, then you know you have a problem and you have to go back and look at it. You have to have a primer negative control where you have not put any DNA. Because if you're getting any of that, some of your reagents might be contaminated.

Manos M:

I think there are some rather inexpensive ways to set up a small room. We were able to put most of our PCR setup in a laminar flow hood in another room. We do not use the fan on the laminar flow hood because that just stirs things up. We use it, essentially, as a still-air box with a UV light. Then the group got too big. That brings me to how to do this inexpensively in a small area. I would either purchase one of those incredibly over-priced still-air tissue culture boxes that have a UV light, small doors that you can work in which are similar to a laminar flow hood, or I would have one made. You could probably have three of them made for the cost of one from a supplier. That can just be put on a bench top in a clean area of the lab where there isn't a lot of traffic. So we now use that as a secondary hood in our regular lab and it's working beautifully. We don't do any PCR product work near that. We keep gloves and all the reagents in there. We have separate cabinets in the room assigned for storage of things for PCR. A certain area of a freezer, part of a cold box, etc. That's working well and I think that's a good option. Again, your gloves need to be stored in there, etc., and you have to have a different grade of paranoia. You don't go answer the phone and then not change your gloves. We go through, I don't want to tell you how many, boxes of gloves per day.

Poiesz B:

I took our people to the operating room. All the molecular biologists. I took them to the OR and showed them what the chief scrub nurse does to anybody that touches their gloves on something that they're not supposed to. How they have to directly re-glove. I think that's it. You can't go taking your gloves and touching everything in the room. You have to act like you're at a surgical specimen. If you pull out of there, take your gloves off and throw the gloves away. Go do your business. Go back in there, put a new set of gloves on. It's that important.

Participant:

You talked about doing some tissue samples and cutting sections and so on. How do you clean your microtome off or how do you prevent contamination of samples?

Manos M:

There are a couple of options for that. If you're rich enough to use disposable microtome blades, and I highly recommend it, which some of our collaborators do. If not, those are cleaned with xylene or ethanol. The thing that you need to be most aware of with that is not just the blade, but you know how a lot of tissues tend to fragment when you slice them. It's those little pieces that come off then, that are flying around the room that you've got to watch out for. You also have to be very careful with the order in which you cut things. For us, we cut the high copy-number positive control last, then clean the whole area. We use toothpicks to pick the slices off and put them in tubes, so those are disposable. Those are the only suggestions that I can give you except that you also have to train the person who is doing the cutting for you.

de la Maza L:

Bernie, you brought up the issue of when you do a multiplex reaction, for some reason, sometimes some of the primers work, and some other ones don't work. Do you have any hints about why sometimes they work and why sometimes they don't work.

Poiesz B:

They work at relatively different efficiencies. We've tried to put the primer pair concentrations in different amounts to make each sample, each virus, or each portion of the virus amplified at relatively the same efficiency. That got to be a little bit mind boggling when we took dose response groups, cause they all had slightly different curves. The clinical samples were occurring with different copy numbers of one virus versus the other. So when we're looking at patients that we know might be multiply infected, or we're looking to see whether a person is positive for four different portions, say of the HTLV-I genome, it became simpler to just accept that they do work at different efficiencies. If there's a lot of one virus in there, we also know. Suppose you came upon a patient with adult T-cell leukemia that had HTLV-I that was also infected with HIV, and they had a 70,000 white count and they're leukemia cells. There are so many copies of HTLV-I. One copy per cell versus the very rare copies of HIV, that a lot of the substrate, depending upon how much you put in and you're willing to spend on it, gets committed to the amplification of the HTLV-I. So, what we found is, if we really want to be rigorous, and it depends on what the purpose is, if we get one of them positive, we have to go back and amplify the others. If they're all negative, that usually means that they weren't there at all, within a certain sensitivity. We know our ability to detect each one of them and each one combined with each other in equal amounts, because we do the standard curves of it. There probably are other reasons that I haven't even gotten into. We haven't even sequenced all the funny little products that get made when you do the PCR.

de la Maza L:

Have you tried to do any quantitation Michele, with the HPVs?

Manos M:

I think the system that we're working with doesn't lend itself very well to quantitation. It's a nice idea, and I think there are some very elegant ways to quantitate back to clinical material. But if you think about what a cervical swab is or what a genital swab is, there is so much randomness in terms of taking that, how many cells you get, whether you hit a lesion or not, whether you hit an infected area of not, etc. I think you can quantitate, but then what does it mean? Possibly quantitating in a cervical lavage would be more useful, but then again, I still think there is this randomness built into what we are doing, that it's not necessarily useful. There is a possibility of doing quantitation on biopsies, and that's something we've been thinking about, but haven't done yet.

de la Maza L:

When you work with your HIV samples, have you found a significant difference in sensitivity between those samples that you do chloroform phenol versus those ones that you just boil?

Poiesz B:

We actually prefer to do a quick lysis, using a quick lysis buffer for DNA and we take an equal amount of cells and use the quick lysis buffer for RNA. Number one, it allows us to put more cells in. When we've put a lot a cells and then chloroform phenol extracted DNA, as we get up to 10, 15, 20 µg, it seems that we pull an inhibitor along with it. Some of the inhibitors are dialyzable. You can dialyze it out. But, others are not, and we're not quite sure what they are. It seems to work easier putting 1 million to 2 million cells into a small volume if we do the quick lysis. I would have to say though, what we've done is to take known clinical samples and cell lines and store chloroform phenol extracted material versus quick lysis prepared material, and bring it out and test it against a standard curve, relative to each other over time. At least for the sequences that we're looking at so far, it seems that the chloroform phenol extracted material stores better. We lose signal over time in material stored in the quick lysis buffer, particularly the RNA. That's all at -70°C. There may be a difference there. If you're going to do it right away, quick lysis seems, at least in our hands, to be a very easy and simple way to do it. Red cell contamination seems to inhibit the PCR reaction. The more red cells you have, whether it's the hemaglobin or some other molecule, is unclear, but red cells seem to dampen your reaction. So, how well your ficoll-hypaque separation of your peripheral blood mononuclears is determines what you get. That was why we either use β-globin or the HLDQ-a signal to normalize HIV signal. Because we can count so many CD4 or PBMs, if you will, but we're not necessarily counting the red cells. Now it looks like it's better if we actually lyse the red cells up front, and then just take your PBMs.

de la Maza L:

Is that also your experience with the HPV, from the point of view of specimens?

Manos M:

I agree with the conclusion that quick prepped material doesn't store as well as clean DNA But in our experience on genital swabs, we were just boiling water or PCR buffer versus doing a quick proteinase digestion in the presence of non-ionic detergent. The amplification signal was one to two orders of magnitude better with the protease digestion. So, we have stuck with that. As far as going through phenol chloroform, again, I am of the "no manipulation of sample unless necessary" school of thought, so we avoid that.

Al-Nakib W (Kuwait University, Kuwait):

I just want to ask whether you include any RNA inhibitors when you are doing reverse transcriptase?

Poiesz B:

We have not, but whether that would improve it or not is unclear. What we would like to do is to test that out to see if it would increase out detection rate and improve the storage capacity of the material, but we haven't done that as of yet.

Al-Nakib W:

We have found that this is crucial for viruses such as rhinoviruses or polioviruses.

Poiesz B:

I wouldn't be surprised. That sounds reasonable.

CLINICAL SEROLOGICAL AND INTESTINAL IMMUNE RESPONSES TO ROTAVIRUS INFECTION OF HUMANS

Ruth Bishop, Jennifer Lund, Elizabeth Cipriani, Leanne Unicomb and Graeme Barnes

Department of Gastroenterology
Royal Children's Hospital
Melbourne 3052, Australia

INTRODUCTION

Rotaviruses are one of the most common infectious agents with the potential to cause disease in humans that are encountered during the normal life span. Infections of the intestinal tract have been documented in newborn babies, in young children, in adults and in the elderly (Bishop, 1986; Hjelt, 1988; Hrdy, 1987; Steinhoff, 1980). The clinical consequences of infection differ in different age-groups (Rodriguez et al. 1987).

In full term, healthy neonates rotavirus infection usually results in asymptomatic or mild disease (Bishop, 1986). For most children the primary rotavirus infection occurs in the first two years of life and may then result in severe disease. Rotaviruses have been shown to be the single most important cause of acute diarrhoea requiring admission to hospital of children aged 6-24 months throughout the world (Bishop, 1986; de Zoysa and Feacham, 1985; Kapikian and Chanock, 1985). In a Washington DC study one child in 272 (3.7/1000 per year) under one year of age and one in 451 (2.2/1000) 12-24 months of age were hospitalized for rotavirus diarrhoea annually (Rodriguez et al. 1980). Rotavirus infection is also common in young children with diarrhoea not requiring admission to hospital (Koopman et al. 1984; Pitson et al. 1986; Rodriguez et al. 1987). Many of these children require outpatient medical supervision in addition to close parental care in their homes.

Rotavirus infection is an equally important cause of diarrheal illness in developing countries. In a comprehensive study of the etiology of severe, often life-threatening diarrheal illness in Bangladesh, rotaviruses were found to be the single most important pathogen in children less than 2 years of age (Black et al. 1980) and rotavirus infection was more likely to result in death than other agents (Black et al. 1981). It was calculated that if fluid replacement therapy were not available, mortality due to rotavirus infection in Bangladesh would reach 2.9 deaths per 1000 children under two years of age per year. A recent epidemiological study in the United States concluded that deaths of young children due to rotavirus infection may be more common in developed coun-

Medical Virology 9
Edited by L.M. de la Maza and E.M. Peterson
Plenum Press, New York, 1990

tries than has been realized, even though treatment by replacement of fluid and electrolyte losses is readily available. Ho et al. (1988) identified approximately 500 deaths per year due to diarrhoea of non-specific aetiology in children less than 5 years of age. The age of children affected, the clinical symptoms and the seasonality (winter) caused them to speculate that many of these deaths may have been due to rotavirus infection.

After 2 years of age the clinical consequences of rotavirus infection usually diminish in severity and most children possess serum antibodies (Gust et al. 1977; Sack et al. 1980; Yolken et al. 1978). Reinfections are common but are usually associated with less severe symptoms (Bishop et al. 1983; Chiba et al. 1986; Grimwood et al. 1983; Gurwith et al. 1981; Mata et al. 1983; Rodriguez et al. 1987). There is a high rate of mild or asymptomatic disease in older children, and in adults in contact with infected children (Grimwood et al. 1983; Ryder et al. 1985; Sack et al. 1980). Adults over 60 years of age appear to have higher attack rates and more severe clinical illness than younger adults (Holzel et al. 1980).

The relatively mild spectrum of clinical illness caused by rotavirus infection in newborn babies and in older children and adults, compared with severe disease observed in children aged 6-24 months, suggests that the major determinant of symptoms is the immune status of the affected individual. Age-related resistance to illness may be partly due to non-immune mechanisms inherent in maturation of the gut (Wolf, 1981) However most studies imply that the age-related acquisition of clinical immunity is determined by previous infection(s).

This review will focus on the development of clinical immunity in humans and the immune responses measurable in serum and in the intestine. Results of studies of immunity in animals will not be reviewed, except where results indicate areas remaining to be studied in humans.

CHARACTERISTICS OF ROTAVIRUSES

The rotavirus genus of the Reoviridae consists of a number of antigenically related and morphologically identical viruses that infect the intestinal tract of man and other mammalian and avian species (Estes et al. 1983; Flores et al. 1986). Rotaviruses were first described as causes of diarrhoea in newborn mice (Adams and Kraft, 1967), calves (Mebus et al. 1969) and young children (Bishop et al. 1973) using electron microscopy of infected intestine.

Under natural conditions these viruses are species specific, although interspecies infectivity can be produced experimentally. Using negatively stained preparations observed by electron microscopy, complete rotavirus particles appear as approximately 70 nm in diameter. They comprise an icosahedral core with an inner capsid and two outer capsids composed of radiating capsomers forming spokes (Latin rota, a wheel). The outer capsid is often missing and single shelled particles with a serrated edge can be seen in most preparations. Only complete double shelled particles are infectious.

Electrophoretic analyses show that all rotaviruses contain 11 segments of double-stranded RNA, and strains can be distinguished on the basis of differences in the migration patterns of the genome segments. Some human viruses exhibit an altered migration of the RNA segments 10 and 11 termed "long" and "short" electropherotypes.

Two major polypeptides, VP_2 and VP_6 (coded by genes 2 & 6 respectively) have been consistently observed in all rotavirus strains analyzed to date. These two proteins have been reported to comprise about 80% of the total virus protein composition and are both located on the inner capsid. The inner shell protein VP_6 comprises the common rotavirus group antigen as well as bearing the epitopes that define the two rotavirus subgroups. The polypeptides of the outer capsid (VP_4, VP_7) both elicit neutralizing antibodies. The combined neutralizing antibodies formed to VP_4 and VP_7 are titrated by plaque neutralization or inhibition of fluorescent cell-forming units. On the basis of differing neutralization titres using hyperimmune sera rotaviruses of human and animal origin are presently classified into at least nine serotypes, of which serotypes 1, 2, 3, 4, 8 and 9 are found in human infections (Hoshino et al. 1984). Serotypes 1 and 2 have been identified only in human infections. Serotypes 3 and 4 infect humans and other animals. Serotype 3 strains appear to be widely distributed in the animal kingdom. The host range of serotypes 8 and 9 is not known at present.

The original classification of rotaviruses into serotypes detected both VP_7 and VP_4 simultaneously. A re-evaluation of serotype classification is being undertaken at present, using monoclonal antibodies specific for either VP_4 or VP_7. Mabs specific for VP_7 of serotypes 1, 2, 3, 4, 8 and 9 human strains have been incorporated in ELISA systems and used to serotype strains directly from stools. Mabs specific for VP_4 have been more difficult to develop and many have proved to be either strain specific or so widely cross-reactive that antigenic differences cannot be demonstrated. It is possible that a binary system of classification of rotavirus strains (similar to that for influenza) will be developed to identify differences in VP_4 and VP_7 separately. At present, identification of VP_7 using specific Mabs shows a close correlation with serotype assigned using absorbed hyperimmune antisera.

Assays for antibodies to VP_4, VP_6 and VP_7 are important in studying immune responses to the virus. ELISA now forms the basis for studies to detect and titrate VP_6 antibodies formed in response to infection and is also adaptable to studies of the class of immunoglobulin (IgG, IgM, IgA, secretory Ig) formed in response to infection. Assays for antibodies to VP_4 and VP_7 separately are being developed at present.

Several "rotavirus-like" organisms have been identified in human and animal infections. These are morphologically identical with rotaviruses but lack any of the rotavirus antigens, including the group antigen VP_6. Their electrophoretic RNA pattern differs from those of conventional rotaviruses. The existence of these "rotavirus-like" particles has caused reclassification of the genera into Groups A, B, C etc. Only Group A rotavirus infection will be discussed in this review. The extent to which Groups B and C infections occur in humans has yet to be determined.

PATHOPHYSIOLOGY

Most studies of pathogenesis of rotavirus infection have involved experimental infection of newborn animals. Where human infection has been studied, the results accord with those seen in animals. Oral infection can be established with very small doses of infectious virus. The minimal infectious dose in pigs using porcine virus has been calculated as 1 PFU/piglet (Graham et al. 1987), in adult humans given homologous virus as a faecal extract as 1-9 PFU/person (Ward et al. 1986) and in mice given EDIM as 1×10^2 PFU (Ramig, 1988). In the latter model the titre of virus identified in the intestine, and the severity of disease induced, increased in parallel with the virus dose (Ramig,

1988). The minimal infectious dose required to establish infection in a heterologous host (for example with a simian rotavirus strain fed to human infants) appears to be much higher, ie 10^4 PFU (Rennels et al. 1987). Heterologous virus given to mice required doses of 10^5-10^6 PFU to establish infection (Ramig, 1988).

Establishment of infection by rotaviruses in the homologous host is followed by replication of rotavirus in mature enterocytes lining the upper two thirds of the villi of the small intestine. Initially infection occurs in the duodenum and upper jejunum and then progresses distally along the small intestine when progeny virions are released. Damage to epithelial cells leads to increased extrusion and stripping of cells from the villi (Estes et al. 1983). Rotaviruses have also been shown to infect ileal dome M cells in natural infections of piglets (Buller and Moxley, 1988). Pathological changes are restricted to the small intestine and include a reduction in villus length, and marked stunting of villi that can progress to changes resembling coeliac disease in young children (Davidson and Barnes, 1979). Lesions in experimental animals are similar and are not uniform throughout the small intestine (Pearson and McNulty, 1977). The lamina propria becomes infiltrated with mononuclear and polymorphonuclear cells. Mild changes have been observed in stomach and large intestine in piglets (Pearson and McNulty, 1977) but virus has not been demonstrated at these levels.

Virus multiplication is restricted to mature villous epithelial cells and no systemic invasion occurs in the immunocompetent host. Rotavirus has been detected in blood in immuno-incompetent children (Saulsbury et al. 1980). Immunofluorescence has revealed rotavirus particles in cells in the lamina propria but it is unlikely that replication occurs in these cells or in cells of the regional lymph nodes.

Pathological changes in the small intestine are associated with a reduction in alkaline phosphatase and lactase (Collins et al. 1988; Davidson and Barnes, 1979) until normal villus structure is restored with regrowth of mature epithelial cells.

In newborn babies and young children rotavirus particles are shed in abundance before onset and after cessation of symptoms (Cameron et al. 1978a, 1978b; Pickering et al. 1988). Particles may commonly be detected in faeces 8-10 days after onset of symptoms (Davidson et al. 1975). The amount of rotavirus shed in faeces may be influenced by the pre-existing immune state of the host. In young children experiencing a primary rotavirus infection with severe clinical symptoms, rotaviruses are shed in amounts approaching 10^8 or more particles per ml faeces (Davidson et al. 1975). Older children and adults undergoing even symptomatic reinfection rarely excrete amounts of rotavirus detectable by current direct diagnostic techniques. Infection is often only detected serologically (Rodriguez et al. 1987). Overall, these observations imply that virus replication is inhibited by active immunity but not by passively acquired maternal antibodies.

CLINICAL SYMPTOMS OF ROTAVIRUS INFECTION

Clinical descriptions concentrate mainly upon severely affected individuals, but there is a wide spectrum of severity of disease (Hamilton and Gall, 1982). Symptoms associated with rotavirus infection are likely to be influenced by the interplay of numerous factors relating to virulence of the infecting strain and pre-existing immunity (both passive and active). There is no evidence for a characteristic "rotavirus syndrome" and symptoms include those characteristic of gastrointestinal infection due to a variety of agents.

The illness is characterized by the abrupt onset of vomiting and diarrhoea, associated with upper respiratory symptoms in 20-75% of patients (Steinhoff, 1980). Controlled studies indicate similar rate of upper respiratory symptoms in children with non-rotavirus gastroenteritis (Hjelt et al, 1987b). Fever is present in the majority of patients. Mild to moderate dehydration, usually isotonic, is a frequent finding. Peripheral white cell counts are rarely elevated and faecal leucocytes uncommon. The course of the disease is usually short with fever and vomiting resolving in 1-2 days. Inadequate or delayed rehydration poses a threat to life and is still a major cause of death of young children in developing countries. Diarrhoea often lasts for 6-8 days and may become persistent and associated with malabsorption of ingested disaccharides.

All human serotypes (1, 2, 3, 4, 8, 9) have been observed to cause acute life-threatening diarrhoea. It is likely that strains vary in virulence but at present there is no means of estimating virulence of human strains. Attempts have been made to associate characteristic clinical features with subgroup I and II strains. Two such attempts yielded diametrically opposed results (Steele et al. 1988; Uhnoo and Svensson, 1986). It is more likely that the differences in clinical symptoms observed during the two studies were due to chance differences in virulence of the four strains, rather than to inherent differences in virulence associated with subgroup antigen.

Severity of infection as judged by clinical symptoms is influenced by age (see later) but the spectrum of clinical symptoms remains the same. Adults generally show diarrhoea and vomiting but lack fever and respiratory symptoms (Rodriguez et al. 1987).

Persistent infections have been established *in vitro* and suggest a potential for viral persistence in nature (Estes et al. 1983). Prolonged excretion, often over many months, has been recorded in children with severe immune deficiencies (Saulsbury et al. 1980). This may be due to constantly renewed infection of newly maturing enterocytes rather than to persisting infection in permissive cells.

CLINICAL IMMUNITY TO ROTAVIRUS DISEASE IN HUMANS

The need for immunoprophylaxis against rotavirus disease has been recognized and candidate rotavirus vaccines derived from heterologous animals (bovine and simian strains) have already undergone field trials (Bishop, 1988; Edelman, 1987; Kapikian et al. 1986). In order to devise an effective strategy it is necessary to determine whether immunity is homotypic or heterotypic and whether either are effective in preventing reinfection and/or disease.

There is no appropriate animal model of human disease in which the existence of homotypic or heterotypic immunity can be proven. Evidence must be deduced from longitudinal observation of infection in young children. Reinfection has been shown to be a common occurrence in children in developed and developing countries (Bartlett et al. 1988; Bishop et al. 1983; Black et al. 1982a,b; Chiba et al. 1986; Eiden et al. 1988; Friedman et al. 1988; Gurwith et al. 1981; Hjelt et al. 1987b; Linhares et al. 1989; Sack et al. 1980; Yamaguchi et al. 1985). When symptoms from sequential episodes have been graded for severity, second and subsequent infections have been asymptomatic or less severe than the initial infection (Bartlett et al. 1988; Bishop et al. 1983; Chiba et al. 1986; Eiden et al. 1988; Friedman et al. 1988; Hjelt et al. 1987b) or have shown similar symptoms to the initial infection (Eiden et al. 1988 , Friedman et al. 1988; Gurwith et al. 1981; Linhares et al. 1989). Three of the studies determined antigenicity of the viruses from sequential infections, and included an outbreak caused by serotype 3 followed by an outbreak with serotype 1 (Friedman et al. 1987), sequential symptomatic reinfections when a subgroup II infection preceded a subgroup I infection (Linhares et al. 1989), and asymptomatic reinfection with serotype 3 following previously symptomatic infection with serotype 3 (Chiba et al. 1986). There are no reports of severe life-threatening symptoms in children experiencing rotavirus reinfection.

Although inconclusive, the results of these surveys tend to support the belief that when sequential rotavirus infections occur it is the first infection that is likely to be the most severe. Homotypic protection has been documented, and partial clinical immunity appears to occur, even between rotaviruses of different serotypes.

Longitudinal studies in Melbourne support this belief. During the course of longitudinal studies conducted on 231 children, recruited as newborn babies and followed by passive surveillance for 12-36 months, we have noted numerous sequential infections. Characteristic patterns of sequential infection are shown in Figure 1 for 11 patients undergoing primary infection with serotypes 1, 2, 3 and 4. In all patients primary infections were associated with moderatesevere symptoms using the scoring system devised by Riepenhoff-Talty et al. 1981. All patients undergoing reinfections 6-28 months later were either asymptomatic or showed mild symptoms. Reinfections were detected serologically using paired sera, so rotavirus strains responsible for reinfections could not be identified. The serotypes predominant in hospitalised children during the period of reinfections are noted in the Figure. Eight out of 12 reinfections may have been due to homologous strains, although four occurred (patients 2,

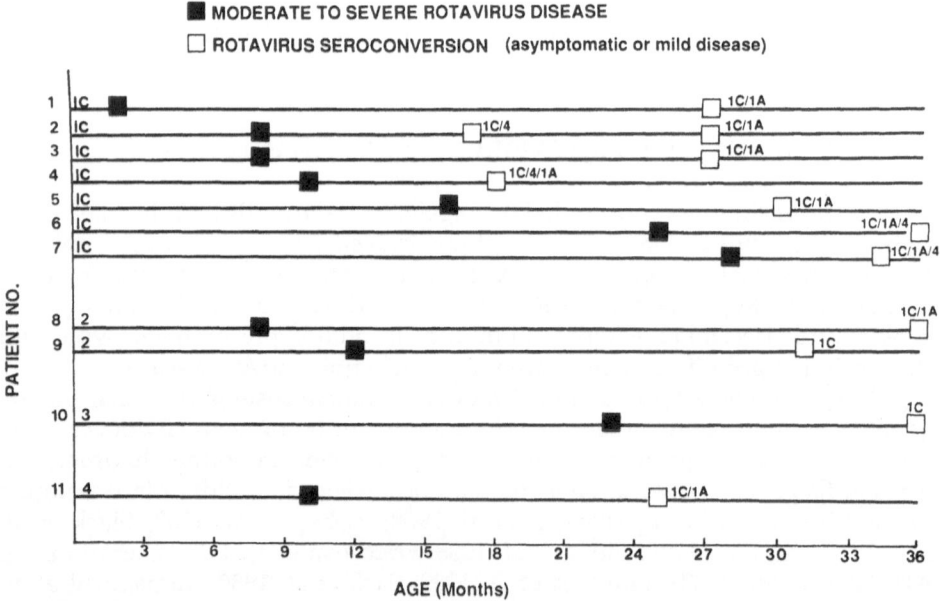

Figure 1. Sequential rotavirus infection in 11 children with primary rotavirus infection due to serotypes 1C, 2, 3 or 4. The serotypes of the predominant community strains during periods of rotavirus reinfection are indicated as 1C/1A/4 (Coulson, 1987).

4, 6, 7) when heterologous serotype 4 strains were common in the community. Strains 1C and 1A must be regarded as homologous since both are serotype 1 strains. Four reinfections (patients 8, 9, 10, 11) are likely to have been due to heterologous serotypes.

These results, together with results of other studies, support the belief that homologous and heterologous clinical immunity may exist in young children after primary infection with serotypes 1, 2, 3 and 4. Such clinical immunity may be only partially protective in that children (at least those in developed countries) may suffer mild-moderate symptoms on reinfection. It seems likely that severe life-threatening illness may only occur in young children experiencing a primary infection.

SEROLOGICAL RESPONSE TO ROTAVIRUS INFECTION

Serological responses to rotavirus infection, either with wild-type human viruses or to rotavirus vaccine candidates comprising animal rotavirus strains, have been repeatedly demonstrated. The responses are influenced by age and by the serological test used. Results from young children, older children and adults, and from newborn babies will be discussed separately.

Primary Infection in Young Children

Most sera have been obtained from young children hospitalised with severe acute rotavirus diarrhoea. A significant rise in rotavirus antibodies to group antigen (VP_6) in convalescent sera (obtained approximately four weeks after onset of symptoms) can be consistently demonstrated using a wide variety of serological techniques including CF, IAHA, and ELISA (Gust et al. 1977; Steinhoff 1980). ELISA has been used extensively to study isotype-specific immunoglobulin responses (Grimwood et al, 1988; Hjelt, 1988; Yolken et al. 1978).

The acute antibody response follows the classical pattern with early appearance of specific IgM followed by IgA and IgG (Abe and Inoye, 1979; Angeretti et al. 1987; Davidson et al. 1983; Grimwood et al. 1988; Hjelt et al. 1985, 1986; Riepenhoff-Talty et al. 1981; Zheng et al. 1988). IgG subclasses show an increase in IgG_1 and IgG_3. IgG_3 antibodies reach peak concentration one week after rotavirus excretion begins, whereas the peak concentration of IgG_1 occurs after two months (Grauballe et al. 1986). Approximately 25% of children show IgG_4 responses, but no IgG_2 responses have been recorded. The time course of development of a survey of acute serum antibody in 44 children admitted to hospital with acute rotavirus diarrhoea (caused by serotypes 1 or 4) is shown in Figure 2 (Grimwood et al. 1988). IgM was initially detected after 2-3 days, with peak incidence at 10 days. Occasional positive results were recorded after 30 days from onset. IgG titres were seldom detected in acute sera (obtained within 7 days of onset of symptoms), were detectable in all patients (more than 3 months old) after 30 days, and continued to rise and were still detectable 6-12 months later. IgA antibodies were initially detected at 8-10 days post onset, the peak incidence rose to peak being recorded at 30 days from onset. IgA antibodies were still detectable at 100 days post onset. Secretory immunoglobulin (ScIg) was detected in a few children at low levels approximately 8 days post onset. The findings are in agreement with most other surveys, however Hjelt et al. (1985) have reported ScIg to be present in most of their patients at 7-14 days post-onset.

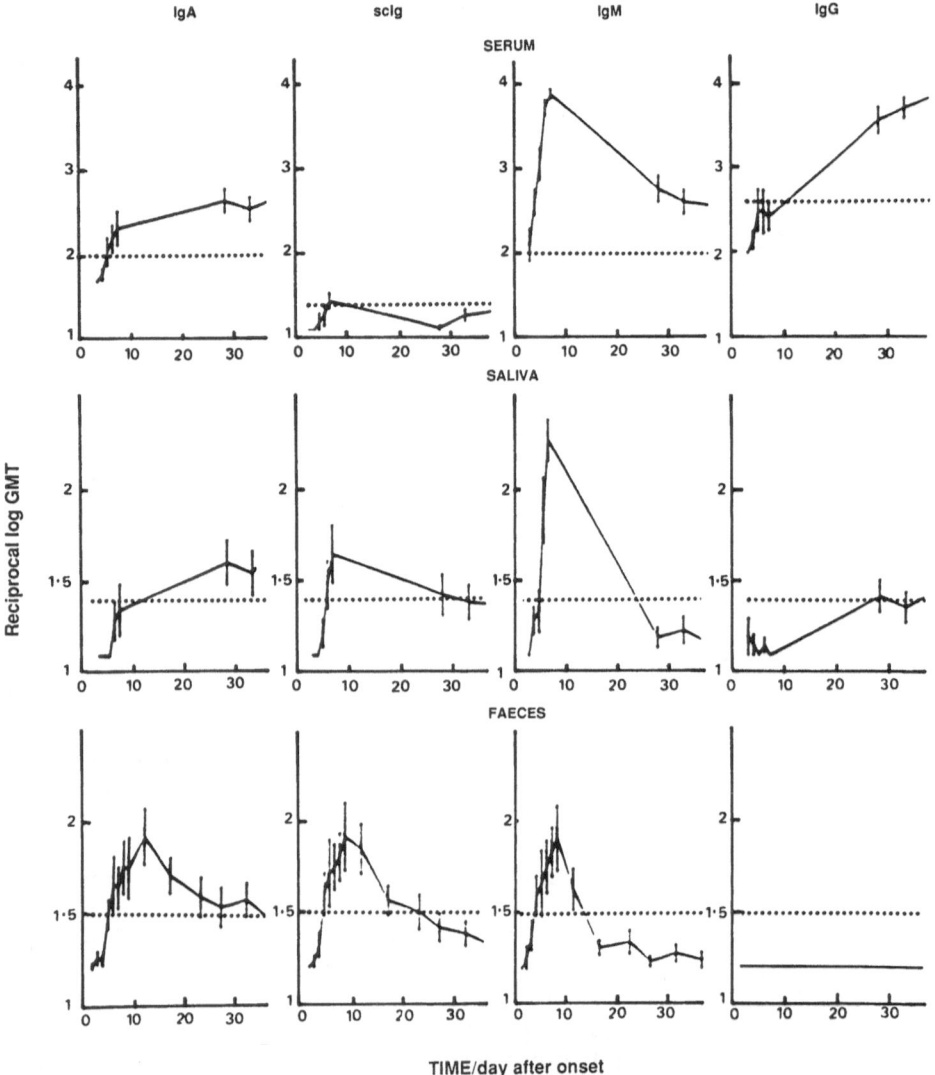

IgA scIg IgM IgG

SERUM

SALIVA

Reciprocal log GMT

FAECES

TIME/day after onset

Figure 2. Time course of development of antirotavirus IgA, ScIg, IgM, IgG in serum, saliva and faeces from 44 young children admitted to hospital with rotavirus diarrhoea.

Serum antibody responses in IgM, IgA, IgG are not as consistently found in children given rotavirus candidate vaccines comprising animal strains (Delem and Vesikari, 1987; Wright et al. 1987). Vaccine trials have indicated that response to infection with heterologous strains (judged by seroconversion) is dose dependant.

Neutralising antibody responses. Serotype-specific responses are measured by neutralization assays, either plaque-reduction neutralization or inhibition of fluorescent focus cell-forming units. Both assays estimate the sum total of anti-VP_4 and anti-VP_7 antibodies. Assays to differentiate the responses to these two neutralizing antigens will be discussed later.

Homotypic and heterotypic serological responses have been observed in rotavirus infections of seronegative children (Brussow et al. 1988; Chiba et al. 1986; Clark et al. 1985; Gerna et al. 1984, Puerto et al. 1987). Serotype 1 or 2 infections have been associated with serotype specific responses (Gerna et al. 1984; Zheng et al. 1985); serotype 4 with a heterotypic response to serotype 1 in addition to the homotypic response (Gerna et al. 1984; Puerto et al. 1987); serotype 3 infections with responses to serotypes 1, 3 and 4 (Chiba et al. 1986). Animal candidate vaccine strains given orally to young children have shown only homotypic neutralizing antibody (Christy et al. 1988; Losonsky et al. 1988) except where pre-existing serotypic cross-reactions of virus strains used have been demonstrated in vitro (Clark et al. 1986).

The induction of heterotypic neutralizing antibodies after natural rotavirus infection may be explained, in part, by the existence of neutralizing epitopes shared between serotypes (Coulson et al. 1986). In addition since neutralizing epitopes are found on the two surface viral polypeptides VP_4 and VP_7 (Hoshino et al. 1985), it is likely that many of the heterotypic serological responses are due to the use of assays that measure both responses simultaneously.

Protein specific responses. Primary and secondary responses to individual rotavirus proteins have been estimated in a limited number of specimens to date. Techniques used include competitive EIA using monoclonal antibodies against VP_7 (Beards & Desselberger, 1989; Shaw et al. 1987), plaque neutralization assays using reassortants containing either VP_4 or VP_7 of human rotaviruses (Flores et al. 1989; Ward et al. 1988), radioimmunoprecipitation assay (RIPA) (Svensson et al. 1987a,b) or immunoblotting (Brussow et al. 1988; Ushijima et al. 1989).

In primary human rotavirus infections of four seronegative children, serum responses to VP_1, VP_2, $VP_{3/4}$, VP_6 and a non-structural protein NS2 were found in convalescent sera (Svensson et al. 1987a). No serum responses to VP_7 of subgroups I or II were identified. All four children developed neutralizing antibodies (estimated by plaque neutralization), and it was concluded that the major component of the neutralizing antibody response must have been due to the VP_4 antigen. In another study, one child with no pre-existing rotavirus antibodies was shown to develop VP_2 and VP_6 antibodies of IgM, IgA class, and VP_7 antibodies of IgG class after a rotavirus infection (Ushijima et al. 1989).

Immune responses to animal rotavirus infections in humans present different responses. RIT4237 (bovine) and RRV (simian) infections induced responses mainly against the inner capsid polypeptides VP_2 and VP_6 (Svensson et al. 1987b). Both failed to induce detectable antibody responses to VP_7 in RIPA assays. Some weak reactions with $VP_{3/4}$ were detected. None of the vaccinated children developed detectable responses to non-structural proteins, indicating possibly the very limited replication of these strains in the heterologous host. However, competitive binding studies using Mabs specific for VP_7 and VP_4 and sera from young children vaccinated with RRV showed serum neutralizing antibody rises homotypic to VP_7 of RRV (72%), and heterotypic to VP_4 of RRV in 56% (Shaw et al. 1987). Fewer responses to VP_7 and VP_4 were detected in a similar assay on sera obtained from children vaccinated with RIT-4237. Data supported a lack of heterotypic immune response to VP_7 using animal rotaviruses. Flores et al. (1989) used human RRV reassortants as vaccine strains to detect polypeptide responses in vaccinated Venezuelan children. Homotypic VP7 responses were observed (in 3-5 month old infants) in 9/23 (39%) infants given serotype 1 reassortant RRV and 52% of infants given serotype 2 reassortant RRV. The children also developed seroresponses to RRV in 48-65%. These findings indicate that the neutralizing activity induced by the reassortants is directed to both VP4 and VP7 components of the virus, but that stronger and more frequent responses were obtained to VP4.

This preliminary evidence that VP4 may be a larger component of the immune response than VP7 in natural and in vaccine induced infection may partly explain the heterologous responses frequently observed with neutralization assays.

Responses to Reinfection

Whereas young children's serum antibody levels, with elevations of serum IgM, are typical of a primary rotavirus infection, the serum antibodies detected in older children and adults post-infection indicate an immune response characteristic of reinfection. Serum IgG rises rapidly to high titres but IgM may be difficult to detect (Haug et al. 1978). In a study of adults following rotavirus diarrhoea no rotavirus specific IgM was detected but significant elevations were found in specific antibodies of the IgG and IgA classes (Sheridan et al. 1981).

Secondary infection with either human rotavirus serotypes occurring naturally or with animal rotaviruses in vaccine trials, is associated with a boost in neutralizing antibody titres to serotypes already experienced (Clark et al. 1986; Kapikian et al. 1983a,b).

Human rotavirus re-infection in children and adults suggests that the response is predominently to VP3/4 (Svensson et al. 1987a, Ward et al. 1988). RIPA of acute and convalescent sera from young children with pre-existing rotavirus antibodies showed a variable number of polypeptides immunoprecipitated. Acute sera immunoprecipitated up to five polypeptides VP1, VP2, VP3, VP4 and VP6. Three of the four convalescent sera precipitated VP7 in addition to these five polypeptides using both subgroup I (DS1) and II (Wa) strains. The same three convalescent sera reacted in addition with NS2, and two reacted in addition with NS3.

Cell Mediated Immune Responses

The involvement of cell-mediated immunity in clearance of rotavirus infection in humans can be deduced from the numerous reports of persistent excretion, (sometimes associated with persistent diarrhoea), observed in children with immune deficiencies (Dolan et al. 1985; Eiden et al. 1985; Pedley et al. 1984; Saulsbury et al. 1980; Wood et al. 1988), and the observation that persistent infection cleared after a bone marrow transplant in one child (Pedley et al. 1984).

Experimental infection of mice has shown the importance of cytotoxic T lymphocytes as possible mediators of heterotypic immunity (Offit and Dudzik, 1988, 1989). However the relevance of these animal models to human infection remains to be established.

Only limited studies have been published to date of cell-mediated immune responses to rotavirus infection in humans. All describe responses in adults, and hence are not describing responses to primary infections. Elderly adults, and a kidney transplant patient with rotavirus diarrhoea developed transient rotavirus-induced lymphoproliferative responses that peaked at 9-16 days post-infection and declined by 30-37 days (Totterdell et al. 1988a, 1988b). By contrast, healthy adults (with pre-existing serum antibody) given a human rotavirus vaccine strain (RIT 4375) or a bovine vaccine (RIT 4237) showed a lymphoproliferative response in only one of twelve subjects. Systemic lymphoproliferative responses to rotavirus have been shown to be absent during rotavirus infection in two young children with decreased numbers of T cells (Totterdell et al. 1988a; Wood et al. 1988).

Other evidence of the potential involvement of cell mediated mechanisms in recovery from rotavirus disease in humans is the demonstration that rotavirus-infected cells stimulate human leucocytes to produce a cytokine that enhances natural killer cytotoxicity (Kohl et al. 1983). Natural killer cells can be mobilized quickly during infection and respond rapidly to lymphokines such as interleukin -2 and interferon. It is likely that these cells play an important part in the immune response to viral infections (Ritz 1989), including primary infections.

Responses to Neonatal Infection

Unlike young children with severe (primary) symptomatic rotavirus infection, newborn babies have been reported to show no or limited serological responses (Bryden et al. 1982; Davidson et al. 1983; Delem and Vesikari, 1987; Gust et al. 1977; Jayashree et al. 1988a), although rotavirus specific mucosal antibody has been demonstrated in saliva from infected newborn babies (Jayashree et al. 1988a).

At birth, cord blood antibodies reflect antibody titres in maternal sera (McLean and Holmes, 1980). Titres decline during succeeding weeks in all babies, even in those infected with rotaviruses in the neonatal period. We obtained maternal and cord sera at birth from 48 babies, together with follow up sera at 1-23 days, and at approximately 3 months post onset of neonatal rotavirus excretion. ELISA IgM rotavirus antibodies, measured using a capture assay (Coulson et al. 1989), were detected in 30 of the 48 babies (62.5%) in sera obtained 1-23 days post onset of rotavirus excretion. The seroconversion rate was not influenced by mode of feeding since 22/32 breast fed babies (68.8%) with rotavirus infection developed serum IgM antibodies compared with 8/12 (66.7%) fed artificial milk formulae. IgM ELISA antibodies were detected as early as 6 days of age and 3 days post onset of rotavirus excretion. No IgM antibodies were identified in maternal sera, cord sera or in infant sera obtained at the age of three months. Peak occurrence of IgM appeared to occur 9-23 days after onset of excretion (Figure 3). Pre-existing serum antibody in cord sera may have influenced the seroconversion rates (Figure 4) since median levels for ELISA units of maternal antirotavirus IgG were lower (800 units) in infants with an IgM response than in infants who did not respond (median 1800 units). In fact, of the 13 infants with maternal anti-rotavirus IgG levels of more than 1500 units (equivalent to a CF antibody titre of more than 1:16) only 5 (38%) showed an IgM response compared with 24 of 30 (80%) with maternal IgG units of less than 1500 (less than CF 1:16).

There was however no absolute correlation between maternal antirotavirus IgG level and development of a serum IgM response. Such a correlation may have been more apparent if serotype 3 specific neutralizing antibody levels in maternal sera against the infecting rotavirus serotype 3 had been estimated. It was also possible that the IgM assay used was influenced by the use of an animal rotavirus strain (SA$_{11}$) as antigen. Substitution of a human rotavirus strain in the assay showed positive IgM results for some sera that had given negative IgM results against SA$_{11}$.

Neutralising antibody to serotype 3 was detected at 3 months of age in 8/12 (69%) of infants infected as newborn babies, compared with 1/10 uninfected infants (Figure 5). No neutralizing antibody to serotype 2 was detected in infants infected with serotype 3 rotavirus. Neutralising antibody titres against serotypes 1 and 4 were detected in 6/12 (50%) and 8/12 (75%) of serotype 3 rotavirus positive infants compared with 3/10 (30%) and 4/10 (40%) of rotavirus negative infants. It is possible that the higher incidence of neutralizing antibody to the heterologous serotype 1 and 4 viruses seen after neonatal infection may have been due to active development of neutralizing antibody to VP$_4$. It is also possible that the results were due to persisting maternal neutralizing antibody, although none had been detected in any of the 22 sera using our ELISA IgG assay.

INTESTINAL RESPONSE TO ROTAVIRUS INFECTION

Most of the studies of response of the gut mucosa have focussed on experimental infection in a variety of newborn animals including mice, calves, lambs, rabbits and pigs. Intestinal antibody, both passively and actively ac-

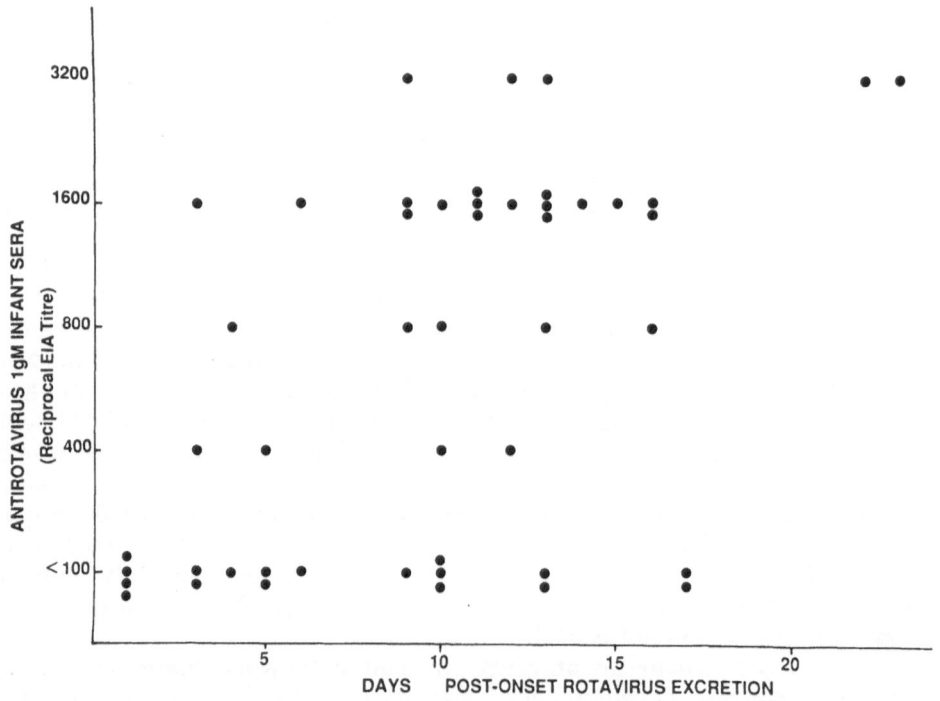

Figure 3. Antirotavirus IgM titres in sera obtained 1-23 days after rotavirus excretion in 48 newborn babies.

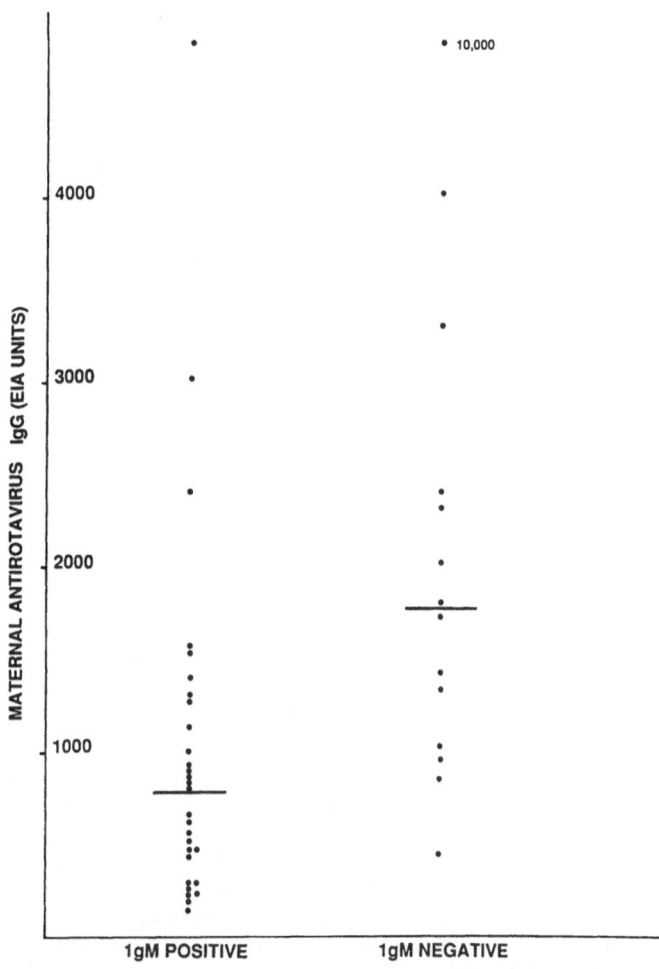

Figure 4. Antirotavirus IgG titres in maternal sera in relation to detection of IgM response in newborn babies infected with rotavirus.

quired has been shown to be primarily involved in protection (Snodgrass and Wells, 1978). The results cannot safely be extrapolated to humans since the animal models do not reproduce human disease at a comparable age.

A limited number of carefully performed studies in young children experiencing primary rotavirus infection, and in adults experimentally infected with a human rotavirus strain have detected antirotavirus IgA in the upper small intestinal secretions obtained by intubation (Bernstein et al. 1989; Davidson et al. 1983; Grimwood et al. 1988). Antibody is detectable approximately 7 days after onset of infection and persists for 3-4 weeks. IgM class antibodies also rise rapidly but are transient (Davidson et al. 1983; Grimwood et al. 1988). IgG class antibodies are only occasionally detected in small intestinal contents and

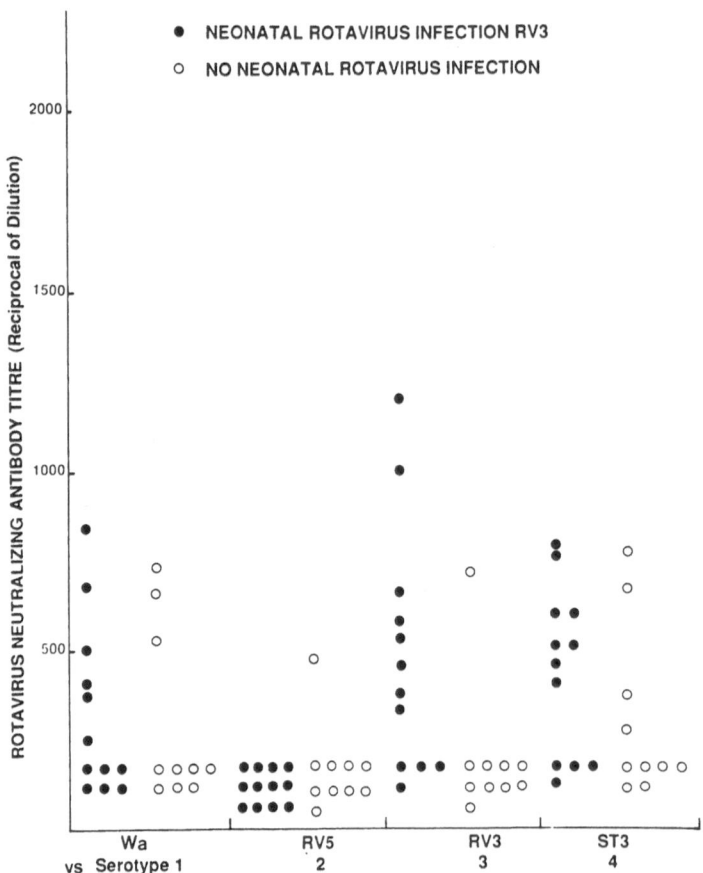

Figure 5. Rotavirus neutralizing antibody titres to serotypes 1, 2, 3 and 4 in infants aged 3 months who did, or did not, excrete serotype 3 rotaviruses during the first 14 days of life.

may derive from high serum antibody levels (Grimwood et al. 1988). Neutralizing antibody has been demonstrated in small intestinal contents in approximately 50% of adults fed a human rotavirus strain (Bernstein et al. 1989).

Numerous studies in children and adults undergoing natural infection, or given rotavirus vaccine animal strains confirm what the mucosa responds by forming specific rotavirus ELISA antibodies. These antibodies are detected in faecal extracts as coproantibodies of the IgA class (Bernstein et al. 1989; Grimwood et al. 1988; Hjelt et al. 1986; Losonsky et al. 1988; Riepenhoff-Talty et al. 1981; Shinozaki et al. 1986a, 1986b; Sonza and Holmes, 1980; Stals et al. 1984; Wright et al. 1987; Yamaguchi et al. 1985). Coproantibody levels peak 10-15 days after a primary infection (Figure 2). The duration of persistence after a primary infection is debatable and has been reported to vary between 3 weeks and 6 months. The comprehensive study conducted by Grimwood et al (1988) using daily specimens for 14 days and then weekly specimens for periods of 1-2 years indicates clearly that IgA coproantibodies sometimes appear only transiently after primary infection (Figure 2). Persisting antibodies reported in other studies may be due to undetected reinfection. Coproantibodies of IgM class and secretory Ig have also been reliably detected after primary infection (Grimwood et al. 1988; Hjelt et al. 1985, 1986). IgA rotavirus antibodies have also been detected in a proportion of nasal washes, throat swabs or saliva (Figure 2) obtained from children with primary infection (Grimwood et al. 1988; Shinozaki et al. 1986a; Stals et al. 1984). Rotavirus antibodies of the IgA class were detected in saliva from 62% of newborn babies sampled 3-12 months after rotavirus infection (Jayashree et al. 1988b).

Anamnestic boosts in ELISA IgA rotavirus coproantibody have been recorded after rotavirus reinfection (Bernstein et al. 1986; Coulson et al. 1989; Wright et al. 1987; Yamaguchi et al. 1985) and imply that memory exists, at least as long as one year after a primary infection (Yamaguchi et al. 1985). IgA coproantibody declines slowly and may persist for long periods after rotavirus reinfection (Coulson et al. 1989; Yamaguchi et al. 1985). Development of coproantibody may be a more reliable index of rotavirus reinfection than serum antibody (Bernstein et al. 1986; Coulson et al. 1989; Losonsky et al. 1988) whereas serum antibody may be a more reliable index of primary infection.

PROTECTION AGAINST ROTAVIRUSES

Immune protective mechanisms against rotavirus infection and disease may be mediated via serum or mucosal antibody (or both), and involve passive or active protection according to age. The intent to which cell-mediated immune mechanisms contribute to protection in humans is unknown.

In newborn and very young children who have not experienced rotavirus infection, immune protection appears to involve participation of maternal antibody present in the gut lumen (derived from breast milk) or in gut mucosa (derived from transplacental transfer of serum antibody). Titres of rotavirus specific antibodies (IgA, neutralizing antibodies) in human colostrum and breast milk vary widely from person to person (Bell et al. 1988; Duffy et al. 1986; Jayashree et al. 1988c; McLean and Holmes, 1980; Rahman et al. 1987; Ringenbergs et al. 1988). Ingestion of rotavirus antibodies in breast milk, in human gammaglobulin or in hyperimmune cow's milk (Barnes et al. 1982; Berger et al. 1983; Ebina et al. 1985) protect against establishment of rotavirus infection and against the clinical effects of infection. Oral antibody appears to act either by preventing infection or by modifying severity of clinical symptoms once infection has occurred. The protective effect is influenced by neutralizing antibody titres, at least when oral vaccines are administered to breast fed infants (Cadranel et al. 1987; Rennels et al. 1987). In natural infection of young infants high levels of placentally acquired serum antibody also appear to be partially protective against rotavirus disease (Riepenhoff-Talty et al. 1986). Once rotavirus infection is established passively administered oral antibodies appear to have no or little influence on recovery from infection (Ebina et al. 1985; Hilpert et al. 1987). The appearance of active antibody is associated with cessation of excretion of rotavirus particles (Stals et al. 1984).

The frequency with which reinfection occurs in all age groups indicates that there is little protection against infection with rotaviruses throughout life. Instead, protection is directed against amelioration of clinical disease once infection has occurred. In adult volunteers challenged orally with a human rotavirus, the prime correlate with resistance was the presence of antirotavirus

IgA in the duodenal lumen (Kapikian et al. 1983a, 1983b). There was no correlation with neutralizing antibody in this study, although another study showed a significant relation between protection and jejunal neutralizing antibody (Ward et al. 1989).

Pre-existing levels of serum antibody resulting from active infection may serve as an index of protection. The presence of high serum IgG or IgA antibody levels (estimated by CF, ELISA) has been associated with protection against rotavirus disease in all age groups (Hjelt et al. 1987a; Ryder et al. 1985; Sack et al. 1980; Sheridan et al. 1981; Ward et al. 1989).

There is evidence that protection is related to pre-existing titre of homotypic antibody against the reinfecting serotype. Chiba et al. (1986) measured serum levels against human serotypes 1-4 in children in a Japanese orphanage who were exposed to consecutive outbreaks of gastroenteritis due to serotype 3 rotaviruses. A serum neutralizing antibody titre to serotype 3 of 1:128 or greater seemed to be protective against infection and disease. Although this study implies that heterotypic neutralizing antibody is not protective, there are too few studies of this nature to draw any firm conclusions at present. Animal studies imply that active and passive heterotypic protection can occur and that it may be mediated via antibody to VP_4 (Hoshino et al. 1988; Matsui et al. 1989; Offit et al. 1986; Taniguchi et al. 1987).

It is possible that serum antibodies, neutralizing or otherwise, are not protective but merely reflect the individual's accumulated exposure to rotavirus. Any correlation of serum antibody level with protection would therefore be coincidental and be linked to other factors such as the capacity to produce gut mucosal antibodies rapidly, or to diffuse anamnestically boosted serum antibody rapidly into the gut lumen.

CONCLUSIONS

Longitudinal microbiological and serological surveillance of children recruited at birth provides convincing evidence that sequential rotavirus infections (with the same or a different serotype) are common during the first 2-3 years of life. Rotavirus infection appears to be an annual event for most young children in developed countries, occurring during the colder months of the year when hospital admissions for rotavirus diarrhoea are at their peak. Reinfections may be even more frequent in tropical climates with endemic rotavirus disease.

These epidemiological observations imply that single or repeated rotavirus infection(s) do not confer lasting protection against reinfection with homologous or heterologous viruses. It remains possible that recent infection could confer transient protection against reinfection. For example, antirotavirus antibody persisting in the gut lumen could prevent infection, as it does when administered orally to infants.

Surveillance studies also suggest that primary rotavirus infection (with serotypes 1, 2, 3, 4, 8 or 9) can protect against severe clinical symptoms on reinfection. Severe life-threatening disease, although not a common consequence of rotavirus infection may only occur during primary infection. Scoring of severity of clinical symptoms associated with sequential infections observed in longitudinal surveillance studies provides evidence that many reinfections (with the same or a different serotype) are associated with mild to moderate symptoms or are asymptomatic. In the absence of homotypic or heterotypic clinical protection it could be anticipated that at least a proportion of children would experience sequential severe disease, particularly as different serotypes frequently co-exist, or supplant each other, in the same community.

It is necessary to understand the humoral and cellular immune responses to rotavirus infection in order to identify the components of the immune response that are involved in clinical protection. Primary rotavirus infection of neonates and young children results in rapid development (2-3 days) of IgM antibody in serum and gut lumen. IgM antibodies rarely persist longer than 30 days and are not detected after reinfection.

Specific IgA, IgG (IgG_1,IgG_3) antibodies appear in serum of young children approximately ten days after primary infection. Neutralizing antibody appears in serum during the same period of time after both neonatal and post-neonatal infection. "Homotypic" serum neutralizing antibodies appear after infection with serotypes 1, 2, 3 and 4, and "heterotypic" antibodies also appear in some previously seronegative children. Assays that detect responses to individual polypeptides in serum reveal that the major responses are to non-neutralizing polypeptides (VP_1,VP_2,VP_6,NS2). There is preliminary evidence that stronger and more frequent responses occur to the neutralizing polypeptide VP_4 than to VP_7. Anamnestic responses in serum IgA, IgG and neutralizing antibody occur after reinfection and are associated with boosts in neutralizing antibody titres to serotypes already experienced. The predominant response again appears to be against VP_4 polypeptide.

Specific IgA antibodies appear in small intestinal contents and in faeces and are detected transiently (for 2-4 weeks) as coproantibodies after primary infection. Anamnestic boosts, indicating that immune memory exists, occur after reinfection, and may then persist for long periods. Boosts of coproantibody levels may be a more reliable index of rotavirus reinfection than serum antibody. There is no information relevant to detection of homotypic, heterotypic or individual polypeptide responses in gut lumenal contents.

On the basis of experiments with mice, it is likely that rotavirus-specific cytotoxic T lymphocytes in the gut mucosa play an important part in immune response to rotavirus infection. We are still largely ignorant of the role of cell-mediated immunity in infection of young children.

Present gaps in understanding the immune response in children, particularly in relation to serum and mucosal response to individual polypeptides make it impossible to decide which component of the immune response is critical to clinical protection. Assuming that severity of disease is at least partly determined by amount of intestinal epithelial cell damage, then an immune mechanism that restricts virus multiplication and spread could provide at least partial clinical protection. Specific antibody present in the gut lumen theoretically could protect against re-infection. The frequency with which re-infection occurs in nature indicates that this defense mechanism plays little part in prevention of human disease. Instead a more persisting immune response, or one capable of rapid mobilization (within 24-48 hours) would be required to restrict virus multiplication on reinfection. Identification of this response would greatly simplify the selection of candidate rotavirus vaccines for human use.

ACKNOWLEDGEMENTS

Some of the results reported here have resulted from team efforts within the Department of Gastroenterology. In particular we wish to acknowledge the origin of Figure 2 which was compiled from data described in the M.D. thesis presented by Dr. Keith Grimwood. During the course of our research we have benefitted from discussions with Dr. Barbara Coulson, and from the dedicated management of patients by the nursing staff of the Royal Children's Hospital. We are grateful to Mrs. Jane Lee and Miss Val Williams for secretarial assistance. The work has been supported by NHMRC and the Royal Children's Hospital Research Foundation.

REFERENCES

Abe Y, Inouye S (1979) Complement-fixing immunoglobulin M antibody response in patients with infantile gastroenteritis. J Clin Microbiol 9:284-286.

Adams WR, Kraft LM (1967) Electron microscopic study of the intestinal epithelium of mice infected with the agent of epizootic diarrhoea of infant mice (EDIM) virus. Am J Pathol 51: 39-60.

Angeretti A, Magi MT, Merlino C, Ferrara B, Ponzi AN (1987) Specific serum IgA in rotavirus gastroenteritis. J Med Virol 23:345-349.

Barnes GL, Hewson PH, McLellan JA, Doyle LW, Knoches AML, Kitchen WH, Bishop RF (1982) A randomized trial of oral gammaglobulin in low-birth-weight infants infected with rotavirus. Lancet II:1371-1373.

Bartlett III AV, Reves RR, Pickering LK (1988) Rotavirus in infant-toddler day care centers: Epidemiology relevant to disease control strategies. J Paediatr 113:435-441.

Beards GM, Desselberger V (1989) Determination of rotavirus serotype-specific antibodies in sera by competitive enhanced enzyme immunoassay. J Virol Methods 24:103-110.

Bell LM, Clarke HF, Offit PA, Horton Slight P, Arbeter AM, Plotkin SA (1988) Rotavirus serotype-specific neutralizing activity in human milk. Am J Dis Child 142:275-278.

Berger R, Hadjiselimovic F, Just M, Reigel P (1983) Effect of feeding human milk on nosocomial infections in an infants ward. Develop biol Standard 53:219-228.

Bernstein DI, Ziegler JM, Ward RL (1986) Rotavirus fecal IgA antibody response in adults challenged with human rotavirus. J Med Virol 20:297-304.

Bernstein DI, McNeal MM, Schiff GM, Ward RL (1989) Induction and persistence of local rotavirus antibodies in relation to serum antibodies. J Med Virol 28:90-95.

Bishop RF (1986) Epidemiology of diarrhoeal disease caused by rotavirus. In Holmgren J, Lindberg A, Mollby R (ed.) Development of vaccines and drugs against diarrhoea. Stockholm:Studentlitteratur,pp 158-170.

Bishop RF (1988) The present status of rotavirus vaccine development. Southeast Asian J Trop Med Pub Hlth 19:429-435.

Bishop RF, Barnes GL, Cipriani E, Lund JS (1983) Clinical immunity after neonatal rotavirus infection. A prospective longitudinal study in young children. N Engl J Med 309:72-76.

Bishop RF, Davidson GP, Holmes IH, Ruck BJ (1973) Virus particles in the epithelial cells of duodenal mucosa from children with acute non-bacterial gastroenteritis. Lancet ii:1281-1283.

Black RE, Merson MH, Rahman ASMM, Yunus M, Alim ARMA, Huq I, Yolken RH, Curlin GT (1980) A two year study of bacterial, viral and parasitic agents associated with diarrhea in rural Bangladesh. J Infect Dis 142:660-664.

Black RE, Merson MH, Huq I, Alim ARA, Yunus MD (1981) Incidence and severity of rotavirus and Escherichia coli diarrhoea in rural Bangladesh. Implications for vaccine development. Lancet i:141-143.

Black RE, Brown KH, Becker S, Yunus M (1982a) Longitudinal studies of infectious diseases and physical growth of children in rural Bangladesh II Incidence of diarrhea and association with known pathogens. Am J Epidemiol 115:315-324.

Black RE, Greenberg HB, Kapikian AZ, Brown KH, Becker S (1982b) Acquisition of serum antibody to Norwalk virus and rotavirus and relation to diarrhea in a longitudinal study of young children in rural Bangladesh. J Infect Dis 145:483-489.

Brussow H, Werchau H, Lerner L, Mietens C, Liedtke W, Sidothi J, Sotek J (1988) Seroconversion patterns to four human rotavirus serotypes in hospitalised infants with acute rotavirus gastroenteritis. J Infect Dis 158:588-595.

Bryden AS, Thouless MF, Hall CJ, Flewett TH, Wharton BA, Mathew PM, Craig I (1982) Rotavirus infections in a special-care baby unit. J Infect 4:43-48.

Buller CR, Moxley RA (1988) Natural infection of porcine ileal dome M cells with rotavirus and enteric adenovirus. Vet Pathol 25:516-517.

Cadranel S, Zeglache S, Jonckheer T, Zissis G, Andre FE, Bogaerts H, Delem A (1987) Factors affecting antibody response of newborns to repeated administrations of the rotavirus vaccine RIT4237. J Pediatr Gastroent Nutr 6:525-528.

Cameron DJS, Bishop RF, Veenstra AA, Barnes GL (1978a) Noncultivable viruses and neonatal diarrhea: a fifteen-month survey in a newborn special care nursery. J Clin Microbiol 8:93-98.

Cameron DJS, Bishop RF, Veenstra AA, Barnes GL, Holmes IH, Ruck BJ (1978b) Pattern of shedding of two noncultivable viruses in stools of newborn babies. J Med Virol 2:7-13.

Chiba S, Nakata S, Urasawa T, Urasawa S, Yokoyama T, Morita Y, Taniguchi K, Nakao T (1986) Protective effect of naturally acquired homotypic and heterotypic rotavirus antibodies. Lancet ii:417-421.

Christy C, Madore HP, Pichichero ME, Gala C, Pincus P, Vosefski D, Hoshino T, Kapikian A, Dolin R, Elmwood M, Panorama Paediatric Groups (1988) Field trial of rhesus rotavirus vaccine in infants. Pediatr Infect Dis J 7:645-650.

Clark HF, Dolan KT, Horton-Slight P, Palmer J, Plotkin SA (1985) Diverse serologic response to rotavirus infection of infants in a single epidemic. Pediatr Infect Dis 4:626-631.

Clark HF, Furukawa T, Bell LM, Offit PA, Perrella PA, Plotkin SA (1986) Immune response of infants and children to low-passage bovine rotavirus (strain WC3). Am J Dis Child 140:350-356.

Collins J, Starkey WG, Wallis TS, Clarke GJ, Worton KJ, Spencer AJ, Haddon SJ, Osborne MP, Candy DCA, Stephen J (1988) Intestinal enzyme profiles in normal and rotavirus-infected mice. J Pediatr Gastroent Nutr 7:264-272.

Coulson BS (1987) Variation in neutralization epitopes of human rotaviruses in relation to genomic RNA polymorphism. Virology 159:209-212.

Coulson BS, Grimwood K, Bishop RF, Barnes GL (1989) Evaluation of endpoint titration, single dilution and capture enzyme immunoassays for measurement of antirotaviral IgA and IgM in infantile secretions and serum. J Virol Methods 26:53-66.

Coulson BS, Tursi JM, McAdam WJ, Bishop RF (1986). Derivation of neutralizing monoclonal antibodies to human rotaviruses and evidence that an immunodominant neutralization site is shared between serotypes 1 and 3. Virology 154:302-313.

Davidson GP, Barnes GL (1979) Structural and functional abnormalities of the small intestine in infants and young children with rotavirus enteritis. Acta Paediatr Scand 68:181-186.

Davidson GP, Bishop RF, Townley RRW, Holmes IH, Ruck BJ (1975) Importance of a new virus in acute sporadic enteritis in children. Lancet i:242-251.

Davidson GP, Hogg RJ, Kirubakaran CP (1983) Serum and intestinal immune response to rotavirus enteritis in children. Infect Immun 40:447-452.

Delem A, Vesikari T (1987) Detection of serum antibody responses to RIT4237 rotavirus vaccine by ELISA and neutralization assays. J Med Virol 21:231-238.

de Zoysa I, Feachem RG (1985) Intervention for the control of diarrhoeal diseases among young children: rotavirus and cholera immunization. Bull WHO 63:569-583.

Dolan KT, Twist EM, Horton-Slight P, Forrer C, Bell IM, Plotkin SA, Clark HF (1985) Epidemiology of rotavirus electropherotypes determined by a simplified diagnostic technique with RNA analysis. J Clin Microbiol 21:753-758.

Duffy LC, Riepenhoff-Talty M, Byers TE, LaScolea LJ, Zielenzny MA Dryja DM, Ogra PL (1986) Modulation of rotavirus enteritis during breast feeding. Implications on alterations in the intestinal bacterial flora. Am J Dis Child 140:1164-1168.

Ebina T, Sato A, Umezu K, Ishida N, Ohyama S, Oizumi A, Aikawa K, Katagiri S, Katsushima N, Imai A, Kitaoka S, Suzuki H, Konno T (1985) Prevention of rotavirus infection by oral administration of cow colostrum containing antihumanrotavirus antibody. Med Microbiol Immunol 174:177-185.

Edelman R (1987) Perspective on the development and deployment of rotavirus vaccines. Pediatr Inf Dis J 6:704-710.

Eiden J, Losonsky GA, Johnson J, Yolken RH (1985) Rotavirus RNA variation during chronic infection of immunocompromised children. Pediatr Infect Dis 4:632-637.

Eiden J, Verleur DG, Vonderfecht SL, Yolken RH (1988) Duration and pattern of asymptomatic rotavirus shedding by hospitalized children. Pediatr Infect Dis 7:564-569.

Estes MK, Palmer EL, Obijeski JF (1983) Rotaviruses: A review. Curr Top Microbiol Immunol 105:123-184.

Flores J, Nakagomi O, Nakagomi T, Glass R, Gorziglia M, Askaa J, Hoshino Y, Perez-Schael I, Kapikian AZ (1986). The role of rotaviruses in pediatric diarrhea. Ped Infect Dis 5:553-562.

Flores J, Perez-Schael I, Blanco M, Vilar M, Garcia D, Perez M, Daoud N, Midthun K, Kapikian AZ (1989) Reactions to and antigenicity of two human-rhesus rotavirus reassortant vaccine candidates of serotypes 1 and 2 in Venezuelan infants. J Clin Microbiol 27:512-518.

Friedman MG, Galil A, Sarov B, Margaleth M, Katzir G, Midthun K, Taniguchi K, Urasawa S, Kapikian AZ, Edelman R , Sarov I (1988) Two sequential outbreaks of rotavirus gastroenteritis: evidence for symptomatic and asymptomatic reinfections. J Infect Dis 158:814-822.

Gerna G, Battaglia M, Milenesi G, Passarani N, Percivalle E, Cataneo E (1984)

Serotyping of cell culture-adapted subgroup 2 human rotavirus strains by neutralization. Infect Immun 43:722-729.

Graham DY, Dufour GR, Estes MK (1987) Minimal infective dose of rotavirus. Arch Virol 92:261-271.

Grauballe PC, Hornsleth A, Hjelt K, Krasilnikoff PA (1986) Detection by ELISA of immunoglobulin G subclass-specific antibody responses in rotavirus infections in children. J Med Virol 18:277-281.

Grimwood K, Abbott GD, Fergusson DM, Jennings LC, Allen JM. (1983) Spread of rotavirus within families: a community based study. Br Med J 287:575-577.

Grimwood K, Lund JCS, Coulson BS, Hudson IL, Bishop RF, Barnes GL (1988) Comparison of serum and mucosal antibody responses following severe acute rotavirus gastroenteritis in young children. J Clin Microbiol 26:732-738.

Gurwith M, Wenman W, Hinde D, Feltham S, Greenberg H (1981) A prospective study of rotavirus infection in infants and young children. J Infect Dis 144:218-224.

Gust ID, Pringle RC, Barnes GL, Davidson GP, Bishop RF (1977) Complement-fixing antibody response to rotavirus infection. J Clin Microbiol 5:125-130.

Hamilton JR, Gall DG (1982) Pathopysiological and clinical features of viral enteritis. In Tyrrell D, Kapikian AZ (ed) Virus infections of the gastrointestinal tract. New York: Marcel Dekker, pp227-238.

Haug KW, Orstavik I, Kvelstad G (1978) Rotavirus infections in families. A clinical and virological study. Scand J Infect Dis 10:265-269.

Hilpert H, Brussow H, Mietens C, Sidoti J, Lerner L, Werchau H (1987) Use of bovine milk concentrate containing antibody to rotavirus to treat rotavirus gastroenteritis in infants. J Infect Dis 156:158-166.

Hjelt K (1988) Acute rotavirus gastroenteritis in children. Danish Med Bull 35:222-236.

Hjelt K, Grauballe PC, Schiotz PO, Andersen L, Krasilnikoff PA (1985) Intestinal and serum immune response to a naturally acquired rotavirus gastroenteritis in children. J Pediatr Gastroent Nutr 4:60-66.

Hjelt K, Grauballe PC, Anderson PO, Howitz P, Krasilnikoff PA (1986) Antibody response in serum and intestine in children up to six months after a naturally acquired rotavirus gastroenteritis. J Pediatr Gastroent Nutr 5:74-80.

Hjelt K, Grauballe PC, Paerregaard A, Nielsen OH, Krasilnikoff PA (1987a) Protective effect of preexisting rotavirus-specific immunoglobulin A against naturally acquired rotavirus infection in children. J Med Virol 21:39-47.

Hjelt K, Nielsen OH, Paerregaard A, Grauballe PC, Krasilnikoff PA (1987b) Acute gastroenteritis in children attending day-care centres with special reference to rotavirus infections II Clinical manifestations. Acta Paediatr Scand 76:763-768.

Ho M, Glass RI, Pinsky PF, Anderson LJ (1988) Rotavirus as a cause of diarrheal morbidity and mortality in the United States. J Infect Dis 158:1112-1116.

Holzel H, Cubitt DW, McSwiggan DA, Sanderson PJ, Church J (1980) An outbreak of rotavirus infection among adults in a cardiology ward. J Infect 2:33-37.

Hoshino Y, Wyatt RG, Greenberg HB, Flores J, Kapikian AZ (1984) Serotypic similarity and diversity of rotaviruses of mammalian and avian origin as studied by plaque reduction neutralisation. J Infect Dis 149:694-702.

Hoshino Y, Sereno MM, Midthun K, Flores J, Kapikian AZ, Chanock RM (1985)

Independent segregation of two antigenic specificities (VP3 and VP7) involved in neutralization of rotavirus infectivity. Proc Natl Acad Sci USA 82:8701-8704.

Hoshino Y, Saif LJ, Sereno MM, Chanock RM, Kapikian AZ (1988) Infection immunity of piglets to either VP3 or VP7 outer capsid protein confers resistance to challenge with a virulent rotavirus bearing the corresponding antigen. J Virol 62:744-748.

Hrdy DB (1987) Epidemiology of rotaviral infection in adults. Rev Infect Dis 9:461-469.

Jayashree S, Bhan MK, Kumar R, Bhandari N, Sazawal S (1988a) Protection against neonatal rotavirus infection by breast milk antibodies and trypsin inhibitors. J Med Virol 26:333-338.

Jayashree S, Bhan MK, Kumar R, Raj P, Glass R, Bhandari N (1988b) Serum and salivary antibodies as indicators of rotavirus infection in neonates. J Infect Dis 158:1117-1120.

Jayashree S, Bhan MK, Raj P, Kumar R, Svensson L, Stintzing G, Bhandari N (1988c) Neonatal rotavirus infection and its relation to cord blood antibodies. Scand J Infect Dis 20:249-253.

Kapikian AZ, Chanock RM (1985) Rotaviruses. In Fields BN, Knipe DM, Chanock RM, Melnick JL, Roizman B, Shope RE (ed.), Virology. New York: Raven Press, pp 863-906.

Kapikian AZ, Wyatt RG, Levine MM, Yolken RH, Vankirk DH, Dolin R, Greenberg HB, Chanock RM (1983a) Oral administration of human rotavirus to volunteers: induction of illness and correlates of resistance. J Infect Dis 147:95-106.

Kapikian AZ, Wyatt RG, Levine MM, Black RE, Greenberg HB, Flores J, Kalica AR, Hoshino Y, Chanock RM (1983b) Studies in volunteers with human rotaviruses. Devel Biol Stand 53:209-218.

Kapikian AZ, Flores J, Hoshino Y, Glass RI, Midthun K, Gorziglia M, Chanock RM (1986) Rotavirus: the major etiologic agent of severe infantile diarrhoea may be controllable by a "Jennerian" approach to vaccination. J Infect Dis 153: 815-822.

Kohl S, Harmon MW, Ping Tang JY (1983) Cytokine-stimulated human killer cytotoxicity: response to rotavirus-infected cells. Pediatr Res 17:868-872.

Koopman JS, Turkish VJ, Monto AS, Gouvea V, Srivastava S, Isaacson RE (1984) Patterns and etiology of diarrhoea in three clinical settings. Am J Epidemiol 119:114-123.

Linhares AC, Gabbay YB, Freitas RB, da Rosa EST, Mascarenhas JDP, Loureiro ECB (1989) Longitudinal study of rotavirus infections among children from Belem, Brazil. Epidem Infect 102:129-145.

Losonsky GA, Rennels MB, Lim Y, Krall G, Kapikian AZ, Levine MM (1988) Systemic and mucosal immune responses to rhesus rotavirus vaccine MMU 18006. Pediatr Infect Dis J 7:388-393.

Mata L, Simhon A, Urrutia JJ, Kronmal RA, Fernandez R, Garcia L (1983) Epidemiology of rotaviruses in a cohort of 45 Guatemalan Mayan Indian children observed from birth to the age of three years. J Infect Dis 148:452-461.

Matsui SM, Offit PA, Vo PT, Mackow ER, Benfield DA, Shaw RD, Padilla-Noriega L, Greenberg HB, (1989) Passive protection against rotavirus-induced diarrhea by monoclonal antibodies to the heterotypic neutralization domain of VP7 and the VP8 fragment of VP4. J Clin Microbiol 27:780-782.

Mebus CA, Underdahl NR, Rhodes MB, Twiehaus MJ (1969) Calf diarrhoea (scours): reproduced with a virus from a field outbreak. Bull Neb Agric Exp Station 233:1-16.

McLean B, Holmes IH (1980) Transfer of antirotaviral antibodies from mothers to their infants. J Clin Microbiol 12:320-325.

Offit PA, Shaw RD, Greenberg HB (1986) Passive protection against rotavirus-induced diarrhea by monoclonal antibodies to surface proteins VP_3 and VP_7. J Virol 58:700-703.

Offit PA, Dudzik KI (1988) Rotavirus-specific cytotoxic T lymphocytes cross-react with target cells infected with different rotavirus serotypes. J Virol 62:127-131.

Offit PA, Dudzik KI (1989) Rotavirus-specific cytotoxic T lymphocytes appear at the intestinal mucosal surface after rotavirus infection. J virol 63:3507-3512.

Pearson GR, McNulty MS (1977) Pathological changes in the small intestine of neonatal pigs infected with a pig reovirus-like agent (rotavirus). J Comp Pathol 87:363-375.

Pedley S, Hundley F, Chrystie I, McCrae MA, Desselberger U (1984) The genomes of rotaviruses isolated from chronically infected immune deficient children. J Gen Virol 65: 1141-1150.

Pickering LK, Bartlett III AV, Reves RR, Morrow A (1988) Asymptomatic excretion of rotavirus before and after rotavirus diarrhoea in children in day care centres. J Pediatr 112:361-365.

Pitson GA, Grimwood K, Coulson BS, Oberklaid F, Hewstone AS, Jack I, Bishop RF, Barnes GL (1986) Rotavirus and other enteric pathogens associated with acute diarrhoea in children in Melbourne Australia: comparison between children treated at home and those requiring hospital admission. J Clin Microbiol 24:395-399.

Puerto FI, Padilla-Noriega L, Zamora-Chavez A, Briceno A, Puerto M, Arias CF (1987) Prevalent patterns of serotype-specific seroconversion in Mexican children infected with rotavirus. J Clin Microbiol 25:960-963.

Rahman MM, Yamauchi M, Hanada N, Nishikawa K, Marishima T (1987) Local production of rotavirus specific IgA in breast tissue and transfer to neonates. Arch Dis Child 62:401-405.

Ramig RF (1988) The effects of host age, virus dose and virus strain on heterologous rotavirus infection of suckling mice. Microbiol Pathogenesis 4:189-202.

Rennels MB, Losonksy GA, Shindledecker CL, Hughes TP, Kapikian AZ, Levine MM, The Clinical Study Group. (1987) Immunogenicity and reactogenicity of lowered doses of rhesus rotavirus vaccine strain MMU18006 in young children. Pediatr Infect Dis 6:260-264.

Riepenhoff-Talty M, Bogger-Goren S, Li P, Carmody PJ, Barrett HJ, Ogra PL (1981) Development of serum and intestinal antibody response to rotavirus after naturally acquired rotavirus infection in man. J Med Virol 8:215-222.

Ringenbergs M, Albert MJ, Davidson GP, Goldsworthy W, Haslam R (1988) Serotype-specific antibodies to rotavirus in human colostrum and breast milk and in maternal and cord blood. J Infect Dis 158:477-479.

Ritz J (1989) The role of natural killer cells in immune surveillance. N Engl J Med 320:1748-1749.

Rodriguez WJ, Kim HW, Brandt CD, Bize B, Kapikian AZ, Chanock RM, Curlin G, Parrott RH (1980) Rotavirus gastroenteritis in the Washington DC area: incidence of cases resulting in admission to the hospital. Am J Dis Child 134: 777-779.

Rodriguez WJ, Kim HW, Brandt CD, Schwartz RH, Gardner MK, Jeffries B, Parrott RH, Kaslow RA, Smith JI, Kapikian AZ (1987) Longitudinal study of rotavirus infection and gastroenteritis in families served by a pediatric medical practice: clinical and epidemiologic observations. Pediatr Infect Dis 6:170-176.

Ryder RW, Singh N, Reeves WC, Kapikian AZ, Greenberg HB, Sack RB (1985) Evidence of immunity induced by naturally acquired rotavirus and Norwalk virus infection on two remote Panamanian islands. J Infect Dis 151:99-105.

Sack DA, Gilman RH, Kapikian AZ, Aziz KMS (1980) Seroepidemiology of rotavirus infection in rural Bangladesh. J Clin Microbiol 11:530-532.

Saulsbury FT, Winkelstein JA, Yolken RH (1980) Chronic rotavirus infection in immunodeficiency. J Pediatr 97:61-65.

Shaw RD, Fong KJ, Losonsky GA, Levine MM, Maldonado Y, Yolken R, Flores J, Kapikian AZ, Vo PT, Greenberg HB (1987) Epitope specific immune responses to rotavirus vaccination. Gastroent 93:941-947.

Sheridan JF, Aurelian L, Barbour G, Santosham M, Sack RB, Ryder RW (1981) Traveller's diarrhoea associated with rotavirus infection: analysis of virus-specific immunoglobulin classes. Infect Immun 39:917-927.

Shinozaki T, Araki K, Ushijima H, Fujii R (1986a) Diagnostic significance of specific IgA coproconversion in rotavirus infection. Eur J Pediatr 145:581-582.

Shinozaki T, Araki K, Ushijina H, Kim B, Tajima T, Fujii R, Minamitani M (1986b) Coproantibody response to rotavirus in an outbreak in a day-care nursery. Eur J Pediatr 144:515-516.

Snodgrass DR, Wells PW (1978) Passive immunity in rotavirus infections. JAVMA 173:565-568.

Sonza S, Holmes IH (1980) Coproantibody response to rotavirus infection. Med J Aust 2:496-499.

Stals F, Walther FJ, Bruggeman CA (1984) Fecal and pharyngeal shedding of rotavirus and rotavirus IgA in children with diarrhoea. J Med Virol 14:333-339.

Steele AD, Bos P, Alexander JJ (1988) Clinical features of acute infantile gastroenteritis associated with human rotavirus subgroups I and II. J Clin Microbiol 26:2647-2649.

Steinhoff MC (1980) Rotavirus: The first five years. J Pediatr 96:611-622.

Svensson L, Sheshberadaran H, Vene S, Norrby E, Grandien M, Wadell G (1987a) Serum antibody responses to individual viral polypeptides in human rotavirus infections. J gen Virol 68:643-651.

Svensson L, Sheshberadaran H, Vesikari T, Norrby E, Wadell G (1987b) Immune response to rotavirus polypeptides after vaccination with heterologous rotavirus vaccines (RIT4237,RRV-1). J gen Virol 68:1993-1999.

Taniguchi K, Morita Y, Urasawa T, Urasawa (1987) Cross-reactive neutralization epitopes on VP3 of human rotavirus: analysis with monoclonal antibodies and antigenic variants. J Virol 61:1726-1730.

Totterdell BM, Patel S, Banatvala JE, Chrystie IL (1988a) Development of a lymphocyte transformation assay for rotavirus in whole blood and breast milk. J Med Virol 25:27-36.

Totterdell BM, Banatvala JE, Chrystie IL, Ball G, Cubitt WD (1988b) Systemic lymphoproliferative responses to rotavirus. J Med Virol 25:37-44.

Uhnoo I, Svensson L (1986) Clinical and epidemiological features of acute infantile gastroenteritis associated with human rotavirus subgroups 1 and 2. J Clin Microbiol 23:551-555.

Ushijima H, Honma H, Ohnoda H, Mukoyama J, Oyanagi H, Araki K, Shinozaki T, Morikawa S, Kitamura T (1989) Detection of anti-rotavirus IgG, IgM and IgA antibodies in healthy subjects, rotavirus infections and immunodeficiences by immunoblotting. J Med Virol 27:13-18.

Ward RL, Bernstein DI, Young EC, Sherwood JR, Knowlton DR, Schiff GM (1986) Human rotavirus studies in volunteers: determination of infectious dose and serological response to infection. J Infect Dis 154:871-880.

Ward RL, Knowlton DR, Schiff GM, Hoshino Y, Greenberg HB (1988) Relative concentrations of serum neutralizing anitbody to VP3 and VP7 proteins in adults infected with a human rotavirus. J Virol 62:1543-1549.

Ward RL, Bernstein DI, Shukla R, Young EC, Sherwood JR, McNeal MM, Walker MC, Schiff GM (1989) Effects of antibody to rotavirus on protection of adults challenged with a human rotavirus. J Infect Dis 159:79-88.

Wolf JL, Cukor G, Blacklow NR, Dambrauskas R, Trier JS (1981) Susceptibility of mice to rotavirus infection: effects of age and administration of corticosteroids. Infect Immunity 33:565-574.

Wood DJ, David TJ, Chrystie IL, Totterdell B (1988) Chronic enteric virus infection in two T-cell immunodeficient children. J Med Virol 24:435-444.

Wright PF, Tajima T, Thompson J, Kokubun K, Kapikian A, Karzon DT (1987) Candidate rotavirus vaccine (rhesus rotavirus strain) in children: an evaluation. Pediatrics 80:473-479.

Yamaguchi H, Inouye S, Yamauchi M, Morishima T, Matsuno S, Isomura S, Suzuki S (1985) Anamnestic response in fecal IgA antibody production after rotaviral infection of infants. J Infect Dis 152:398-400.

Yolken RH, Wyatt RG, Kim HW, Kapikian AZ, Chanock RM (1978) Immunological response to infection with human reovirus-like agent: measurement of anti-human reovirus-like agent immunoglobulin G and M levels by the method of enzyme-linked immunosorbent assay. Infect Immun 19:540-546.

Zheng B, Han S, Yan Y, Liang X, Ma G, Yang Y, Ng MH (1988) Development of neutralizing antibodies and Group A common antibodies against natural infections with human rotavirus. J Clin Microbiol 26:1506-1512.

DISCUSSION

Riepenhoff-Talty M (Children's Hospital, Buffalo, NY):

I know that the active vaccine program is going on. What do you think of the possibility, at least in developed countries, of a passive approach?

Bishop R (Royal Children's Hospital, Melbourne, Australia):

What Dr. Riepenhoff-Talty is referring to is that it is known from animal studies that if enough antibody is placed in the gut lumen it will prevent rotavirus infection and hence, will protect against disease. This also works in young children. There are some published studies where human gammaglobulin or bovine colostrum with rotavirus antibody of high titer has been used. The results show that in this way, you can prevent infection, but it's a short term prevention. As soon as passive antibody administration ceases, you are back to the stage of having an unprotected child. I think passive protection may be valuable in certain circumstances as a prophylactic measure. For example, with children in whom you want to prevent nosocomial infection, who are in a vulnerable age group, and in the hospital for a procedure such as cardiac surgery in an intensive care nursery.

Warford A (Kaiser Permamente, North Hollywood, CA):

Do you have any experience with rotavirus infections in immunocompromised patients?

Bishop R:

No, I don't. There are several reports that in immunoincompetent children, rotavirus infection is very persistent. This obviously implies that cell mediated immunity has got some role to play in control of rotavirus infection. It is considered that the T cell system is involved in active immunity in young children. This is an area that very definitely needs study. Animal work implies that cell mediated immune system can respond very rapidly to reinfection and that the response is heterologous.

Malherbe H (Salt Lake City, UT):

Dr. Bishop has shown very clearly the degradation of the villi in the gut as a result of rotavirus infection. That, probably in most instances, just results in a relatively mild watery diarrhea. Surely it exposes those villi to infection by other agents. Especially, in underdeveloped countries. Would it not be necessary then to have a vaccine that is entirely administered in the gut. A relatively avirulent strain of rotavirus that can be delivered orally to produce antibody in the gut? In the developing countries, breast feeding is predominantly practised, but as we've seen in South Africa, as the rate of development to the people increases, so breast feeding falls away, and bottle fed infants are more commonly seen. It's quite possible, I should imagine, that the rotavirus prevalence there might lead to more serious diarrhea in those bottle fed infants, and it needs antibody administered locally in the gut.

Bishop R:

I am certainly in favor of breast feeding, say for at least six months of life, in any setting. In the developing countries, it often goes on longer than that. In a developed country, breast feeding will decrease the frequency of rotavirus infection, but it doesn't appear to eliminate it entirely. Obviously, the amount of specific neutralizing antibody present in the gut lumen and the imbalance with the dose that is received is critical. Breast feeding is basically protective. On the other hand, bottle fed children aren't necessarily vulnerable to rotavirus disease. We've seen many children in our intensive care nurseries who have had rotavirus infection in the first three or four days of life, who were fed artificially with formula and yet, had asymptomatic infections. It may have been that the virus itself was avirulent. It may also be that the serum antibodies acquired via the placenta modified the symptoms of disease.

MEDICAL VIROLOGY OF SMALL ROUND GASTROENTERITIS VIRUSES

Neil R. Blacklow

University of Massachusetts Medical School
Division of Infectious Diseases,
Department of Medicine
Worcester, Massachusetts, 01655, USA

SPECTRUM OF VIRAL GASTROENTERITIS AGENTS

Despite the medical importance of viral diarrheas, the responsible etiologic agents were totally unknown until the discovery of Norwalk virus in 1972 (Kapikian et al. 1972). The main reason for this delay in their recognition was the inability to cultivate in cell culture etiologic agents from stool specimens. This was in marked contrast to the success in cultivating the many viruses causing respiratory tract disease, measles, mumps and other common viral syndromes. It was only when electron microscopic (EM) and antigen detection techniques were applied to the study of stool specimens that viral gastroenteritis agents were discovered and their medical importance subsequently ascertained.

Since 1972, a number of etiologic agents of human viral gastroenteritis have been discovered (Cukor and Blacklow, 1984). They are summarized in Table 1. Rotavirus is the best known and studied pathogen, and it also is the agent most responsible for medically important diarrheal disease, being the principal cause of gastroenteritis of infants and young children worldwide. A few strains of morphologically indistinguishable rotaviruses lack a common viral group antigen and are therefore called non-group A rotaviruses. To date, these have been associated with disease only in China. Another category of pathogens is the enteric adenoviruses, which are the only adenovirus strains (serotypes 40 and 41) that are clearly associated with diarrheal disease. They appear to be a distant second in importance to rotaviruses as a cause of endemic diar-rhea of infants and young children in temperate climates. Enteric coronaviruses have been described in stool samples from both ill and well individuals, particularly in those from third world countries, and their role if any, in human diarrhea remains uncertain.

The remainder of the human gastroenteritis viruses listed in Table 1 are small (20 to 40 nm in diameter), and generally round in shape. They form the topic of this chapter, and have often been referred to as the "small round gastroenteritis viruses." In addition to their morphological similarities, they share other characteristics in common. These agents have been extremely refractory to *in vitro* cultivation, and, to date, only the astroviruses have been adapted to growth in cell culture. This is in contrast to many field isolates of rotavirus and enteric adenovirus, which now can be cultivated *in vitro*, albeit inefficiently in specialized or manipulated cell lines. Furthermore, several of the small round gastroenteritis viruses are closely linked to epidemic outbreaks of acute gastroenteritis. There is also evidence, either direct or in-direct, that most of these agents possess single stranded RNA. The dif-ficulties in studying the biology of these agents have also led to their consideration in common; however, this should not imply that these viruses are necessarily related viro-

TABLE 1.   Human Gastroenteritis Viruses

Virus	Virion Diameter (nm)	Nucleic Acid Type	Epidemiology	Medical Importance Demonstrated
Rotavirus	70-75	dsRNA[a]	Major pathogen of young children and infants in winter months; adults typically asymptomatic, occasionally ill.	Yes
Non-Group A Rotavirus	70-75	dsRNA	Diarrheal outbreaks in China, often in adults, often waterborne.	Only in China
Enteric Adenovirus	70-80	dsDNA	Endemic diarrhea of infants and young children in temper-ate climates.	Yes
Enteric Coronavirus	100-150	ssRNA[b]	Uncertain, (?) role in necro-tizing enterocolitis of infants.	No
Norwalk virus	27-32	Not known	Epidemics in older children and adults; often spread from contaminated water or under-cooked shellfish.	Yes
Norwalk-like Viruses (Other "SRSV")[c]	27-40	Not known	Probably similar to Nor-walk, with Snow Mountain, Hawaii as examples.	Partially
Calicivirus	27-38	ssRNA	Usually young children; shellfish-borne spread occurs in adults.	Partially
Astrovirus	27-32	ssRNA	Some outbreaks described in young children.	Partially
Small Round Featureless Agents	20-26	Not known	Weak evidence for disease, with serological proof of recent infection lacking (e.g. W, Ditchling, cockle agents).	No

[a]   ds = double-stranded
[b]   ss = single stranded
[c]   SRSV = Small round structured viruses

logically and, indeed, it is already known for example, that the astroviruses and caliciviruses are quite distinct from one another.

COMPARATIVE MEDICAL VIROLOGY OF SMALL ROUND GASTROENTERITIS VIRUSES

The first of these viruses to be studied was Norwalk virus (Blacklow et al. 1972; Kapikian et al. 1972); it can be considered the prototype agent also because more is known about its medical virology than the other small round gastroenteritis viruses. In a careful series of comparative EM studies, Caul has attempted to categorize all of the small viruses morphologically and his efforts provide a useful basis for discussing these agents (Caul and Appleton, 1982; Caul, 1988). Norwalk virus and similar agents form a first category, as shown in Table 2, namely, small round structured viruses (SRSV's) or Norwalk-like viruses as they have also been called (Christensen, 1989). These agents all share an amorphous surface structure with a feathery ragged outline that lacks geometric symmetry. They also share characteristics of density in cesium chloride (range 1.34-1.42 gm/cm^3) and their derivations from epidemics or family outbreaks of gastroenteritis. Indeed, for the most part they have been named after the location of the outbreak (eg, Norwalk, Ohio; Hawaii; Snow Mountain, Colorado; Montgomery County, Maryland; Taunton, England; Otofuke and Sapporo, Japan). As shown in Table 2, at least three of these viruses are distinct, based on immune electron microscopy (IEM) studies of their immunological relatedness. These three agents, Norwalk, Hawaii, and Snow Mountain, have also induced disease in human volunteers (Dolin et al. 1971, 1982; Schreiber et al. 1974), thereby providing the necessary human clinical materials for development of immunoassays to detect each of these viruses in stools and their antibodies in sera (Blacklow et al. 1979; Dolin et al. 1986; Greenberg et al. 1978; Treanor et al. 1988b). Little clinical and immunological information is available about the other small round structured viruses listed in Table 2, as they have not been administered to volunteers or had immunoassays developed.

The second category of viruses outlined in Table 2 is encompassed by the caliciviruses and the astroviruses. These viruses have a definitive classical surface structure as visualized by EM. The virion surface of caliciviruses has cup-shaped indentations or hollows (Latin calix: cup) which may form a six pointed star ("Star of David"). The astroviruses, by contrast, have a five or six-pointed star appearing on their surface which consists of a continuous unbroken rounded structure unlike the caliciviruses for which the surface structure is broken by the hollows. These morphologic features of the caliciviruses and astroviruses are described thoroughly elsewhere (Madeley, 1979b). Immunoassays have recently been developed for the detection of caliciviruses and astroviruses (Herrmann et al. J. Infect. Dis., In Press; Nakata et al. 1988) and will provide the means to assess fully their medical importance as human pathogens, in the same way that immunoassay permitted the recognition of the medical importance of Norwalk virus. It is already known that, despite their morphologic differences, Norwalk virus and two strains of human calicivirus share some immunologic relatedness based on human immune responses to Norwalk virus developing in patients with calicivirus gastroenteritis (Cubitt et al. 1987).

The third category of small round viruses said to be associated with gastroenteritis is comprised of certain small round featureless viral particles (Table 2). Their structure lacks discernible surface features and a sharply delineated outer

113

TABLE 2. Clinical and Immunological Relationships Among Small Round Gastroenteritis Viruses

Virus	Category	Induction of Disease in Adult Volunteers	Immunoassay Developed	Immunological Relationships Studied
Norwalk	SRSV[a] (Norwalk-like)	Yes	Yes	Distinct from Hawaii, Snow Mountain; related to Montgomery County, calicivirus
Hawaii	SRSV (Norwalk-like)	Yes	Yes	Distinct from Norwalk, Snow Mountain, Montgomery County
Snow Mountain	SRSV (Norwalk-like)	Yes	Yes	Distinct from Norwalk, Hawaii, Montgomery County
Montgomery County	SRSV (Norwalk-like)	Yes	No	Related to Norwalk (identical?)
Taunton (UK)	SRSV (Norwalk-like)	Not Done	No	Distinct from Astrovirus
Other "SRSV's"	SRSV (Norwalk-like)	Not Done	No	No
Otofuke and Sapporo	SRSV (Norwalk-like)	Not Done	No	Otofuke and Sapporo are related, but not related to Norwalk
Calicivirus	Classically structured	Not Done	Yes	Related to Norwalk. Distinct from Astrovirus
Astrovirus	Classically structured	Infection induced but not disease	Yes	Distinct from Norwalk, Hawaii, Snow Mountain, calicivirus
Marin County	Classically structured	Infection induced but not disease	Yes	Is Astrovirus Type 5
W, Ditchling, cockle, Paramatta	Small round featureless (parvoviruses?)	Not Done	No	Distinct from Norwalk

[a] SRSV: Small round structured viruses.

edge, in contrast to the SRSV's which do possess a delineated outline, albeit feathery and ragged. Also, the small round featureless particles seem to be smaller than 27-40nm sized SRSV's, caliciviruses and astroviruses (Tables 1 and 2), being reported as 20-26nm in diameter (Caul, 1988). These smaller agents include Ditchling, W (Wollan), cockle and Paramatta (Caul, 1988), all of which have been seen in stools of individuals experiencing epidemic gastroenteritis. However, the etiological relationship of these small agents to gastroenteritis is not clear, because serological evidence of recent infection with most of these agents is lacking, unlike the case for the SRSV's, astroviruses and caliciviruses. It has been suggested that these 20-26nm particles represent parvoviruses in feces that are passengers and bear no clear relationship to gastroenteritis (Caul, 1988).

With this background information about the comparative medical virology of small round gastroenteritis viruses in mind, Norwalk virus, other SRSV's, caliciviruses and astroviruses will now be discussed individually. The smaller 20-26nm Ditchling, W, cockle and Paramatta agents will not be covered due to their lack of clear association with gastroenteritis (see above).

NORWALK VIRUS

Norwalk virus was the first viral agent to be definitively associated etiologically with human gastroenteritis (Blacklow et al. 1972; Kapikian et al. 1972). It was first captured for laboratory study by human volunteer experiments (Dolin et al. 1971), and then detected by IEM in the stools of these volunteers (Kapikian et al. 1972). Because of an extensive series of human experimental studies performed with Norwalk virus during the 1970's, clinical material (stools and paired sera) became available for the development of immunoassays to recognize the virus in stools and its antibodies in human sera (Blacklow et al. 1979; Greenberg et al. 1978). These immunoassays have been the prerequisite for our understanding of the medical importance of Norwalk virus as a major cause of epidemics of gastroenteritis (Greenberg et al. 1979b, Kaplan et al. 1982). It is interesting that, despite our understanding of Norwalk virus epidemiology, this virus has still to be cultivated in cell culture, despite intensive efforts. Moreover, none of the Norwalk-like viruses (SRSV's) have been cultivated to date, and far fewer human volunteer studies have been performed with the Norwalk-like agents than with Norwalk virus. As a result, there are limited defined human clinical materials with which to set up immunoassays for the SRSV's and to perform widespread epidemiological investigations to ascertain their medical importance. Thus, little is known about the medical importance of the SRSV's in contrast to Norwalk virus, and, in order to discuss these agents, we of necessity must concentrate on the prototype agent, Norwalk virus.

Biological Characteristics

As viewed by IEM, Norwalk virus is a round-shaped, non-enveloped particle, 27nm in its average diameter, with a feathery, ragged outline that is difficult to recognize in feces by direct EM (Kapikian et al. 1972). This is probably because the virus is shed in relatively low titer in feces (in contrast to rotavirus, for example) and is also small.By IEM reaction with human convalescent sera possessing Norwalk antibodies, particles are aggregated and are then able to be visualized. No clearcut substructure is apparent, perhaps due to its being obscured by antibody in the IEM reaction; however, a suggestion of small indentations on the virion surface has been described, although these are not as pronounced as the cupshaped depressions of caliciviruses (Kapikian and Chanock, 1985). Also, most IEM studies of Norwalk virus have been performed on   -70°C stored stool specimens, which has been reported to obscure charac-teristic calicivirus morphology seen in fresh stool samples (Cubitt et al. 1987). It has not been possible to visualize Norwalk virus particles in EM studies of small intestinal mucosa biopsy specimens from infected volunteers.

Because Norwalk virus has not been cultivated in cell culture and is also shed in relatively small amounts in feces, the nature of its nucleic acid has not been determined. However, sufficient amounts of virus have been purified from feces to determine its protein composition. The virion possesses a single structural protein of molecular weight 59,000 daltons (Greenberg et al. 1981a). The only defined group of animal viruses known to possess one major virion structural protein of reported similar molecular weight is the calicivirus group. Although Norwalk virus lacks the characteristic calicivirus cup-shaped depressions on the surface of the virion, it has shown hints of surface indentations and has mostly been examined in stool samples that have been frozen and thawed, as mentioned above.

Clinicopathological Findings

The name, Norwalk virus, is derived from an outbreak of "winter vomiting disease" that occurred in the community of Norwalk, Ohio in 1968 (Dolin et al. 1971). A bacteria- and toxin-free filtrate derived from the stool of an affected adult in the outbreak reproduced the naturally occurring disease syndrome when administered orally to healthy adult volunteers (Dolin et al. 1971). Subsequent volunteer studies helped to delineate the clinical findings and pathophysiology of Norwalk virus infection. Approximately half of inoculated volunteers developed illness, with a spectrum of symptoms ranging from diarrhea without vomiting in some, to vomiting without diarrhea in others, and to both diarrhea and vomiting in additional individuals. Thus, it seems that the descriptive syndromes of "winter vomiting disease" and "viral diarrhea" may be produced by the same agent. Illness develops following an 18 to 48 hour incubation period and is self-limited, generally lasting for 12 to 48 hours.

Norwalk virus gastroenteritis is accompanied by a mucosal lesion of the proximal small intestine that is characterized by damage to villous absorptive cells, infiltration of the lamina propria by polymorphonuclear leukocytes and mononuclear cells, and villous shortening with crypt hypertrophy (Schreiber et al. 1973). These changes are associated with malabsorption of d-xylose, lactose and fat, and revert to normal within two weeks of the onset of illness. By contrast, there is no alteration of the morphology of the rectal or gastric mucosa during illness, fecal leukocytes are not present, and gastric secretion of hydrochloric acid, pepsin, and intrinsic factor are not altered. Thus, the term "enteritis" may seem appropriate for this clinical entity. However, there is a marked delay in gastric emptying, and this abnormal gastric motor function is likely responsible for the nausea and vomiting that often occurs with Norwalk illness (Meeroff et al. 1980). Based on this observation, it is proper to categorize this clinical entity as "gastroenteritis." Oral administration of bismuth subsalicylate reduced the severity and duration of abdominal cramps in ill volunteers and shortened gastrointestinal symptoms from 20 to 14 hours (Steinhoff et al. 1980).

Diagnosis

Norwalk virus can be detected in stools by IEM (Kapikian et al. 1972), and a general rating of serum antibody activity can be made by its ability to clump the virus in an IEM reaction (Parrino et al. 1977). However, the IEM technique is obviously cumbersome and not suited for routine or rapid diagnosis. Norwalk viral antigen was first detected in stools by radioimmunoassay (RIA) and serum antibody to the virus measured by an RIA blocking test (Blacklow et al. 1979; Greenberg et al. 1978). These RIA tests, and all subsequent immunoassays that have been developed rely upon the use of stools and sera from human volunteers as critical reagents. As a result, all diagnostic assays for Norwalk virus and its antibodies are restricted to the few research laboratories that possess the necessary human reagents. There are no commercially available assays for Norwalk virus. The virus has not been able to be purified sufficiently from human stool to permit preparation of the hyperimmune animal serum reagents needed for diagnostic purposes.

An enzyme-linked immunosorbent assay (ELISA) to detect Norwalk virus antigen in stools has been developed that is more sensitive than the RIA test (Herrmann et al. 1985, 1986). It is also more convenient than RIA as it does not require radioactivity and expensive laboratory equipment; instead it uses horseradish peroxidase as its label. A biotin-avidin ELISA test has also been developed that measures both Norwalk antigen and antibody (Gary et al. 1985). IgM antibody responses to Norwalk virus can be detected and indicate recent infection, but the IgM assays also require reagents derived from human volunteers (Cukor et al. 1982; Erdman et al. 1989).

Epidemiology

The epidemiology of Norwalk virus as a medically important pathogen has been well defined using the immunoassays described above. Disease usually occurs in an epidemic fashion. Forty-two percent of 74 acute gastroenteritis outbreaks studied in the United States were associated with Norwalk virus, and an additional 23% were provisionally associated (a minority of patients with seroconversions) with the virus or related viruses (Kaplan et al. 1982). These outbreaks typically occur in certain settings: recreational camps, cruise ships, contaminated drinking or swimming water, ingestion of raw or poorly cooked shellfish, nursing homes, schools (elementary to college), and community or family locations. Outbreaks occur during all seasons, affecting older children and adults, but sparing infants and young children. Antibody prevalence levels by age in the United States are very low in childhood, and rapidly rise during adolescence and early adulthood, reaching about 60% of the population by middle age. In developing tropical nations such as Thailand, Bangladesh, Ecuador and the Philippines, antibody prevalence levels rise at the earlier ages of 2 to 6 years (Cukor et al. 1980; Greenberg et al. 1979a). Infection has been conclusively shown to be spread by the fecal-oral route based on volunteer experiments (Blacklow et al. 1972; Dolin et al. 1971). There are some epidemiological data to suggest air-borne transmission of a Norwalk-like virus hospital outbreak in which it was hypothesized that the virus became airborne as a result of explosive diarrhea and vomiting or by movement of contaminated laundry (Sawyer et al. 1988). This hypothesis is consistent with the reported detection of Norwalk virus in vomitus (Greenberg et al. 1979c). It also fits logically with the extremely rapid secondary spread of Norwalk virus infection that is often observed.

Clinical Immunity

Our understanding of the clinical immunity to Norwalk virus infection is derived primarily from human volunteer studies. When a group of 12 volunteers was inoculated with Norwalk virus and rechallenged 27 to 42 months later, the same six volunteers who became ill on the first challenge again became ill on rechallenge (Parrino et al. 1977). Significant serum antibody rises, usually from baselines of pre-existing antibody, occurred with each illness. The six volunteers who were clinically well on the initial challenge remained well on rechallenge. Moreover, serological responses and prechallenge Norwalk antibody were for the most part absent in these non-ill volunteers. Short-term resistance to illness was observed as most previously ill volunteers remained well when rechallenged 4 to 14 weeks later (Blacklow et al. 1972, 1979; Parrino et al. 1977).

These studies indicate that the rise in serum antibody seen after Norwalk illness appears to be a marker for infection in susceptible individuals but lacks a protective role. Also, Norwalk illness commonly occurs in the presence of serum antibody. In contrast, those who resist illness usually have low or absent levels of antibody both prior to and after exposure to the virus. Intestinal antibody levels follow the same pattern as serum antibody (Blacklow et al. 1979; Greenberg et al. 1981b). The explanation for this pattern of clinical immunity is unclear. One unproven suggestion relies on a genetic control of Norwalk virus resistance or susceptibility, perhaps at the level of intestinal receptor sites. Another possible explanation is that repetitive exposures to the virus are needed to generate eventual illness and immune response. This would indicate that the resistant volunteers have had fewer naturally occurring previous exposures to the virus than susceptible subjects. Regardless, it is quite clear that immunity to Norwalk virus is not long lasting and that repeated bouts of illness throughout life would seem possible. This is consistent with the frequent role of Norwalk virus in nursing home outbreaks of gastroenteritis.

Immunological Relatedness To Caliciviruses

Human immune responses to Norwalk virus detected by RIA have developed in patients experiencing gastroenteritis caused by two well-defined strains of human caliciviruses (Cubitt et al. 1987). Human caliciviruses can be divided into several serologically distinct strains, using IEM (Cubitt et al. 1987). In two gastroenteritis outbreaks, due to the UK4 strain of calicivirus, 16 of 20 (80%) symptomatic patients developed seroconversions by IEM to the etiologic agent UK4. Twelve of these 20 (60%) patients also developed antibody rises by immunoassay to Norwalk virus. In another two gastroenteritis outbreaks due to the UK2 strain of calicivirus, 8 of 8 (100%) symptomatic patients developed seroconversions by IEM to the etiologic agent, UK2. Two of these 8 (25%) patients also developed antibody rises by RIA to Norwalk virus. These data, relying upon naturally occurring human immune responses, support the concept that Norwalk virus belongs to the family Caliciviridae which is also suggested by the morphological and biochemical characteristics of the virus outlined above in the section on "Biological Characteristics".

NORWALK-LIKE VIRUSES

The Norwalk-like viruses (SRSV's) share with Norwalk virus the morphological and biophysical characteristics described in the "Comparative Virol-

118

ogy" section above. They are also derived from epidemic or family outbreaks of gastroenteritis and are listed in Table 2. Much less information is available for each of these agents than for Norwalk virus. They are discussed individually, below.

Hawaii

The Hawaii agent was first described in a family diarrheal outbreak and is an antigenically distinct agent from Norwalk virus and the Montgomery County agent based on cross challenge studies in volunteers (Wyatt et al. 1974). It produces clinically indistinguishable disease from Norwalk virus, and also induces a similar reversible proximal small intestinal lesion (Schreiber et al. 1974). The virus particle is able to be visualized in feces by IEM, and is immunologically distinct by IEM testing from Norwalk, Montgomery County and Snow Mountain agents (Dolin et al. 1982; Thornhill et al. 1977). As is the case for all of the Norwalk-like viruses, limited human clinical materials are available for the establishment of immunoassays for widespread use which would permit an understanding of Hawaii's epidemiological and biological properties. Recently, an ELISA test that has detected Hawaii antigen in stools of volunteers and seroconversion in these subjects has been described (Treanor et al. 1988b). The ELISA antigen test has also confirmed the antigenic distinctiveness of Hawaii from the Norwalk and Snow Mountain agents.

Snow Mountain

The Snow Mountain agent (SMA) is derived from a gastroenteritis outbreak at a resort camp in Colorado and was first identified in stools of volunteers by IEM (Dolin et al. 1982). Like Norwalk virus, it possesses a single structural protein of molecular weight 62,000, which is compatible with the calicivirus group (Madore et al. 1986a). Both RIA and ELISA tests have been developed to detect SMA in stools and seroconversions to the virus (Dolin et al. 1986; Madore et al. 1986b). Widespread epidemiological studies have not been performed. SMA is antigenically distinct from Hawaii and Montgomery County agents by IEM and immunoassay testing. It is interesting, however, that 2 of 21 patients in a SMA outbreak also seroconverted to Norwalk virus by RIA testing, suggesting that some antigenic determinant(s) are shared between these viruses (Guest et al. 1987).

Of particular interest has been the recent production of a monoclonal antibody against the Snow Mountain agent by *in vitro* immunization of murine spleen cells (Treanor et al. 1988a), a technique which has also been used to prepare monoclonal antibodies to cell cultivated astrovirus (Herrmann et al. 1988). This technique permits the use of small quantities of antigen for immunization, an important issue in attempting to raise antibodies to the currently noncultivatable Norwalk-like viruses. A monoclonal antibody of the IgM class has been prepared that reacts with purified SMA and its single structural protein but not with Norwalk or Hawaii viruses. The antibody is less efficient than standard polyclonal antibody in detecting SMA in stools by immunoassay. However, this antibody should facilitate further purification and characterization of SMA. Preparation of additional monoclonal antibodies to Norwalk-like viruses should greatly facilitate future extensive diagnostic and seroepidemiologic studies.

Montgomery County

Derived from a family gastroenteritis outbreak in Maryland, the Montgomery County agent has produced illness in volunteers, and appears similar to Norwalk virus based on cross challenge studies in volunteers (Wyatt et al. 1974). It also seems to be identical to Norwalk virus based on IEM tests (Thornhill et al. 1977) and is now considered to be Norwalk virus.

Taunton

This agent is derived from a hospital gastroenteritis outbreak in the United Kingdom (Caul et al. 1979; Lewis et al. 1988). Its relationship to other SRSV's has not been reported. By IEM testing, an outbreak has been recently uncovered at a nursing home in Florida (Glass et al. 1989).

Other SRSV's

A few other agents have been described by EM or IEM and derived from outbreaks of gastroenteritis (Appleton, 1987; Caul and Appleton, 1982). Little is known about these agents such as whether they actually may be Norwalk virus, SMA or Hawaii agent.

Otofuke and Sapporo

The Otofuke and Sapporo agents from Japan are related to one another by IEM, but not to Norwalk virus (Kogasaka et al. 1981). The Japanese agents are reported to be 33 to 40nm in diameter, in contrast to the 27nm diameter reported as characteristic for Norwalk, Hawaii, Montgomery County, and Snow Mountain viruses. However, precise measurements of these agents are difficult and may vary from one laboratory to another (Caul and Appleton, 1982). For example, the Norwalk and Hawaii viruses have been estimated at 30 to 35nm diameter by British workers, and, indeed, Norwalk virus has been measured by its discoverers to be 32nm in its longest diameter (Caul, 1988; Kapikian and Chanock, 1985).

CALICIVIRUS

Human calicivirus is a classically structured small round gastroenteritis virus (Table 2). Other caliciviruses are found in animals such as swine, cats, sea lions, cattle, pigs, chicken and dogs, but gastroenteritis is not a prominent feature of some of these infections. However, these animal caliciviruses have been characterized virologically to a much greater extent than the human diarrheal agent which has yet to be cultivated *in vitro*, and this information has been useful as a model for studying human calicivirus (Cubitt, 1987; Schaffer, 1979). The animal caliciviruses contain positive single stranded RNA and were formerly classified as a genus of the family, Picornaviridae. However, with the recognition of their distinct morphology, single major polypeptide composition, and differing genome strategy, they have been reclassified as a separate family, Caliciviridae.

Human calicivirus is detected in diarrheal stools by direct EM, and does not require IEM for its recognition. Its characteristic morphology (see

"Comparative Medical Virology" section above) by EM has sufficed for use as a diagnostic tool. Development of more convenient assays has been slow and human volunteer studies have not been performed. An RIA to detect human calicivirus in stools has been developed using a Japanese strain, and relies upon a hyperimmune guinea pig serum raised against the virus purified from stool (Nakata et al. 1983). The assay has also been adapted to measure serum antibody (Nakata et al. 1985a). More recently, the RIA has been converted to an ELISA format to detect the virus and its antibody (Nakata et al. 1988).

At least three serologically distinct strains can be demonstrated by IEM, (UK1, UK2 and Japan) and evidence exists for two additional strains, UK3 and UK4 (Cubitt et al. 1987). The RIA for calicivirus antigen detects UK1, UK2, UK4 and Japan in stools, indicating a group-specific antigen for these viruses; however, the RIA antibody test has not detected seroconversions in patients with the two UK strains that have been studied. As mentioned above under "Norwalk virus," some patients in calicivirus UK2 and UK4 outbreaks have seroconverted to Norwalk virus, which supports the concept that Norwalk virus belongs to the family, Caliciviridae. The Norwalk virus and calicivirus strains described above are antigenically distinct, however, as they do not cross-react by IEM.

Antibody age prevalence studies indicate the acquisition of antibody during early childhood in a manner similar to that for Group A rotavirus but different from Norwalk virus (Cubitt and McSwiggan, 1987; Nakata et al. 1985a, 1988). By school age, most young children possess serum antibodies.

Calicivirus disease has been reported in two main epidemiological patterns (Cubitt, 1987). The primary and most frequent pattern is that of gastroenteritis in infants and young children (Cubitt and McSwiggan, 1981; Suzuki et al. 1979), which occurs in the general pediatric population as well as in schools, orphanages and in nosocomial settings. Symptoms are described as indistinguishable from rotavirus (Cubitt and McSwiggan, 1981). Three percent of children with diarrhea in day care centers in the U.S. have been reported to have their illness due to calicivirus (Matson et al. 1989), an incidence rate half that noted for rotavirus and higher than that noted for Campylobacter, Shigella, and Salmonella. One EM study of pediatric viral gastroenteritis in Buffalo, New York has found calicivirus in 2% of 304 cases in contrast to rotavirus (76%), adenovirus (10%), and astrovirus and other small round viruses (7%) (Riepenhoff-Talty et al. 1983). Another pediatric EM study in London showed calicivirus in 5%, rotavirus in 71%, adenovirus in 14%, astrovirus in 7%, and Norwalk-like agents in 3% (Cubitt, 1987). Data collected by Cubitt suggest that the UK1 and Japan strains of human calicivirus are particularly associated with pediatric gastroenteritis.

The second epidemiological pattern for calicivirus illness is indistinguishable from Norwalk virus disease. Attack rates are high in adults, and outbreaks are often associated with consumption of contaminated shellfish, water, and cold foods (Cubitt et al. 1981, 1987). This form of illness has been associated with the UK3 and UK4 strains (Cubitt, 1987), and it is particularly interesting that individuals experiencing UK4 disease have been those who have also seroconverted to Norwalk virus.

Little is known about the immune response to calicivirus. One study indicates that the presence of serum antibody to the Japan strain in infants correlates with resistance to illness (Nakata et al. 1985b). There are no comparable data reported for the UK4 strain which is the virus known to bear some similarities to Norwalk virus.

ASTROVIRUS

Human astrovirus, like calicivirus, can be categorized as a classically structured round gastroenteritis virus (Table 2). Other astroviruses are found that produce diarrhea in animals such as sheep, cattle, cats, dogs, pigs, turkeys and ducks. These animal agents are unrelated immunologically to the human astrovirus. The human astrovirus, unlike calicivirus, can be cultivated in HEK or LLCMK2 cell cultures treated with trypsin (Herrmann et al. 1988; Lee and Kurtz, 1981), and therefore a body of information, albeit sparse, exists about the virology of the agent. It contains positive single stranded RNA that codes for four structural proteins (Kurtz, 1989; Kurtz and Lee, 1987) of similar molecular weight as is found in picornaviruses; however, the distinctive star-shaped morphology of astrovirus distinguishes it from the picornaviruses.

Five serotypes of human astrovirus are described (Kurtz and Lee, 1984). These include the Marin County agent, which was uncovered in a gastroenteritis outbreak that occurred in a convalescent home for the elderly (Oshiro et al. 1981). The difficulties in categorizing correctly small round gastroenteritis viruses are exemplified by the Marin County agent which was originally described as a Norwalk-like virus (Oshiro et al. 1981). However the Marin County agent is now known to be an astrovirus type 5 that possesses some cross-reactivity with type 1 (Herrmann et al. 1987). Astrovirus is distinct immunologically from Norwalk virus, Hawaii virus, SMA and calicivirus (Herrmann et al. 1988; Konno et al. 1982).

Human astrovirus can be detected in diarrheal stools by direct EM, and this has served, until recently, as the only diagnostic tool for this agent. All published studies on the clinical and epidemiological features of infection have relied entirely on EM, and therefore, a complete understanding of the medical and epidemiological importance of this agent is lacking because of the cumbersome nature of EM. Recently, however, astrovirus-specific monoclonal antibodies have been prepared that react with a common group antigen shared by all astrovirus serotypes including the Marin County agent (Herrmann et al. 1988). These monoclonal antibodies have now been employed in an ELISA format to detect astroviruses in stools with a high degree of sensitivity and specificity (Herrmann et al. J. Infect. Dis., In Press). Thus, a simple and convenient means now exists to facilitate studies on the importance of astroviruses in gastroenteritis.

Antibody age prevalence studies, using IEM or immunofluorescence, indicate the acquisition of antibody in over 70% of British children by 3 to 4 years of age (Kurtz and Lee, 1978) and the presence of antibody in 50% of young adults in Japan (Konno et al. 1982). Recently, an RIA to detect IgM and IgG antibodies to astrovirus has been developed and has established that antibodies to astrovirus type 1 are present in half of British children 6 to 23 months of age (Wilson and Cubitt, 1988).

Descriptions of astrovirus clinical illness are limited in scope. Watery diarrhea is prominent and vomiting is said to be less common (Kurtz, 1989; Kurtz and Lee, 1987). Illness is most frequent in young children from infancy up to 7 years of age (Ashley et al. 1978; Konno et al. 1982; Kurtz et al. 1977), and appears to be less severe than rotavirus diarrhea (Kurtz et al. 1977). Eighty percent of 79 babies infected with astrovirus had diarrhea (Madeley, 1979a). Adult infection is usually asymptomatic or mild (Konno et al. 1982; Kurtz et al. 1977; Oshiro et al. 1981), but outbreaks in adults are described in residential facilities for the elderly (Gray et al. 1987; Oshiro et al. 1981).

When administered to eight adult volunteers, an astrovirus containing inoculum produced disease in one subject who also shed large amounts of virus (Kurtz et al. 1979). Nine subsequent volunteers were given a fecal filtrate derived from the volunteer with diarrhea, and astrovirus shedding subsequently occurred in two of them without illness. Virus specific antibody rises were noted in 13 of 16 inoculated volunteers studied. It thus appears that the astrovirus strain studied causes a transmissible infection that is of low pathogenicity for adults. These findings are also consistent with the usual mild or asymptomatic infection pattern felt to occur in adults; however, it is also clear from the Marin County agent outbreak that illness can be prominent in adults (Oshiro et al. 1981). In regard to clinical immunity, it is clear that in the presence of serum antibody, infection of volunteers did not result in diarrhea (Kurtz et al. 1979).

REFERENCES

Appleton, H (1987) Small round viruses: classification and role in food-borne infections. In Novel Diarrhoea Viruses. CIBA Foundation Symposium 128, Chichester: John Wiley and Sons, pp. 108-119.

Ashley CR, Caul EO, Paver WK (1978) Astrovirus-associated gastroenteritis in children. J Clin Pathol 31:939-943.

Blacklow NR, Dolin R, Fedson DS, DuPont H, Northrop RS, Hornick RB, Chanock RM (1972) Acute infectious nonbacterial gastroenteritis: etiology and pathogenesis. Ann Intern Med 76:993-1008.

Blacklow NR, Cukor G, Bedigian MK, Echeverria P, Greenberg HB, Schreiber DS, Trier JS (1979) Immune response and prevalence of antibody to Norwalk enteritis virus as determined by radioimmunoassay. J Clin Microbiol 10:903-909.

Caul EO (1988) Small round human fecal viruses. In Pattison JR (ed), Parvoviruses and Human Disease. Boca Raton: CRC Press, pp. 139-163.

Caul EO, Appleton H (1982) The electron microscopical and physical characteristics of small round human fecal viruses: an interim scheme for classification. J Med Virol 9:257-265.

Caul EO, Ashley CR, Pether JVS (1979) "Norwalk-like" particles in epidemic gastroenteritis in the UK. Lancet 2:1292.

Christensen ML (1989) Human viral gastroenteritis. Clin Microbiol Rev 2:51-89.

Cubitt WD (1987) The candidate caliciviruses. In Novel Diarrhoea Viruses. CIBA Foundation Symposium 128. Chichester: John Wiley and Sons, pp. 157-179.

Cubitt WD, McSwiggan DA (1981) Calicivirus gastroenteritis in North West London. Lancet 2:975-977.

Cubitt WD, McSwiggan DA (1987) Seroepidemiological survey of the prevalence of antibodies to a strain of human calicivirus. J Med Virol 21:361-368.

Cubitt WD, Pead PJ, Saeed AA (1981) A new serotype of calicivirus associated with an outbreak of gastroenteritis in a residential home for the elderly. J Clin Pathol 34:924-926.

Cubitt WD, Blacklow NR, Herrmann JE, Nowak NA, Nakata S, Chiba S (1987) Antigenic relationships between human caliciviruses and Norwalk virus. J Infect Dis 156:806-814.

Cukor G, Blacklow NR (1984) Human viral gastroenteritis. Microbiol Rev 48:157-179.

Cukor G, Blacklow NR, Echeverria P, Bedigian MK, Puruggan H, Basaca-Sevilla V (1980) Comparative study of the acquisition of antibody to Norwalk virus in pediatric populations. Infect Immun 29:822-823.

Cukor G, Nowak NA, Blacklow NR (1982) Immunoglobulin M responses to the Norwalk virus of gastroenteritis. Infect Immun 37:463-468.

Dolin R, Blacklow NR, DuPont H, Formal S, Buscho RF, Kasel JA, Chames RP, Hornick R, Chanock RM (1971) Transmission of acute infectious nonbacterial gastroenteritis to volunteers by oral administration of stool filtrates. J Infect Dis 123:307-312.

Dolin R, Reichman RC, Roessner KD, Tralka TS, Schooley RT, Gary W, Morens D (1982) Detection by immune electron microscopy of the Snow Mountain agent of acute viral gastroenteritis. J Infect Dis 146:184- 189.

Dolin R, Roessner KD, Treanor JJ, Reichman RC, Phillips M, Madore HP (1986) Radioimmunoassay for detection of the Snow Mountain agent of viral gastroenteritis. J Med Virol 19:11-18.

Erdman DD, Gary GW, Anderson LJ (1989) Development and evaluation of an IgM capture enzyme immunoassay for diagnosis of recent Norwalk virus infection. J Virol Meth 24:57-66.

Gary GW Jr, Kaplan JE, Stine SE, Anderson LJ (1985) Detection of Norwalk virus antibodies and antigen with a biotin-avidin immunoassay. J Clin Microbiol 22:274-278.

Glass RI, Monroe SS, Stine S, Madore P, Lewis D, Cubitt D, Grohmann G, Ashley C (1989) Small round structured viruses: the Norwalk family of agents. In Farthing, MJG (ed.), Viruses and the Gut. London: Swan Press Ltd., pp. 87-90.

Gray JJ, Wreghitt TG, Cubitt WD, Elliot PR (1987) An outbreak of gastroenteritis in a home for the elderly associated with astrovirus type 1 and human calicivirus. J Med Virol 23:377-381.

Greenberg HB, Wyatt RG, Valdesuso J, Kalica AR, London WT, Chanock RM, Kapikian AZ (1978) Solid-phase microtiter radioimmunoassay for detection of the Norwalk strain of acute nonbacterial epidemic gastroenteritis virus and its antibodies. J Med Virol 2:97-108.

Greenberg HB, Valdesuso J, Kapikian AZ, Chanock RM, Wyatt RG, Szmuness W, Larrick J, Kaplan J, Gilman RH, Sack DA (1979a) Prevalence of antibody to the Norwalk virus in various countries. Infect Immun 26:270-273.

Greenberg HB, Valdesuso J, Yolken RH, Gangarosa E, Gary W, Wyatt RG, Konno J, Suzuki H, Chanock RM, Kapikian AZ (1979b) Role of Norwalk virus in outbreaks of nonbacterial gastroenteritis. J Infect Dis 139:564-568.

Greenberg HB, Wyatt RG, Kapikian AZ (1979c) Norwalk virus in vomitus. Lancet 1:55.

Greenberg HB, Valdesuso JR, Kalica AR, Wyatt RG, McAuliffe VJ, Kapikian AZ, Chanock RM (1981a) Proteins of Norwalk virus. J Virol 37:994-999.

Greenberg HB, Wyatt RG, Kalica AR, Yolken RH, Black R, Kapikian AZ, Chanock RM (1981b) New insights in viral gastroenteritis. Perspect Virol 11:163-187.

Guest C, Spitalny KC, Madore HP, Pray K, Dolin R, Herrmann JE, Blacklow NR (1987) Foodborne Snow Mountain Agent gastroenteritis in a school cafeteria. Pediatrics 79:559-563.

Herrmann JE, Nowak NA, Blacklow NR (1985) Detection of Norwalk virus in stools by enzyme immunoassay. J Med Virol 17:127-133.

Herrmann JE, Kent GP, Nowak NA, Brondum J, Blacklow NR (1986) Antigen detection in the diagnosis of Norwalk virus gastroenteritis. J Infect Dis 154:547-548.

Herrmann JE, Hudson RW, Blacklow NR, Cubitt WD (1987) Marin County agent, an astrovirus. Lancet 2:743.

Herrmann JE, Hudson RW, Perron-Henry DM, Kurtz JB, Blacklow NR (1988) Antigenic characterization of cell-cultivated astrovirus serotypes and development of astrovirus-specific monoclonal antibodies. J Infect Dis 158:182-185.

Kapikian AZ, Chanock RM (1985) Norwalk group of viruses. In B.N. Fields (ed.), Virology, New York: Raven Press, pp. 1495-1517.

Kapikian AZ, Wyatt RG, Dolin R, Thronhill TS, Kalica AR, Chanock RM (1972) Visualization by immune electron microscopy of a 27-nm particle associated with acute infectious nonbacterial gastroenteritis. J Virol 10:1075-1081.

Kaplan JE, Goodman RA, Schonberger LB, Lippy EC, Gary GW (1982) Gastroenteritis due to Norwalk virus: an outbreak associated with a municipal water system. J Infect Dis 146:190-197.

Kogasaka R, Nakamura S, Chiba S, Sakuma Y, Terashima H, Yokoyama T, Nakao T (1981) The 33- to 39-nm virus-like particles, tentatively designated as Sapporo agent, associated with an outbreak of acute gastroenteritis. J Med Virol 8:187-193.

Konno T, Suzuki H, Ishida N, Chiba R, Mochizuki R, Tsunoda A (1982) Astrovirus-associated epidemic gastroenteritis in Japan. J Med Virol 9:11-17.

Kurtz JB (1989) Astrovirus. In Farthing MJG (ed.), Viruses and the Gut. London: Swan Press Ltd., pp. 84-87.

Kurtz J, Lee T (1978) Astrovirus gastroenteritis age distribution of antibody. Med Microbiol Immunol 166:227-230.

Kurtz JB, Lee TW (1984) Human astrovirus serotypes. Lancet 2:1405.

Kurtz JB, Lee TW (1987) Astroviruses: human and animal. In Novel Diarrhoea Viruses. CIBA Foundation Symposium 128. Chichester: John Wiley and Sons, pp. 92-101.

Kurtz JB, Lee TW, Pickering D (1977) Astrovirus associated gastroenteritis in a children's ward. J Clin Pathol 30:948-952.

Kurtz JB, Lee TW, Craig JW, Reed SE (1979) Astrovirus infection in volunteers. J Med Virol 3:221-230.

Lee TW, Kurtz JB (1981) Serial propagation of astrovirus in tissue culture with the aid of trypsin. J Gen Virol 57:421-424.

Lewis DC, Lightfoot NF, Pether JVS (1988) Solid phase immune electron microscopy with human immunoglobulin M for serotyping of Norwalk-like viruses. J Clin Microbiol 26:938-942.

Madeley CR (1979a) Viruses in the stools. J Clin Pathol 32:1-10.

Madeley CR (1979b) Comparison of the features of astroviruses and caliciviruses seen in samples of feces by electron microscopy. J Infect Dis 139:519-523.

Madore HP, Treanor JJ, Dolin R (1986a) Characterization of the Snow Mountain agent of viral gastroenteritis. J Virol 58:487-492.

Madore HP, Treanor JJ, Pray KA, Dolin R (1986b) Enzyme-linked immunosorbent assays for Snow Mountain and Norwalk agents of viral gastroenteritis. J Clin Microbiol 24:456-459.

Matson DO, Estes MK, Glass RI, Bartlett AV, Penaranda M, Calomeni E, Tanaka T, Nakata S, Chiba S (1989) Human calicivirus-associated diarrhea in children attending day care centers. J Infect Dis 159:71-78.

Meeroff JC, Schreiber DS, Trier JS, Blacklow NR (1980) Abnormal gastric motor function in viral gastroenteritis. Ann Intern Med 92:370-373.

Nakata S, Chiba S, Terashima M, Sakuma Y, Kogasaka R, Nakao T (1983) Microtiter solid-phase radioimmunoassay for detection of human calicivirus in stools. J Clin Microbiol 17:198-201.

Nakata S, Chiba S, Terashima H, Nakao T (1985a) Prevalence of antibody to human calicivirus in Japan and Southeast Asia determined by radioimmunoassay. J Clin Microbiol 22:519-521.

Nakata S, Chiba S, Terashima H, Yokoyama T, Nakao T (1985b) Humoral immunity in infants with gastroenteritis caused by human calicivirus. J Infect Dis 152:274-279.

Nakata S, Estes MK, Chiba S (1988) Detection of human calicivirus antigen and antibody by enzyme-linked immunosorbent assays. J Clin Microbiol 26:2001-2005.

Oshiro LS, Haley CE, Roberto RR, Riggs JL, Croughlan M, Greenberg H, Kapikian A (1981) A 27-nm virus isolated during an outbreak of acute infectious nonbacterial gastroenteritis in a convalescent hospital: a possible new serotype. J Infect Dis 143:791-796.

Parrino TA, Schreiber DS, Trier JS, Kapikian AZ, Blacklow NR (1977) Clinical immunity in acute gastroenteritis caused by Norwalk agent. N Engl J Med 297:86-89.

Riepenhoff-Talty M, Saif LJ, Barrett HJ, Suzuki H, Ogra PL (1983) Potential spectrum of etiological agents of viral enteritis in hospitalized infants. J Clin Microbiol 17:352-356.

Sawyer LA, Murphy JJ, Kaplan JE, Pinsky PF, Chacon D, Walmsley S, Schonberger LB, Phillips A, Forward K, Goldman C, Brunton J, Fralick RA, Carter AO, Gary WG, Glass RI, Low DE (1988) 25- to 30-nm virus particle associated with a hospital outbreak of acute gastroenteritis with evidence for airborne transmission. Amer J Epid 127:1261-1271.

Schaffer FL (1979) Caliciviruses. Comp Virol 14:249-284.

Schreiber DS, Blacklow NR, Trier JS (1973) The mucosal lesion of the proximal small intestine in acute infectious non-bacterial gastroenteritis. N Engl J Med 288:1318-1323.

Schreiber DS, Blacklow NR, Trier JS (1974) The small intestinal lesion induced by Hawaii agent acute infectious nonbacterial gastroenteritis. J Infect Dis 129:705-708.

Steinhoff MC, Douglas RG Jr, Greenberg HB, Callahan DR (1980) Bismuth subsalicylate therapy of viral gastroenteritis. Gastroenterology 78:1495-1499.

Suzuki H, Konno T, Kutsuzawa T, Imai A, Tazawa F, Ishida N, Katsushima N, Sakamoto M (1979) The occurrence of calicivirus in infants with acute gastroenteritis. J Med Virol 4:321-326.

Thornhill TS, Wyatt RG, Kalica AR, Dolin R, Chanock RM, Kapikian AZ (1977) Detection by immune electron microscopy of 26 to 27 nm virus-like particles associated with two family outbreaks of gastroenteritis. J Infect Dis 135:20-27.

Treanor J, Dolin R, Madore HP (1988a) Production of a monoclonal antibody against the Snow Mountain agent of gastroenteritis by in vitro immunization of murine spleen cells. Proc Nat Acad Sci USA 85:3613-3617.

Treanor JJ, Madore HP, Dolin R, (1988b) Development of an enzyme immunoassay for the Hawaii agent of viral gastroenteritis. J Virol Meth 22:207-214.

Wilson SA, Cubitt WD (1988) The development and evaluation of radioimmune assays for the detection of immune globulin M and G against astrovirus. J Virol Meth 19:151-160.

Wyatt RG, Dolin R, Blacklow NR, DuPont H, Buscho R, Thornhill TS, Kapikian AZ, Chanock RM (1974) Comparison of three agents of acute infectious nonbacterial gastroenteritis by cross-challenge in volunteers. J Infect Dis 129:709-714.

DISCUSSION

Wright J (Gull Laboratories, Salt Lake City, UT):

Any estimates on the size of the genome of either the astroviruses or caliciviruses?

Blacklow N:

What has been reported on this has been very minimal. John Kurtz at Oxford has preliminary data that human retrovirus has a positive-strand RNA genome of about 7,500 nucleotides. I am not aware of any data on human calcicivirus.

Bishop R (Royal Children's Hospital, Melbourne, Australia):

I was very interested that you have developed an ELISA as a monoclonal. In the past, one of the problems with identifying astros was being sure of the actual appearance. The star-shape looks lovely when you see it, but you don't always see it. I'm just wondering, you had a very good correlation between EM and an ELISA. Would you, in every case that you saw them in EM, have picked them as an astro if you hadn't had the ELISA? How often, in other words, did you see the star on those positive ones?

Blacklow N:

These samples were looked at by EM by two different laboratories, blinded. The first was by David Cubitt in London, and the second was in our laboratory by Dorothy Henry. So, the specimens were doubly confirmed as being EM positive. In our studies to date, samples have been first looked at by EM, and then by ELISA.' We have subjected other round viral particles appearing by electron microscopy to the assay, and they are negative. These include human calicivirus strains from David Cubitt, Norwalk-like specimens that we've had, Snow Mountain agent from Rochester, and others, and these have all been negative in the astrovirus monoclonal based ELISA.

Riepenhoff-Talty M (Children's Hospital, Buffalo, NY):

I wonder if you recall Maria Szymanski and Peter Middleton and their mini reo. Would you comment on what you think mini reo is?

Blacklow N:

I don't know what it represents. It definitely is not related to the Reoviridae. Most workers in the field have not detected or described mini-reovirus in stools. What has been your experience with that?

Riepenhoff-Talty M (Children's Hospital, Buffalo, NY):

We have not been able to detect it either.

Lennette E (California Public Health Foundation, Berkeley, CA):

What about the toro viruses? You have made no reference to those, but can you say anything about the characteristics and their potential importance?

Blacklow N:

These are interesting agents which are not small round gastroenteritis viruses, they're shaped differently, being enveloped, pleomorphic and 130 mm in diameter. The Berne and Breda agents described by Flewett have been noted in a couple of individual cases of diarrhea. Tom Flewett in the United Kingdom has performed extensive seroepidemiologic studies for these so called toro viruses, and has not found that they seem to have any significant role at all in human diarrhea.

RABIES - NEW CHALLENGES BY AN ANCIENT FOE

Richard W. Emmons

Viral and Rickettsial Disease Laboratory
Division of Laboratories
California State Department of Health Services
Berkeley, California, 94704, USA

INTRODUCTION

It has only been four years since Dr. Jean Smith presented to the 5th session of this series of symposia an excellent talk entitled "The Changing World of Rabies" (Baer and Smith, 1986). That very timely report discussed new techniques in laboratory diagnosis, new knowledge of the pathogenesis of rabies and the host's immune response, the exciting developments in monoclonal antibody analysis for strain differentiation, and hopeful trends in prevention and treatment of rabies. But it is worthwhile to check in again on the progress that is being made - and the challenges that still remain. I'll need to repeat some of the topics covered in 1985, but will highlight some other new information as well; and anyway, this audience has changed; and the fascination and challenge of rabies are as great as ever.

So turn for a few minutes from the challenge of "new" viruses such as HIV-1, human gastroenteritis viruses, HHV-6, and other important ones to consider this most ancient virus - and still among the most feared and most important in the world. Rabies was described as early as 2200 B.C. in the Babylonian code of Eshnunna; in 400 B.C. by Democritus; and in 200 B.C. by Aristotle (Robinson, 1984). The classic descriptions and basic knowledge of this horrible disease, recorded then and up to the beginning of the 20th century A.D., are largely valid today. It may seem that little progress has been made in its control. Despite considerable success in most so-called "developed" countries in preventing canine and human cases, the global problem is probably getting worse. Although the human disease is highly preventable, and animal sources of rabies could be greatly suppressed, if not eliminated, in many places, if sufficient social concern, will, medical organization, and money were available and applied, much of the time these resources are diverted elsewhere; and rabies, the threat of rabies, and/or the costs of rabies prevention remain major problems in both industrialized and developing countries in most of the world. A 1981 WHO survey revealed 20,482 reported human deaths per year (Baer et al. 1988), and such surveys are widely recognized as greatly underestimating the true problem. In fact, in India alone, special surveys estimated that some 20,000 cases occur annually (some say twice this), and over 250,000 people are given antirabies treatment each year. (Baer, 1988; Warrell and Warrell, 1988). Thailand, the Philippines, China, other Asian and African countries also report large numbers of cases. Even in the United States of America, where human cases are extremely rare, some 600,000 to 2,000,000 dog bites occur annually and an estimated 20,000 to 30,000 people are given post-exposure anti-rabies treatment annually (Robinson, 1984). The costs of these treatments and of dog and cat vaccination in the U.S.A. amount to over $300,000,000 per year. The costs of the fear of contracting rabies cannot be calculated. Compare these data with the 90,990 AIDS cases, 52,435 of them fatal, up to March 31, 1989 in the U.S.; and some 100,000 cases of AIDS recorded in the world since 1983.

Indigenously-acquired human rabies is now very rare in the U.S.A. In the past decade there were only seven such cases; the most recent one in California was in 1987 (Bartlow et al. 1988); the previous indigenous California case was in 1969. But the threat of exposure is always with us; and extensive intercontinental travel and the "global village" concept are realities. We owe it to our fellow humans and fellow creatures, as well as for our own self-interest, to try to rid the entire world of this scourge. It is hard to see that rabies virus has any need to exist, or plays any useful role in the world.

Despite this gloomy picture, and the antiquity and persistence of this terrible foe, there are some new findings and new tools to help meet some of the challenges rabies presents. The total subject is too large to cover comprehensively in this short time, and I recommend a number of excellent, recent reviews for background information (Baer, 1975; Baer and Smith, 1985; Baer et al. 1988; Johnson, 1989; Kaplan, 1983; Kuwert et al. 1985; Robinson, 1984; Sureau et al. 1988). I will briefly review, mainly from the public health laboratory point of view, just a few topics which may be of special interest to you as clinicians, laboratorians, and public health professionals, which have had some significant, recent research and discussion in the literature: the virus - its properties and ecology; the clinical disease and its management; prevention of rabies; and laboratory diagnosis. Only selected references are provided, not necessarily the initial reports or the most comprehensive ones, but they give entree to the vast literature on rabies and illustrate the main points of this report.

THE VIRUS AND ITS ECOLOGY

Despite early recognition of rabies as a viral disease, until the past decade much remained unknown about the basic virus properties. Significant advances have now been made in our understanding of the molecular biology of the virus and other members of the rhabdovirus group. The rabies virus (RV) particle is approximately 180 nm x 75 nm, often described as "bullet shaped"; and consists of a clearly defined lipid bilayer envelope, studded with 10 nm long peplomers formed of an 80,000 dalton glycoprotein (GP) - the major surface antigen. The two envelope membranes include two non-glycosylated proteins, M1 and M2, external to the nucleocapsid complex which contains a core of tightly structured helix of 30-35 coils - the single stranded RNA molecule of about 4.6×10^6 Da, with negative sense polarity. Also contained in the nucleocapsid is RNA-dependent RNA transcriptase (the enzyme on which the virus depends for duplication) and structural proteins L, N, and NS. Recent cloning and sequencing of the complete genome has revealed its organization and molecular differentiation from other rhabdoviruses. The five structural genes (N, NS, M, G, and L) and the deduced amino acid sequences for these have been almost completely determined. These protein structures have specific biologic and immunologic properties - the transcriptase, the nucleoprotein (NP), non-structural phosphoprotein, matrix protein, and glycoprotein, which is the most extensively analyzed, since it induces and binds viral neutralizing antibodies and is also a determinant and target for virus immune T lymphocytes. The techniques of molecular biology will continue to aid greatly in explaining the immunology, pathogenesis, genetic variability, virus-host interactions, modes of viral replication and transmission, and the unique immunologic and biologic properties of rabies and rabies-like viruses (Baer et al. 1988; Kaplan, 1983; Koprowski, 1988; Wunner et al. 1988). It has been shown, e.g., that loss of pathogenicity by some fixed strains of virus is associated with replacement of arginine at position 333 of the glycoprotein with isoleucine, glutamine, or glutamic acid (Wunner et al. 1988).

Rabies virus strains in nature are remarkably stable antigenically, despite the large number of minor variants that exist, and the relative ease with which variants can be selected *in vitro* with neutralizing monoclonal antibodies. The variants found in nature do not readily undergo the "shifts" or "drifts" seen for influenza viruses, e.g., and have not undergone evolutionary diversification, as have, apparently, many of the arbovirus groups. The virus is basically the same throughout the world, producing the same clinical disease patterns, is identified by the same diagnostic tests, and a single vaccine and immune gamma globulin prevent rabies anywhere in the world. However, one or more rabies-related viruses do pose problems in this regard, and closer analysis of the rabies strain variations that do exist is revealing interesting new information.

The development of rabies monoclonal antibodies (Mabs), following Kohler and Milstein's report in 1975 of the basic method for hybridoma production (Kohler and Milstein, 1975), has helped greatly in the analysis of strain variation, supplementing standard methods such as host-specificity, cross-protection, cross-neutralization, other serologic tests, and cell-culture growth characteristics (Baer and Smith, 1985; Bussereau et al. 1988; Dietzschold et al. 1987, 1988; Flamand et al. 1980a, 1980b; Rupprecht et al. 1987; Smith, 1988; Smith et al. 1984, 1985, 1986; Whetstone et al. 1984; Wiktor and Koprowski, 1978; Wiktor et al. 1980).

Hybridomas secreting a large variety of GP or NP-specific Mabs are produced by fusion of myeloma cells with BALB/C mouse splenocytes from mice which have been immunized against various RV strains. For example, 41 GP-specific, and 33 NC-specific Mabs were prepared for some of the initial studies. For analysis of NP variants, reactivity patterns by indirect immunofluorescence (IIF) are determined on cells infected in suspension, then distributed to wells of plates; 60-80% of cells become infected in 24 h, and can then be stained by IIF. Glycoprotein characterization is done by measuring the virus neutralization (Nt) index for each anti-glycoprotein-specific Mab against each virus isolate. (Dietzschold et al. 1988). Screening of isolates with NC Mabs is relatively easy by IIF or EIA, and results are highly reproducible. In contrast, analysis with GP-specific Mabs is slower and more difficult, since the Nt assay takes 3-4 days, and results are less reproducible. However, both methods are useful; e.g. RV isolates from bats generally are homogeneous by anti-NC analysis, but are readily distinguished from raccoon, fox, skunk, or other strains; but they are quite heterogeneous by GP analysis and show interesting differences between bat species.

With the help of Mab analysis, the taxonomy of the rhabdoviruses has been much better clarified. The family *Rhabdoviridae* includes over a hundred viruses of vertebrates, invertebrates, and plants, all very similar in morphology and other basic physicochemical characteristics. The subgroup or genus

Lyssavirus currently includes four serotypes: rabies virus, Lagos bat virus (from frugivorous bats on Lagos Island, Nigeria, in 1956); Mokola virus (first isolated in 1968 from shrews near Ibadan, Nigeria) and Duvenhage virus (isolated in 1971 from a man of that name who died of a rabies-like illness following a bite by a sick bat in Transvaal, South Africa). Virus strains very similar, but apparently not identical, to Duvenhage virus have been isolated from bats in northern Europe, and are thought responsible for human cases of rabies-like disease; and Mokola virus has also been claimed to cause human illness. Much more study of these rabies-related viruses and their significance in human illness is needed. They show extensive cross-reactions and some cross-protection among themselves and with rabies virus, but are clearly distinct. A virus described in 1958 from the brain of a horse with "staggers", a Nigerian disease suggestive of rabies, has some serologic relationship to rabies virus but has been inadequately studied. Two other viruses (Obodhiang virus, from *Mansonia* mosquitoes in Sudan; and Kotonkan virus from *Culicoides* gnats in Nigeria) which are somewhat related antigenically to rabies virus, but are quite different in other respects, are of no known importance for humans (Bussereau et al. 1988; Dietzschold et al. 1987, 1988; Kaplan 1983; Kuwert et al. 1985; Wiktor and Koprowski 1978; Wiktor et al. 1980).

While these rabies-related viruses are fascinating and must be better understood, RV itself remains challenging enough for our resources. Domestic or semi-domestic dogs, and to a lesser extent cats, constitute the most important hosts in most of the world. Other domestic or farm animals, such as cattle, sheep, and horses, are occasionally involved. Major wildlife reservoir species in the world include the fox in western Europe and Canada; wolf and raccoon dog in eastern Europe; skunks, insectivorous bats, raccoons and foxes in the U.S.A; the arctic fox in northern polar regions; and the mongoose in southeast Asia and some Caribbean islands.

The Mab technique has facilitated analysis of viral strains from these hosts and has helped to explain some mysteries or clarify some uncertainties; for example, to reveal the most probable source of infection in the unexpected human or animal case who had no known bite exposure. A particular NC Mab from ERA/SAD strain of virus is being used to identify vaccine virus in field trials of oral immunization of foxes. Mabs help in pathogenesis studies *in vivo*, or in *in vitro* cell culture studies of genetic stability or variations; in distinguishing vaccine strains from street strains of virus when studying apparent live vaccine-induced clinical rabies in animals, or rare cases of vaccine failure in people or animals; and in describing the topography of the antigenic structure of glycoproteins and nucleocapsid proteins (Dietzschold et al. 1988).

Much has been learned, but there are many unsolved or new questions to be answered. What are the long term trends in wildlife rabies? Are there regular or predictable cycles in various important species which can help control programs be more efficiently directed? Are there changes in basic reservoir hosts in a region, such as the growing importance and northward spread during the past decade of raccoon rabies along the east coast of the U.S., (Jenkins et al. 1988), which can be explained and prevented? Pathogenesis of infection in animals or cell cultures is still not well enough understood. How does the virus remain latent during the often very long incubation period? What triggers the virus to become active and progress from the latent state to spread to and through neural tissue and cause the clinical disease? What is the true nature of the viral receptor? It is not completely accepted by all researchers that it is identical to the acetylcholine receptor. It is curious that many and diverse types of cells in culture can be infected, but rabies virus appears to attach to and infect very limited types of cells *in vivo*, primarily neural cells and salivary gland cells. What actually protects against RV infection? Is it Nt antibody that binds to and neutralizes incoming virus? Or does Nt antibody elicit the host's T cell response? And perhaps viral NP not the GP, is the major target for cytotoxic T cells, since NP or fragments of it can be detected on the surface of infected cells. Clearly, much of importance remains to be learned (Koprowski, 1988). What is the true status and significance of the so-called "chronic" rabies or the asymptomatic "carrier state", claimed or shown under unusual, special circumstances. There are old claims and poorly documented, anecdotal reports in the older literature of such phenomena, often misinterpreted or too readily incorporated into folklore; some have not stood up to careful scrutiny or experimentation and are now doubted, such as the claim that vampire bat rabies involves healthy carrier bats. However, recovery from verified clinical rabies can readily be demonstrated in a proportion of certain animal hosts infected experimentally with certain strains of virus; and at least three, perhaps more, human cases have survived the disease. There are intriguing reports of experiments with African strains of virus that showed, in special circumstances, virus secretion in the saliva of infected dogs, even though the animal remained healthy (Fekadu, 1988). These reports need further verification. It is important that such experiments or claims not be misinterpreted or over-emphasized. I do not know of any records that such a purported "carrier" animal has actually transmitted rabies. The basic rule in rabies prevention remains that if a biting dog remains healthy during the 10-day observation period, it can be released from quarantine, and the person bitten need not be given antirabies treatment.

HUMAN CASES - CLINICAL PICTURE AND MANAGEMENT

Despite much progress in the field of antiviral drugs and intensive care in general, the picture for human rabies remains gloomy. Survival of three human cases, and in one of these apparently complete recovery with no residual damage, was documented up to 1988 (Baer et al. 1988; Hattwick, 1974). A recent short research communication from India mentioned 54 clinically-diagnosed cases over a 17-year period, with 2 survivors, ages 18 and 10 (Gode et al. 1988). Insufficient clinical details and no supporting laboratory data were given to support the validity of this claim, but it adds to the body of anecdotal literature on non-fatal rabies (Hattwick, 1974; Remington et al. 1985; Warrell and Warrell, 1988).

There are three special challenges to be met: (1) We must improve our techniques for early recognition of cases; (2) We must continue to try for case survival and recovery, and also must carefully document, virologically and serologically, that this has occurred, to dispel the pervasive sense of despair and inevitability of death that may lead to a self-fulfilling outcome, if it prevents us from the extraordinary efforts that may be required to cure the disease; and (3) We must learn how to better manage the hospital environment and medical staff caring for the patient. The earliest symptoms of rabies are often nonspecific and suggest some other illness: malaise, headache, anorexia, sore throat, photophobia, gastrointestinal or upper respiratory symptoms, fatigue, and low grade fever. Subtle mental changes and anxiety may suggest a psychiatric disorder, or malingering. Neuralgia or paresthesias, especially at the site of exposure, are the only relatively specific symptoms, if they occur. In the absence of an exposure history, the physician can easily be misled and there may be reluctance to consider or voice suspicion of such a rare and serious diagnosis as rabies. In fact, the patient or family may not be willing to admit to (from fear of the consequences), may not recognize the significance of, may not remember, or may not even have ever been aware of a bite or other significant exposure by a rabid animal within the past months or year. In rare cases, incubation periods of several years appear to be valid, making the exposure history even less reliable and the exposure event even less likely to be recalled. California's most recent case was a 13 year old Filipino boy in San Francisco, who became ill in December, 1987 and died of rabies 12/15/1987. No history of animal bite or significant exposure could be elicited, and he had not been outside of the U.S.A. since he immigrated from the Philippines in 1981. (Bartlow et al. 1988). The virus strain isolated from his brain at autopsy was typed by Dr. Jean Smith at the Centers for Disease Control as very similar to dog strains from the Philippines (personal communication). A recent case in Oregon died 2/3/89, with no known animal bite, but a probable exposure in Mexico at least 11 months earlier (Centers for Disease Control, 1989).

Since 1980 in the U.S.A, 6 of the 10 rabies cases had no reported or known animal bite or exposure (Centers for Disease Control, 1989; Fishbein et al. 1988). In three of these cases, rabies was not suspected and diagnosed until after death. So far, there is no laboratory test to show, during the incubation period, if a person actually had become infected by the virus when bitten by the suspect or proven rabid animal, and if that person is doomed to develop rabies disease or not. However, in many cases during the later prodromal period of clinical illness, it is possible to diagnose rabies and thus trigger appropriate intensive care and case management. Immunofluorescence (IF) staining of frozen sections of skin has been positive in many of the cases where it was tried (Bryceson et al 1975; Warrell et al. 1988). Less frequently, IF staining of corneal epthelial cell smears has shown positive results. Also, antibody in response to rabies infection can be detected by the rapid IIF method or the RFFIT test for Nt antibody, but not until about a week after onset of the first symptoms (Johnson, 1989).

As mentioned earlier, it is important to anticipate the need for verification of the etiology of encephalitis cases, and to collect and properly store the specimens needed - serum, cerebrospinal fluid, saliva, throat washings, corneal smears, skin biopsies, etc. Also, stool samples are not important for rabies studies but should be collected to rule out enteroviruses which may cause encephalitis which mimics rabies. When recovery from rabies does occur, we will want the incontrovertible proof. In the three well-studied cases, antibody titers in sera and CSF have greatly exceeded those due to vaccine stimulation, and are accepted as proof of rabies disease. Virus was not isolated, and it is probable that in the case that is going to survive, virus growth and progression will be halted by the patient's immune response before the virus can reach sites where it can be recovered, such as the peripheral cerebral cortex, the CSF, saliva, skin nerves, or corneal cells. Perhaps new methods such as the polymerase chain reaction, in-situ hybridization, or other probe techniques will be successful.

Treatment of rabies remains as a huge challenge. Results of recent, aggressive attempts to cure the disease have been very disappointing. Trials with ribavirin, interferon, and intravenous hyperimmune gamma globulin have not been conclusive, although some hopeful results have been noted and research in these directions is continuing. (Anderson et al. 1984; Fishbein and Baer, 1988; Merigan et al. 1984; Robinson, 1984; Vodopija, 1988; Warrell and Warrell, 1988).

The management of hospitalized cases of rabies is a topic more for a clinical symposium than this one, and there are many recent, excellent reviews emphasizing the need for intensive care, a climate of optimism, and the minimal hazard to hospital staff so that inappropriate fear, over-use of preventive vaccination, etc. can be avoided (Baer and Smith, 1986; Remington et al. 1985; Warrell and Warrell, 1988).

In the first week or two of clinical illness, virus can be isolated from patient saliva, but the rapidly rising antibody levels soon neutralize it. In fact, no human-to-human transmission has been documented, except by corneal transplant (Kaplan, 1983; Warrell and Warrell, 1988). At least six such cases have been reported, the first by Houff et al. (1979) in a 37-year old woman who received a transplant from a man who had died of an undefined, progressive paralytic disease; a case in France; two cases in Thailand who each received a corneal transplant from the same donor; and most recently, two cases in India (Gode and Bhide, 1988). In another episode in Algiers, in 1981, rabies was diagnosed in the corneal donor soon after the transplant had occurred, and the recipient was promptly flown to Paris, treated with HDCV, HRIG and human leucocyte interferon, and has remained well (Sureau et al. 1981).

PREVENTION

There are not many infectious diseases for which exposure can be recognized and effective treatment instituted after infection has occurred - fortunately, rabies is one of them. Even the old, Semple type vaccines, prepared from animal brain tissue, which are still what most of the world must use if anything at all is available, are quite effective, relatively speaking. The modern post-exposure treatment regimen, consisting of thorough washing of the wound site, immune gamma globulin (preferably HRIG) and a 5 or 6 dose series of HDCV intramuscularly, has been extremely effective. There has been no recorded failure of this regimen to prevent rabies, if it has been properly instituted within 24 h of exposure, and only rare failures at all have been attributed to delay in use, or inadvertent subcutaneous inoculation (Immunization Practices Advisory Committee, 1984; Fishbein and Baer, 1988; Warrell and Warrell, 1988; Wiktor, et al. 1964, 1988).

The major challenge to the world is not new techniques or new principles, but to achieve the political will, social will and organizational and economic means to implement what we already know how to do. Cost is the big impediment. However, new technology and research are promising to help in this regard.

The development of HRIG to replace antiserum produced in equines was an important step, because of the high rate of hypersensitivity reactions and severe serum sickness which occurred with the equine product. Studies of an HRIG produced by ion exchange chromatography for I.V. use in post-exposure treatment have recently been reported (Aoki et al. 1989). However, because HRIG is expensive and difficult to produce, and basically is unavailable to developing countries, there is renewed interest in preparing a safer equine RIG (Wilde et al. 1987, 1989). Similarly, the replacement of nerve tissue vaccines and the safer, but still less than ideal, duck embryo vaccine, by HDCV's requiring only 5 or 6 doses, rather than 14-23 doses, was a major step forward. But these vaccines are too expensive for the countries where vaccination is most needed. A number of lower-cost, purified tissue culture-based or other vaccines have been developed and put into field trials over the past several years. They show much promise, and comparable protection to that provided by HDCV's. These include: vaccines prepared from fetal rhesus monkey cells

(Rabies Vaccine Adsorbed, Michigan Dept. Public Health); primary embryonated duck eggs (Swiss Serum and Vaccine Institute, Berne); primary hamster kidney; primary fetal bovine kidney (FBKC, Pasteur Institute); primary dog kidney; primary chick embryo (PCEC, Behring Werke); primary Japanese quail embryo; a new, purified VERO cell culture vaccine (Institute Merieux, now licensed in France); and others (Baer, 1988; Berlin et al. 1982; Burgoyne et al. 1985; Centers for Disease Control, 1988; Sureau, 1988; Vodopija, 1988; Wiktor et al. 1988). Semple type vaccines are still used in over 50 countries where rabies is a major problem. There is clearly a need for national self-sufficiency in safer, cheaper cell culture vaccine production.

Other innovations are under study or proposed, including: use of multisite intradermal injections, requiring less vaccine and only three clinic visits (Warrell and Warrell, 1988); a "2-1-1" schedule of intramuscular inoculations under study in Zagreb, Yugoslavia, also requiring only three clinic visits (Vodopija, 1988); synthetic peptides as vaccines (thus far disappointing, but needing much more study); subunit vaccines; and anti-idiotype vaccines which might give broader protection against rabies-related viruses as well as rabies.

For pre-exposure immunization of veterinarians, animal control personnel and others at high risk of contact with rabid animals, studies continue on the most effective, convenient and cheapest methods, including many studies of intradermal immunization as a more cost-effective method (Baer et al. 1988). A combination of rabies and tetanus vaccines has even been considered for routine childhood immunization in regions of the world where the risk is particularly high (Lery et al. 1986).

Extensive studies are in progress on better ways to control rabies in the animal host. Control of rabies in dogs and, to a less extent, cats, in the U.S.A. and many industrialized countries has been largely successful, and compendia of approved and required vaccines and schedules are published regularly (National Association of State Public Health Veterinarians, Inc., 1989; Veterinary Public Health Unit, 1988). Again, cost and other governmental priorities have prevented such progress in the countries most in need of it.

Wildlife rabies is a much more difficult problem. Extensive trials over the past decade in Switzerland, Germany and elsewhere in Europe, and more recently in Canada and the U.S.A, have been carried out to attempt to reduce or eliminate wildlife rabies foci by use of oral vaccination with live, attenuated virus; or by recombinant vaccinia vaccines expressing RV glycoprotein. Major concerns about safety from reversion of the attenuated strain to virulence, and transmission and inadvertent pick up of the baits by non-target species, or other problems, have not been realized thus far, but continue to require great caution in implementing such programs. Although apparently effective in decreasing fox rabies in parts of Europe, the difficulty of evaluating the efficacy of such a program, in the face of long-term, cyclic trends of wildlife rabies, must be recognized. There have also been simultaneous significant decreases in fox rabies in some areas of Europe where oral vaccination programs were not being carried out (Baer, 1988; Baer et al. 1988; Esposito et al. 1988; Jenkins et al. 1988; Tolson et al. 1987).

LABORATORY DIAGNOSIS

I will conclude with just a few comments about the laboratory diagnosis of rabies. The standard procedures have been well described and there are many good references, so they need not be reiterated (Baer, 1975; Baer and Smith,

1985; Baer et al. 1988; Johnson, 1989; Sureau et al. 1988). I have already mentioned the role of IF of skin biopsy or corneal smears; IIF and RFFIT serologic tests in early diagnosis of human cases; and the value of Mabs for preparing strain-specific and sensitive reagents. A Mab-based IF conjugate (Centocor) and the BBL reagent (Becton Dickinson Co.) prepared from equine serum are very sensitive and specific, but too expensive for the developing countries where they are most needed. Some way of reducing the cost and providing the staff training and practical application of this procedure is an urgent need.

For example, an enzyme immunoassay has been proposed (Nicholson et al. 1985) as a cheaper method, not requiring a fluorescence microscope. Experimental application of avidin-biotin peroxidase staining (Fekadu et al. 1988) and dot hybridization (Ermine et al. 1988; Vishawapoka, 1988) and a rapid enzyme immunodiagnostic test using homogenized brain specimens (Bourhy et al. 1989; Perrin and Sureau, 1987; Perrin et al. 1986) have also been described. These are promising, but will require more experience and extensive comparative testing with the standard IF method before they can replace it.

Techniques for demonstration by IF staining of rabies antigen in formalin-fixed, paraffin-embedded tissues, ordinarily non-reactive, by trypsin digestion have been helpful in pathologic studies, in retrospective diagnosis of initially unsuspected cases when fresh or frozen tissue was not collected, and in experimental work (Johnson et al. 1980, Umoh and Blenden, 1981; Umoh et al. 1985). Some fascinating "archeological virology" remains to be done, since antigenicity is preserved for a long time, probably for many decades.

Although rabies diagnosis can depend largely on IF staining, there are many situations where we need to have the virus isolate - to verify the diagnosis when IF staining results are equivocal, to analyze the strain by Mab or other means, and so on. Standard mouse inoculation tests are beginning to be replaced for routine diagnosis by cell culture inoculation, which can be quicker and more convenient in some ways. Various cell cultures, such as primary hamster kidney, the BHK-21 hamster kidney cell line, and mouse neuroblastoma cells, have been used (Bourhy et al. 1989; Larghi et al. 1975; Rudd et al. 1980; Smith et al. 1978; Sureau et al. 1988; Umoh and Blenden 1983).

For serologic diagnosis, there is a clear need for a simpler test to replace the IF, mouse, and RFFIT tests, since these are labor-intensive. Much study has been done on enzyme immunoassay tests but they have not become routinely available as yet (Baer, 1975; Baer et al. 1988; Kuwert et al. 1985; Nicholson et al. 1985).

Antibody assays based on hemadsorption (Grandien, 1985), a microneutralization test based on EIA rather than IF (Mannen et al. 1987), and a dot immunobinding assay (Heberline et al. 1987) have also recently been described. It will be important to demonstrate the close correlation of results by these procedures with standard Nt antibody results and with actual immunity and resistance to challenge, if they are to be relied upon to decide the need for vaccine boosters in preexposure preventive vaccination or postexposure treatment situations.

Rabies diagnostic tests in the U.S.A. and perhaps most countries are done in public health laboratories, as I feel they should be. If this practice were ever to change, and clinical hospitals and laboratories began to use newer, simple, diagnostic tests based on enzyme immunoassays, molecular probes, or other techniques, then very strong efforts would have to be made to maintain the quality assurance, reference testing, specimen storage, record-keeping, and all the carefully constructed, interdependence and interaction of the laboratory with public health, animal control organizations, the Centers for Disease Control and the World Health Organization that have evolved over the years. If public health laboratories are to retain this responsibility, then they must continue to be sufficiently funded, staffed and supported so they can act as emergency testing and reporting laboratories and be responsive to the special anxieties, concerns and needs that revolve around rabies.

REFERENCES

Anderson LJ, Nicholson KG, Tauxe RV, Winkler WG (1984) Human rabies in the United States, 1960 to 1979: epidemiology, diagnosis and prevention. Ann Int Med 100:728-735.

Aoki FY, Rubin ME, Friesen AD, Bowman JM, Saunders JR (1989) Intravenous human rabies immunoglobulin for post-exposure prophylaxis: serum rabies neutralizing antibody concentrations and side effects. J Biol Standard 17:91-104.

Baer GM (Ed) (1975) The Natural History of Rabies, Volumes I and II; Academic Press, New York, San Francisco, London.

Baer GM (1988) Research towards rabies prevention: overview. Rev Infect Dis 10 (Suppl 4):S576-S577.

Baer GM, Smith JS (1985) Rabies virus, pp 790-795. In Lennette EH, Balows A, Hausler WJ, Jr., Shadomy HJ (Eds) Manual of Clinical Microbiology 4th Ed. American Society for Microbiology, Washington, D.C.

Baer GM, Smith JS (1986) The changing world of rabies, pp 347-400. In de la Maza LM, Peterson EM (Eds) Medical Virology V. Lawrence Erlbaum Associates, Inc. Hillsdale, NJ.

Baer GM, Bridbord K, Hui FW, Shope RE, Wunner WH, (Guest Editors) (1988) Research towards rabies prevention. A symposium. Washington, D.C., 3-5 November, 1986. Rev Infect Dis 10 (Suppl 4):S573-S815.

Berlin BS, Mitchell Jr, Burgoyne GH, Oleson D, Brown WE, Goswick C, McCullough NB (1982) Rhesus diploid rabies vaccine (adsorbed), a new rabies vaccine. Results of initial clinical studies of preexposure vaccination. JAMA 247(12):1726-1728.

Bourhy H, Rollin PE, Vincent J, Sureau P (1989) Comparative field evaluation of the fluorescent-antibody test, virus isolation from tissue culture, and enzyme immunodiagnosis for rapid laboratory diagnosis of rabies. J Clin Microbiol 27(3):519-523.

Bryceson ADM, Greenwood BM, Warrell DA, Davidson N McD, Pope HM, Lawrie JH, Barnes HJ, Bailie WE, Wilcox GE (1975) Demonstration during life of rabies antigen in humans. J Infect Dis 131:71-74.

Burgoyne GH, Kajiya KD, Brown DW, Mitchell JR (1985) Rhesus diploid rabies vaccine (adsorbed): a new rabies vaccine using FRhL-2 cells. J Infect Dis 152:204-210.

Bussereau F, Vincent J, Coudrier D, Sureau P (1988) Monoclonal antibodies to Mokola virus for identification of rabies and rabies-related viruses (1988) J Clin Microbiol 26:2489-2494.

Centers for Disease Control (1988) Rabies vaccine, absorbed: a new rabies vaccine for use in humans. MMWR 37:217-218,223. Centers for Disease Control, Public Health Service, US Dept Health and Human Services, Atlanta, GA.

Centers for Disease Control (1989) Human Rabies - Oregon, 1989. MMWR 38:335-337. Centers for Disease Control, Public Health Service, US Dept Health and Human Services. Atlanta, GA.

Dietzschold B, Rupprecht CE, Tollis M, Lafon M, Mattei J, Wiktor TJ, Koprowski (1988) Antigenic diversity of the glycoprotein and nucleocapsid proteins of rabies and rabies-related viruses: implications for epidemiology and control of rabies. Rev Infect Dis 10 (Suppl 4):S785-S798.

Dietzschold B, Tollis M, Rupprecht CE, Celis E, Koprowski H (1987) Antigenic variation in rabies and rabies-related viruses: cross-protection independent of glycoprotein-mediated virus-neutralizing antibody. J Infect Dis 156:815-822.

Ermine A, Tordo N, Tsiang H, (1988) Rapid diagnosis of rabies infection by means of a dot hybridization assay. Molec Cell Probes 2:75-82.

Esposito JJ, Knight JC, Shaddock JH, Novembre FJ, Baer GM (1988) Successful oral rabies vaccination of raccoons with raccoon poxvirus recombinants expressing rabies virus glycoprotein. Virology 165:313-316.

Fekadu M (1988) Pathogenesis of rabies virus infection in dogs. Rev Infect Dis 10 (Suppl 4):S678-S683.

Fekadu M, Greer PW, Chandler FW, Sanderlin DW (1988) Use of the avidin-biotin peroxidase system to detect rabies antigen in formalin-fixed paraffin-embedded tissue. J Virol Methods 19:91-96.

Fishbein DB, Baer GM (1988) Animal rabies: implications for diagnosis and human treatment. Ann Intern Med 109:935-937.

Fishbein DB, Dobbins JG, Bryson JH, Pinsky PF, Smith JS (1988) Rabies surveillance, United States 1987. Centers for Disease Control, Public Health Service, US Department of Health and Human Services, Atlanta, GA. CDC Surveillance Summaries, Sept 1988; MMWR 37 (No SS-4):1-17.

Flamand A, Wiktor TJ, Koprowski H (1980a) Use of hybridoma monoclonal antibodies in the detection of antigenic differences between rabies and rabies-related virus proteins. I. The nucleocapsid protein. J Gen Virol 48:97-104.

Flamand A, Wiktor TJ, Koprowski H (1980b) Use of hybridoma monoclonal antibodies in the detection of antigenic differences between rabies and rabies-related virus proteins. II. The glycoprotein J Gen Virol 48:105-109.

Gode GR, Bhide NK (1988) Two rabies deaths after corneal grafts from one donor. Lancet ii:791.

Gode GR, Saksena R, Batra RK, Kalia PK, Bhide NK (1988) Treatment of 54 clinically diagnosed rabies patients with two survivals. Indian J Med Res 88:564-566.

Grandien M, Fridell E Kindmark C-O (1985) Intradermal immunization with reduced doses of human diploid cell strain rabies vaccine: evaluation of antibody response by ELISA and mixed hemadsorption test. Scand J Infect Dis 17:173-178.

Hattwick MW (1974) Human rabies. Publ Health Rev Vol III(3):229-273.

Heberling RL, Kalter SS, Smith JS, Hildebrand DG (1987) Serodiagnosis of rabies by dot immunobinding assay. J Clin Microbiol 25:1262-1264.

Helmick CG (1983) The epidemiology of human rabies postexposure prophylaxis, 1980-1981. JAMA 250:1990-1996.

Houff SA, Burton RC, Wilson RW, Henson TE, London WT, Baer GM, Anderson LJ, Winkler WG, Madden DL, Sever JL (1979) Human-to-human transmission of rabies virus by corneal transplant. N Eng J Med 300:603-604.

Immunization Practices Advisory Committee (1984) Rabies prevention United States, 1984. MMWR 33:393-402. Centers for Disease Control, Public Health Service, US Dept Health and Human Services, Atlanta, GA.

Jenkins SR, Perry BD, Winkler WG (1988) Ecology and epidemiology of raccoon rabies. Rev Infect Dis 10(Suppl 4):S620-S625.

Johnson HN (1989) Rabies virus. Ch 26, pp 893-923. In Schmidt NJ, Emmons RW (Eds). Diagnostic Procedures for Viral, Rickettsial and Chlamydial Infections, 6th Ed, American Public Health Association, Washington, DC.

Johnson KP, Swoveland PT, Emmons RW (1980) Diagnosis of rabies by immunofluorescence in trypsin-treated histologic sections. JAMA 244:41-43.

Kaplan C (1983) Recent studies on rabies. Ch 12, pp 255-262. In Waterson AP (Ed) Recent Advances in Clinical Virology. Number Three. Churchill Livingston, Edinburgh, London, Melbourne, New York.

Köhler G, Milstein C (1975) Continuous cultures of fused cells secreting antibody of predefined specificity. Nature 256:495-497.

Koprowski H (1988) Glimpses into the future of rabies research. Rev Infect Dis 10 (Suppl 4):S810-S813.

Kuwert E, Merieux C, Koprowski H, Bögel K (Eds) (1985) Rabies in the Tropics. Springer-Verlag, Berlin, Heidelberg, New York, Tokyo.

Lery L, Rotivel Y, Trabaud MA, Parvaz P, Mouterde S, Relyveld EH (1986) Combined tetanus-rabies vaccination. Dev Biol Stand 65:209-220.

Mannen K, Mifune K, Reid-Sanden FL, Smith JS, Yager PA, Sumner JW, Fishbein DB, Tong TC, Baer GM (1987) Microneutralization test for rabies virus based on an enzyme immunoassay. J Clin Microbiol 25:2440-2442.

Merigan TC, Baer GM, Winkler WC, Bernard KW, Gibert CG, Chany C, Veronesi R (1984) Human leucocyte interferon administration to patients with symptomatic and suspected rabies. Ann Neurol 16:82-87.

National Association of State Public Health Veterinarians, Inc. (1989) Compendium of Animal Rabies Control, 1989 MMWR 37:789-796. Centers for Disease Control, Public Health Service; US Dept Health and Human Services, Atlanta, GA

Nicholson KG, Warrell MJ, Xueref C, Lee S (1985) pp 334-339. In Kuwert E, Merieux C, Koprowski H, Bögel K (Eds) Rabies in the Tropics. Springer-Verlag, Berlin, Heidelberg, New York, Tokyo.

Perrin P, Rollin PE, Sureau P (1986) A rapid rabies enzyme immuno-diagnosis (RREID): a useful and simple technique for the routine diagnosis of rabies. J Biol Stand 14:217-222.

Perrin P, Sureau P (1987) A collaborative study of an experimental kit for rapid rabies enzyme immunodiagnosis (RREID). Bull.WHO 65:489-493.

Remington PL, Shope T, Andrew J (1985) A recommended approach to the evaluation of human rabies exposure in an acute-care hospital. JAMA 254:67-69.

Robinson PA (1984) Rabies virus. Ch 18, pp 485-511. In Belshe RB (Ed) Textbook of Human Virology. PSG Publishing Co, Inc. Littleton, MA.

Rudd RJ, Trimarchi CV, Abelseth MK (1980) Tissue culture technique for routine isolation of street strain rabies virus. J Clin Microbiol 12:590-593.

Rupprecht CE, Glickman LT, Spencer PA, Wiktor TJ (1987) Epidemiology of rabies virus variants. Differentiation using monoclonal antibodies and discriminant analysis. Amer J Epidemiol 126:298-309.

Smith AL, Tignor GH, Emmons RW, Woodie JD (1978) Isolation of field rabies virus strains in CER and murine neuroblastoma cell cultures. Intervirology 9:359-361.

Smith JS (1988) Monoclonal antibody studies of rabies in insectivorous bats of the United States. Rev Infect Dis 10(Suppl 4):S637-S643.

Smith JS, Reid-Sanden FL, Roumillat LF, Trimarchi C, Clark K, Baer GM, Winkler WG (1986) Demonstration of antigenic variation among rabies virus isolates by using monoclonal antibodies to nucleocapsid proteins. J Clin Microbiol 24:573-580.

Smith JS, Sumner JW, Roumillat LF, Baer GM, Winkler WG (1984) Antigenic characteristics of isolates associated with a new epizootic of raccoon rabies in the United States. J Infect Dis 149:769-774.

Smith JS, Sumner J, Roumillat F, Baer GM, Winkler WG (1985) Epidemiological analysis of street rabies viruses from enzootic areas of the United States. pp 604-610. In Kuwert E, Merieux C, Koprowski H, Bögel K, (Eds) Rabies in the Tropics. Springer-Verlag. Berlin, Heidelberg, New York, Tokyo.

Sureau P (1988) History of rabies: advances in research towards prevention during the last 30 years. Rev Infect Dis 10(Suppl 4):S581-S584.

Sureau P, Lafon M, Baer GM (1988) Rhabdoviridae: Rabies and Vesicular Stomatitis Viruses. Ch 29, pp 571-594. In Lennette EH, Halonen P, Murphy FA (Eds) Laboratory Diagnosis of Infectious Diseases. Principles and Practice. Springer-Verlag. New York, Berlin, Heidelberg, London, Paris, Tokyo.

Sureau P, Portnoi D, Rollin P, Lapresle C, Lapresle C, Chaouni-Berbich A (1981) Prevention de la transmission interhumaine de la rage apres greffe de cornee. C R Acad Sci Paris III 293:689-692.

Umoh JU, Blenden DC (1981) Immunofluorescent staining of rabies virus antigen in formalin-fixed tissue after treatment with trypsin. Bull.WHO 59:737-744.

Umoh JU, Blenden DC (1983) Comparison of primary skunk brain and kidney and raccoon kidney cells with established cell lines for isolation and propagation of street rabies virus. Infect Immun 41:1370-1372.

Umoh JU, Ezeokoli CD, Okoh AEJ (1985) Immunofluorescent staining of trypsinized formalin-fixed brain smears for rabies antigen: results compared with those obtained by standard methods for 221 suspect animal cases in Nigeria. J Hyg Camb 94:129-134.

Veterinary Public Health Unit (1988) Compendium of US Licensed Animal Rabies Vaccines and their application in dogs under the California Rabies Control Program. (September 1988) California State Department of Health Services, Berkeley, CA.

Vishawapoka U, Hemachudha T, Tepsumethanon W, Polsuwan C, Tirawatnpong S, Phanuphak P (1988) Detection of rabies antigen in canine parotid glands by dot-blot technique. Lancet 1:881.

Vodopija I (1988) Current issues in human rabies immunization. Rev Infect Dis 10 (Suppl 4):S758-S763.

Vodopija I, Sureau P, Smerdel S, Lafon M, Baklaic Z, Ljubicic' M, Svjetlicic' M (1988) Interaction of rabies vaccine with human rabies immunoglobulin and reliability of a 2-1-1 schedule application for postexposure treatment. Vaccine 6:283-286

Warrell DA, Warrell MJ (1988) Human rabies and its prevention: an overview. Rev Infect Dis 10 (Suppl 4):S726-S731.

Warrell J, Looareesuwan S, Manatsathit S, White NJ, Phuapradit P, Vejjajiva A, Hoke CH, Burke DS, Warrell DA (1988) Rapid diagnosis of rabies and post-vaccinal encephalitides. Clin Exp Immunol 71:229-234.

Webster WA (1987) A tissue culture infection test in routine rabies diagnosis. Can J Vet Res 51:367-369.

Whetstone CA, Bunn TO, Emmons RW (1984) Use of monoclonal antibodies to confirm vaccine-induced rabies in ten dogs, two cats, and one fox. JAVMA 185:285-288.

Wiktor TJ, Fernandes MV, Koprowski H (1964) Cultivation of rabies virus in human diploid cell strain WI-38. J Immunol 93:353-366.

Wiktor TJ, Flamand A, Koprowski H (1980) Use of monoclonal antibodies in diagnosis of rabies virus infection and differentiation of rabies and rabies-related viruses. J Virol Methods 1:33-46.

Wiktor TJ, Koprowski H (1978) Monoclonal antibodies against rabies virus produced by somatic cell hybridization: detection of antigenic variants. Proc Natl Acad Sci USA 75:3938-3942.

Wiktor T, Plotkin SA, Koprowski H (1988) Rabies vaccine. Ch 22, pp 474-491. In Plotkin SA, Mortimer EA Jr (Eds) Vaccines. WB Saunders Co., Harcourt Brace Jovanovich, Inc. Philadelphia, London, Toronto, Montreal, Sydney, Tokyo.

Wilde H, Chomchey P, Prakonongari S, Punyaratabandhu P (1987) Safety of equine rabies immune globulin (Letter). Lancet 2:1275.

Wilde H, Chomchey P, Prakongsri S, Puyaratabandhu P, Chutivongse S (1989) Adverse effects of equine rabies immune globulin. Vaccine 7:10-11.

Wunner WH, Larson JK, Dietzschold B, Smith CL (1988) The molecular biology of rabies viruses. Rev Infect Dis 10 (Suppl 4):S771-S784.

DISCUSSION

Lennette E (California Public Health Foundation, Berkeley, CA):

As you showed, there was one case of human rabies in 1989 in Oregon and in 1987 in California. Nothing in 1988. So then why is human rabies such an important disease?

Emmons R:

Because of the fear and concern. The diagnosis is not confirmed very often, but it's raised relatively frequently, and it seems like more frequently recently. The State Virus Lab in California gets about a dozen suspect cases, which really have some possible validity to them which we then have to help study quickly. I am sure that this is true in many other states as well, mostly with people who have traveled over seas. It's a scary disease.

Lennette E:

That wasn't a very fair question, actually. As I recall, there were close to 5,000 animal cases reported last year and the year before. Although domestic animals, representing about 10% of that rabid population accounted for about 65% or 70% of human exposures, which had to be treated. That is a considerable number for vaccination. You mentioned animal control of the population. How are we going to achieve that when (this is my pitch), the prevalent opinion in this country is to have everything absolutely risk free. I'm speaking now Dick, of the vaccine developed by Wistar, which they tried to use in a field trial in a couple of islands off the coast of South Carolina. This was turned down on the basis that the genetically engineered vaccinia virus vector was not shown to be safe because the vaccine strain produced encephalitis in humans. That has always been known. So, I don't know how you can rule it out on that basis. Now, they go up to Virginia, and this is now under consideration to see if they can use Paramour Island off the Virginia coast for a field trial. When you have all these obstacles thrown in your way, it becomes very, very difficult, especially when you're dealing with groups who have only one major objective in mind, and the hell with everything else. Sorry, I didn't mean to make a speech. I did what I asked the audience not to do. But, how do you feel about these vaccines for the control of the wildlife population?

Emmons R:

All these things need to be tried. They need to be tried very carefully with all the safeguards. There is concern also about the possible spread and pathogenicity of the vaccinia virus itself, which is little used now. It can be a pathogen in its own right, a reversion of virulence. The difficulty of proving that these control programs are really working when you have these natural fluctuations of rabies in wildlife is a terrible problem. How to reduce wildlife rabies has got to continue to be tackled. Nature is always ahead of us. Population reductions just dampen it for a while and, back it comes. The only place that has really been successful in eliminating rabies is in certain island situations like Japan and Great Britain, where apparently there are no native terrestrial wildlife hosts to re-introduce or keep the virus going. Then, elimination of dog-rabies has apparently permanently eliminated the disease. I don't know the answer. We have to keep trying, but we do have to do it very, very carefully because of the concerns about live vaccines and their possible reversion to virulence. I wish there were a better carrier than vaccinia, it's kind of a nasty virus to be used.

Lennette E:

Dick, the use of the vaccinia virus vector on the basis of possible encephalopathy holds water, because we've known that for years. Perhaps one in 10,000 people who were vaccinated, up to the time vaccines were stopped, developed encephalitis. The highest incidence I know of occurred in the Netherlands about 30 years ago, when one out of 5,000 who received the vaccine developed encephalitis. But, you have to take a risk somewhere.

Lovelace M (Oregon Health Sciences University, Portland, OR):

I had the distinct experience of caring for the Oregon rabies case. I can tell you that I wanted to just echo from a clinician's standpoint, all of the things you have talked about. Number one, this individual came in and was seen in an outlying hospital with mild central nervous system problems, but predominantly abdominal discomfort. He actually underwent abdominal surgery for appendicitis. This gentlemen then, following surgery, had a very slow recovery and had neurologic sequelea that made this case very confusing, clinically. Secondly, as it became clear that he had an encephalitis, the ability to make a diagnosis of some of the unusual encephalitidies complicated his course. Finally, his end-stage, because of a number of conditions, he went very quickly down hill. We didn't have the luxury of time to institute a lot of intensive supportive care. I just wanted to make a point that, in a country where we don't have a lot of rabies cases, and where individuals who are being seen are being seen in predominantly community hospitals, I think one of the big problems that we have is the general lack of awareness of the very non-CNS clinical presentations of rabies. I think that education of the physician populations is really a key problem in the developing countries in instituting the kind of supportive care that has to be done very early in order to really support a patient through this. I really would like to see public health laboratories make available some of the diagnostic tests very early, and encouraging the submission of specimens like skin biopsies and conjunctival biopsies to their public health labs for and as a part of CNS or encephalitis cases as part of a routine work-up. This gentlemen also had about an 11 month time of incubation from the time he left his home in Mexico. Again, this gentlemen probably did not acquire rabies in the United States, but had a long latency from his initial bite which occurred in his hometown in Mexico.

Emmons R:

It is a dilemma, because it would be great if public health labs could be ready to be a fast clinical lab. They probably need to be more than we are, but it is expensive to maintain a seven day a week, 24 hour day staff, ready to do these tests. Most of the time, it proves not to be necessary. There is a hesitation. You don't want every meningitis case and every flu case to be thought of as rabies, not because it triggers extra work for the lab, that's beside the point. That can be paid for, we could make the tests available, but it does trigger a lot of what turns out to be unnecessary alarm and concern on the part of the hospital staff. You like to avoid that if it's not really needed, but you don't want to miss a case of rabies. So, you are on the horns of a dilemma. Maybe if this got to be so routine as an automatic rule out in encephalitis cases, then people would become familiar with it and not be so alarmed by rabies. We've had over and over again, the experience that the word is mentioned, the case is thought of as a possibility, and then, several days of real problems occur within the hospital environment. Theoretically, with universal precautions that are being used more and more now, this can assuage the alarm concerning the hospital staff.

Lovelace M:

One more point is that because we had instituted universal body substance precautions in our hospital, we didn't have the amount of exposure. We rapidly instituted education procedures, resulting in only one or two post exposure vaccinations. Something that I think is very unique in hospital exposure settings. So, again, a pitch for universal body substance precautions. It's very important, I think, in keeping that under control. I agree that the public health laboratories have difficulty crossing over in being available for clinicians. On the other hand, unless we have access to diagnostic techniques that are early, we're going to end up not being able to support these people and being able to add to the number of cases that recover from rabies.

Lennette E:

Cogent to that is, you have to have a high index of suspicion. What you said is very pertinent and apropos of the patient's case presentation. As I recall, the patient that Dr. Emmons mentioned here that Mike Hattwick took care of also was rather confusing initially. And this patient, incidentally if I remember correctly, represents the longest surviving case of rabies in the literature.

Participant:

Just one comment about other vectors. Joe Esposito, in our branch at CDC has got a raccoon pox vector that, in animal studies, is showing good degrees of protection. Since they've just got another isolate of raccoon pox from the swamps of New Jersey, they think that they can show that the virus is already in the environment, and thus, won't be contaminating the environment with it. Still, there will be people who will oppose its use.

HTLV-I INFECTIONS

Yorio Hinuma

Shionogi Institute for Medical Science
Settsu-shi, Osaka 566, Japan

INTRODUCTION

Retroviruses are involved in many naturally occurring neoplasms in various animal species. Their involvement in human neoplasias has been the subject of extensive research, but not until the early 1980s was definitive evidence obtained for casual association of human retroviruses with a certain human disease. A type C retrovirus, named human T-cell leukemia virus (HTLV-I), was isolated in 1980 from cases of sporadic, cutaneous T-cell lymphoma-leukemia in the USA (Poiesz et al. 1980, 1981). Independently, a retrovirus designated as adult T-cell leukemia virus (ATLV) was obtained in 1981 from cases of endemic adult T-cell leukemia (ATL) in Japan (Hinuma et al. 1981; Yoshida et al. 1982). Later, HTLV-I and ATLV were identified to be the same virus (Watanabe et al. 1984). There is now evidence that this retrovirus is etiologically related only to ATL and not to other malignancies, including T-cell lymphoid malignancies other than ATL.

ATL, which was first described in 1977 as a new malignant entity (Takatsuki et al. 1977), is endemic in Japan, especially in the southwest. Clinically, acute type ATL is characterized by rapid progression, poor prognosis, skin lesions, and hypercalcemia (Uchiyama et al. 1977). Recently, ATL cases were also found in other parts of the world such as the Caribbean basin and Taiwan (Blattner et al. 1983; Catovsky et al. 1982; Hinuma et al. 1983; Kuo et al. 1985).

In 1981, an antigen detectable by indirect immunofluorescence (IF) was found in a T-cell line, MT-1, derived from peripheral blood lymphocytes (PBLs) of a patient with ATL (Hinuma et al. 1981). Moreover, C-type retrovirus particles were detected in the same cell line by electron microscopy (EM). The antigen, ATLA, reacted with sera from almost all ATL patients and also with sera from about 25% of the healthy adults tested in areas where ATL is endemic, but with very few sera of subjects from non-endemic areas (Hinuma et al. 1981). Treatment of MT-1 cells with iododeoxyuridine (IUdR) increased their production of HTLV-I particles. Even larger amounts of HTLV-I and ATLA were detected in another T-cell line, MT-2, established by co-cultivation of normal cord-blood lymphocytes with leukemic cells from an ATL patient (Miyoshi et al. 1981). Furthermore, the antigens and virus particles were also shown to be induced when PBL from seropositive subjects were cultured *in vitro*, regardless of whether the subjects were ATL patients. Therefore, healthy seropositive subjects were considered to be HTLV-I carriers (Gotoh et al. 1982; Hinuma, et al.

1982a; Miyoshi, et al. 1982). Immortalization of normal lymphocytes has been achieved quite easily by their cocultivation with X-irradiated MT-2 cells or primary PBL from seropositive individuals (Yamamoto et al. 1982b). In biochemical studies, monoclonal integration of HTLV-I proviral DNA was detected in fresh PBL from ATL patients, but not in PBL from healthy adults (Yoshida et al. 1982). These data strongly indicate that HTLV-I is involved in the leukemogenesis of ATL.

Surprisingly, in 1985, two independent observations in Martinique and Japan noted the presence of antibodies to HTLV-I in the serum and cerebrospinal fluid of patients with chronic myelopathies of unknown origin (Gessain et al. 1985; Osame et al. 1986). These diseases in Martinique and Japan have been called tropical spastic paraparesis (TSP) and HTLV-I associated myelopathy (HAM) respectively. After careful comparison of the two diseases, the TSP and HAM have been found to be a same disease which might be caused by infection with HTLV-I. Occurrence of both ATL and TSP/HAM in a single patient is rare.

PROPERTIES OF HTLV-I

Retroviruses (RNA tumor viruses) have been isolated from many and diverse vertebrate species, including man. All retroviruses share a similar genetic structure and a similar pathway for replication. Retrovirus genomes contain two subunits of 30 to 35S RNA, forming a 60 to 70S RNA complex and reverse transcriptase. Maturation of retroviruses is completed after budding from the membrane of host cells. Retroviruses can be classified on the basis of their morphology as type B, type C, and type D. In addition, some RNA tumor viruses are known to be transmitted through germ cells (endogenous retrovirus). HTLV-I appears to correspond to type C, on the basis of its appearance in electron micrographs of thin sections of ATL cell lines. HTLV-I was shown to have a density of 1.152 to 1.16 g/cm3 by sucrose-density gradient centrifugation (Poiesz et al. 1980; Yoshida et al. 1982).

The ATLA complex detected by IF has been studied biochemically by immunoprecipitation with ATLA-reactive sera and lysates of HTLV-I-producing cells, precipitated HTLV-I, and culture supernatants free of cells and virus (Kalyanaraman et al. 1981; Schneider et al.,1984a,b; Yamamoto and Hinuma, 1982). The ATLA antigen complex was found to consist of three glycopolypeptides which are the env gene products, gp46, gp67, and gp68; the viral core polypeptides p28, p24, p19, and p15, and about eight intracellular polypeptides of 40 to 70K, some of which are precursors of viral structural polypeptides. But the ATLA antigen complexes in different ATLA-positive cell lines were found to vary to some extent (Koyanagi et al., 1984; Sugamura et al., 1984). Although primary leukemic cells contain HTLV-I genomes, these antigens are detectable only after *in vitro* culture of the cells (Gotoh et al., 1982; Hinuma, et al., 1982a).

An HTLV-I provirus integrated into fresh leukemic cells of an ATL patient was cloned, and its complete nucleotide sequence was determined (Seiki et al. 1983). The proviral genome of HTLV-I is composed of 9,032 nucleotides, and the genes includes, gag, pol, env, and pX. The pX sequence, where three proteins with molecular weights of 21, 27 and 38K could be encoded (Seiki et al. 1985), was implicated as being the transforming gene (onc gene) of HTLV-I. However, unlike all known retrovirus onc genes, this region was not homologous to the proto-onc genes in normal cellular DNA.

Analysis of DNA from fresh leukemic cells of ATL patients for HTLV-I provirus by Southern blotting showed that in all ATL cases examined, at least one HTLV-I genome was integrated into the leukemic cell DNA, but that its site of integration varied (Yamaguchi et al. 1984; Yoshida et al. 1984). Defective proviruses were found in fresh leukemic cells and shown to retain the env, pX, and 3'LTR regions, suggesting that these regions, and especially the pX region of the HTLV-I provirus, are important for initiation or maintenance of monoclonal proliferation of ATL cells.

HTLV-II was isolated in 1982 from a cell line established from the spleen of a patient with a T-cell type of hairy cell leukemia (Kalyanaraman et al. 1982). This virus was related to, but distinct from, the prototype of HTLV-I. Recently HTLV-II was reported to be quite frequently isolated from i.v. drug abusers in USA, demonstrating double infection with the human immunodeficiency virus type-1 (HIV-1). The etiological role of the HTLV-II in a specific human disease is not obvious at present.

HIV-1, also known as lymphadenopathy-associated virus (LAV)/HTLV-III/ARV) is the causative agent of the acquired immunodeficiency syndrome (AIDS). HIV-1 preferentially infects a subset of T-cells with the CD4 surface marker, and its reverse transcriptase shows preference for $Mg2+$ over $Mn2+$ for maximal activity. In these characteristics it appears to be somewhat like HTLV-I. However, HIV-1 has cytopathic effects such as causing syncytium formation, but it does not cause cell transformation. HIV-1 belongs to the lentivirus group of retroviruses, whereas HTLV-I belongs to the oncovirus group.

Viruses different from, but very similar to HTLV-I have been found in Old World monkeys. One of the viruses is called simian T-lymphotropic virus (STLV) (Hayami, 1986; Hayami et al. 1983). Infection of man with virus from monkeys or vice versa is very unlikely, as judged from the results of seroepidemiological and biochemical studies.

PATHOGENESIS OF HTLV-I

Cocultivation of normal lymphocytes from adults or newborn babies with HTLV-I-positive cell lines (e.g., MT-2) or primary leukemic cells of patients with ATL resulted in transformation and continuous growth of the recipient cells (Hoshino et al. 1983; Miyoshi et al. 1981; Popovic et al. 1983; Yamamoto et al. 1982b). All the cell lines established in this way gave positive reactions for ATLA and released HTLV-I into the medium, as revealed by electron microscopy. Most of these cell lines contained cells with the T-cell surface-marker CD4, which were very similar to those exhibited by leukemic cells of ATL patients. Although most of the cell lines established by coculture of normal lymphocytes with either primary leukemic cells or cell lines carrying HTLV-I are of T-cell lineage, the host range of this virus is very wide with respect to infectivity (Yamamoto and Hinuma, 1985); the virus can infect a variety of human and mammalian cells, such as those of rabbit, rat, monkey, and feline origin, by the coculture method. Moreover, it can also infect non-lymphoid cells, including fibroblasts, endothelial cells, lung cells, sarcoma cells, and cancer cells. Interestingly, after viral infection, only T-cells seem to be transformed. These data suggest that non-T-cells may play a role as a reservoir of the virus *in vivo*, although they do not become malignant. Indeed, continuous Epstein-Barr virus-genome-positive B cell lines expressing ATLA were established from PBL or lymph node biopsy specimens from ATL patients without the aid of interleukin 2 (IL-2) or cocultivation (Yamamoto et al. 1982a).

Several outstanding characteristics of ATL leukemogenesis by HTLV-I are known (Hunsmann and Hinuma, 1987; Yamamoto and Hinuma 1985): 1) random integration of provirus DNA of HTLV-I into DNA of the host cells; 2) monoclonal proliferation of leukemic cells; 3) the absence of viral proteins in tumor cells *in vivo*; 4) a long latency; and 5) a low frequency of occurrence (1/1,000 to 1/3000), but high frequency of establishment of infected cells *in vitro*. These characteristics could be explained as follows: HTLV-I initially infects various types of cells, including CD4+, CD8+ T cells and B cells, polyclonally. Expression of virus genomes, especially pX genomes, in T4 cells leads to proliferation of the cells. However, these cells may also express viral structural antigens on their surface and so be eliminated by the host immune surveillance system. This period of infection may last a long time, corresponding to the long latent period in healthy carriers. Then, selection of a cell carrying HTLV-I provirus but no viral antigens takes place for some as yet unknown reason, leading to monoclonal proliferation of leukemic cells. Although HTLV-I is essential for the occurrence of ATL, an additional factor(s) must be needed for selection of a single cell and initiation of its monoclonal growth from the population of cells that are polyclonally infected with HTLV-I. Thus, it appears that ATL leukemogenesis results from the interaction of HTLV-I with various additional factors.

MODE OF TRANSMISSION AND SEROEPIDEMIOLOGY

There are many healthy, seropositive individuals in endemic areas of ATL, their incidence apparently correlates with the frequency of occurrence of ATL (Hinuma et al. 1981; 1982b; Shimoyama et al. 1982). A nationwide survey of anti-ATLA in volunteer blood donors in Japan (Maeda et al. 1984) indicated that the incidence of seropositive donors was high in Kyushu (8%) with lower incidence of 0.3 to 1.2% in other areas. The incidence of seropositive donors was found to increase with age. Moreover, the virus is prevalent in certain families in ATL endemic areas (Miyoshi et al. 1982; Tajima et al. 1982); when the mother of a family is seropositive, the chance of her children also being seropositive is greater than when the mother is seronegative. ATLA-positive lymphocytes can readily be detected by short-term culture of peripheral lymphocytes of seropositive individuals; this indicates that most of these persons are healthy carriers of HTLV-I. Vertical genetic transmission of viruses, known as endogenous retroviruses, is rather common in animals. However, this route of transmission has been ruled out in the case of HTLV-I. Several possibilities for mother-to-child infection either vertically or horizontally have been considered, including transplacental and intracervical infections and perinatal infections by HTLV-I-contaminated cervical secretions or breast milk. Of these, milk from the virus-carrying mother is considered to be the most probable, since virus-carrying lymphocytes can be detected in the milk (Nakano et al. 1986). Examination of married couples showed that the virus may also be transmitted horizontally from husband to wife. In this regard it is important to note that the semen from seropositive adults contains lymphocytes that are positive for HTLV-I provirus (Nakano et al. 1984).

Blood transfusion is also a very important route of horizontal transmission of HTLV-I, since, of the several possible routes, this iatrogenic infection was found to be the major cause of increase in the frequency of affected persons (Okochi et al. 1984). Patients who received at least one unit of whole blood or packed erythrocytes or platelet concentrates from donors carrying antibodies to HTLV-I produced anti-ATLA. In contrast, no anti-ATLA was detected in any of the recipients of fresh frozen plasma prepared from seronegative donors or cell-free blood components even from seropositive donors.

Patients who had received whole blood or blood components containing cells derived from seropositive donors usually became positive for antibodies to HTLV-I within 50 days. This seroconversion was most probably due to primary infection with HTLV-I, because the recipients initially showed an IgM antibody response. The sequence of HTLV-I proviral DNA was also detected in the peripheral blood of these recipients (Sato and Okochi, 1986).

Individuals who showed seroconversion after blood transfusion did not show any symptoms indicative of acute infection with HTLV-I. In this regard, ATL seems to differ from AIDS, in which infectious mononucleosis-like symptoms have been observed as a result of acute viral infection. It is important to note that TSP/HAM patients having experiences of blood transmission are not rare. The ages of patients with TSP/HAM is general-ly younger than those with ATL and the number of patients with TSP/HAM is larger in females than in males, in contrast to ATL.

CLINICAL FEATURES OF ATL

The incidence of ATL is slightly higher in men than women. The average age of the patients is 55 years. The most characteristic finding is rapid proliferation of abnormal T-lymphocytes with lobulated, deformed nuclei in the blood and bone marrow. Most of the patients have lymphadenopathy and hepatomegaly or splenomegaly, or both. No mediastinal tumors are observed. About one-third of the patients have various cutaneous lesions without severe itching or excoriation. Blood chemical analyses frequently show hypercalcemia, dysfunction of the liver, and hypoproteinemia. Remission is rarely achieved by common antileukemic chemotherapy and is frequently interrupted by lethal complications, presumably owing to deficiency of cell-mediated immunity (Katsuki et al. 1986, 1987). Several immune deficiencies are notable in ATL patients and virus-infected individuals (Asou et al. 1986). These include general lymphadenopathy; chronic lung diseases (e.g., interstitial pneumonitis, bronchopneumonia, and lung fibrosis); opportunistic infections of the lung; cancer in various sites (e.g., the liver, stomach, skin and vagina); M proteinemia; chronic renal failure; skin candidasis; and some parasitic diseases such as strongyloidiasis. The median survival time is about three months from the onset of treatment.

ATL is a clinical entity that includes various spectra of disease. It is subdivided into five groups (Takatsuki, 1989). 1) Acute ATL - this is the most common disease pattern; however, there may be no fundamental difference between this type and the crisis type described below, which develops from chronic or smoldering ATL. 2) Chronic ATL some cases of this type of ATL were previously diagnosed as having T-cell chronic lymphatic leukemia. When such patients have anti-HTLV-I antibody as well as proviral DNA integrated into leukemia cells, they should be regarded as having chronic ATL. 3) Smoldering ATL in this type of ATL, there is no increase in the leukocyte number, but abnormal lymphocytes are definitely present in the peripheral blood for a rather long time. 4) Crisis-acute conversion from chronic or smoldering ATL. 5) Lymphoma type ATL in this type of ATL no leukemic cells are detectable in the peripheral blood, and the prognosis is usually bad. Conversion from the lymphoma type to leukemia is not unusual. The proportions of these types of ATL are about 54% acute type, 17% chronic type, 11% smoldering type, 7% crisis type, and 11% lymphoma type.

In addition, it should be noted that five patients with clinically typical ATL without HTLV-I have recently been reported (Shimoyama et al. 1986). Thus, HTLV-I may not be involved in all patients with ATL: some factors other than HTLV-I may be able to cause a clinical picture that is indistinguishable from that of typical ATL without the involvement of HTLV-I.

LABORATORY DIAGNOSIS OF HTLV-I INFECTION

Specimen Collection

Since HTLV-I can be isolated only from cultured cells, whole blood or tissue samples are processed for separation of mononuclear cells (lymphocytes) by the Ficoll-Conray method. In the case of whole blood, heparinized blood is used for this purpose. For collection of mononuclear cells, samples should be separated immediately or kept at room temperature during transport. Plasma can also be obtained for antibody tests before application of the blood to a Ficoll-Conray gradient to separate mononuclear cells. When sera are to be tested for antibody only, they can be shipped and stored at ambient temperatures, unless they are contaminated with microorganisms or will be in transit for a long period.

Detection of HTLV-I

There is evidence that HTLV-I genomes are expressed only slightly *in vivo*. This is probably owing to the *in vivo* selection of leukemic cells that do not express most viral information, especially membrane antigens on which immune attack occurs. Thus, the strategies for detection of virus and virus antigens depend mainly on the expression and amplification of viral genomes after culture. However, proviral DNA can be detected directly in cells without their cultivation.

Nucleic Acid Hybridization Method. Provirus DNA of HTLV-I integrated into cells is detectable with complementary DNA (cDNA) to the viral genome labeled with radioactive materials (Reitz et al. 1981, 1983; Yoshida et al. 1982). The Southern-blot and dot-blot methods can both be used for this purpose. For the Southern-blot procedure, lymphocyte preparations containing leukemic cells are isolated from fresh peripheral blood or lymph nodes of patients. A provirus clone such as 1 ATK-1, containing HTLV-I provirus, is used as a source of the probe for detecting the proviral sequence, since the total nucleotide sequence of this clone has been determined (Seiki et al. 1983).

Induction of HTLV-I In Vitro. PBLs are separated by Ficoll-Conray gradient centrifugation from the peripheral blood of ATL patients or healthy virus carriers. They are then cultured with or without normal PBLs from cord blood or peripheral blood. Cells are also cultured either with or without IL-2. Cocultivation of provirus-containing cells with normal lymphocytes results in transformation immortalization and establishment of cell lines at high frequencies. Control cultures consist of normal PBLs alone. The cultures are examined at least once a week for cell transformation, which is defined as the appearance of scattered aggregations and subsequent increase in their size and number. Transformation is confirmed by observation of continuous growth of the cells during serial subcultivation. Without cocultivation, however, even leukemic cells from patients can usually grow only transiently. When transformed or growing, these cells are used for studies on the virus or viral information detected by IF, enzyme-linked immunosorbent assay (ELISA), Western blot (WB), radioimmunoprecipitation (RIP), EM, or nucleic acid hybridization.

Detection of HTLV-I Antibodies

IF has been used as a reliable and standard method for measuring HTLV-I antibodies. ELISA (Taguchi et al. 1983) and particle agglutination (PA) (Ikeda et al. 1984) were developed for screening for anti-HTLV-I antibodies in donor blood. Kits for the ELISA and PA procedures are available commercially such as EITEST-ATL (Eisai Inc., Tokyo, Japan) and Serodia-ATLA (Fuji-rebio Inc., Tokyo, Japan), IF, RIP, and WB have also been used as confirmatory tests. Concerning IF, ELISA and PA but not WB and RIP, some key points are mentioned as follows.

Immunofluorescence. As described in a previous section, indirect IF method has contributed greatly to the discovery of ATLV (HTLV-I), its association with the disease, and seroepidemiological studies of the virus (Hinuma et al. 1981; Maeda et al. 1984; Tajima and Hinuma, 1985). The IF is very reliable and specific and can be used for both confirmation and screening of ATLA antibodies.

As a target, MT-1 cells (Miyoshi et al. 1979) expressing ATLA in 2 to 50% of the total cell population are used widely, although other HTLV-I-carrying cells could also be used. When the sera are positive for ATLA antibodies, mono- or multinuclear giant cells show specific fluorescence. A great advantage in use of MT-1 cells for the IF procedure is this apparent morphological feature of positive cells. Moreover, it is easy to determine whether the test sera are positive for antibodies, since many of the MT-1 cells do not contain antigen and so serve as negative controls. The antibody titer is expressed as the reciprocal of the end dilution of the serum that gives positive fluorescence. FITC-labeled anti-human IgM can also be used as a second antibody to detect IgM antibody to HTLV-I.

Enzyme-linked Immunosorbent Assay. In ELISA for HTLV-I antibody (Saxinger and Gallo, 1983; Taguchi et al. 1983), antigens are prepared from extracts of MT-2 cells or other cells that produce much HTLV-I by treatment of the cells with detergent. The assay kit consists of antigen-coated cups, the reaction solution, the enzyme substrate, and solution for stopping the enzyme reaction. The assay is carried out in microcups.

Particle Agglutination Test. The reagent for the PA test is prepared from gelatin particle carrier sensitized with ATLA on the principle that these sensitized particles are agglutinated by ATLA antibody in serum or plasma (Ikeda et al. 1984; Kobayashi et al. 1988). Antigen is prepared by concentrating the culture fluid of a virus-producing cell line, subjecting it to sucrose-gradient centrifugation, collecting the virus fraction corresponding to a density of about 1.16 g/cm3, and finally disrupting the purified HTLV-I with detergent. The PA assay kit consists of several reagents, including reconstituting solution, serum diluent, sensitized particles, unsensitized particles, and positive control serum. Plastic microplates are used for the PA test. The test is performed for either qualitative (screening) or quantitative purposes. The PA test is carried out at the room temperature. Usually two hours are satisfactory for reading the results because complete agglutination with antibody positive serum occurs quickly.

This assay has the following advantages. 1) As a microtiter technique, the test procedure is very simple and is particularly suitable for mass-screening of test samples. 2) The test is rapid: results can be assessed by the naked eye after about 2 h. 3) The PA test involves the use of a newly developed artificial carrier, Fuji particles, that do not show the nonspecific agglutination usually observed with use of other carriers. 4) In the single PA test, IgM antibody as well as IgG antibody is detectable.

Problems Associated with Screening for HTLV-I Antibody. As described already, ATL is a malignancy that has a long latent period after viral infection. The onset of leukemia probably results from the interaction of the virus with other factors. Thus, for control of ATL, it is very important to prevent viral infection itself. For preventing HTLV-I infection from transfused blood, virus-infected blood must be identified and removed from the blood bank.

Three procedures - PA, ELISA and IF - can be used to screen for HTLV-I antibody in the serum of donor blood, but each procedure has both advantages and disadvantages. For the PA test, a rather low dilution of serum, such as 1:16, is generally used. However, this could result in nonspecific agglutination due to undetermined factors and lead to misreading of the results. Moreover, the results for serum samples with low PA titers, such as 16 and 32 are not highly reproducible. However, all IF-positive samples are also positive by PA, and, therefore, PA may be suitable for screening sera of donor bloods. Since sera that are PA-positive but IF-negative usually have PA titers of 16 or 32, if these PA titers are regarded as negative, the coincidence rate of PA and IF is increased. But if this is done, some IF-positive samples with PA titers of 16 and 32 may be overlooked. Thus for the purpose of screening donor bloods, it seems better to regard sera with PA titers of more than 16 as positive. This idea is also supported by the fact that some sera that are PA-positive but IF-negative have been shown to be positive for antibody to an env gene product, gp68 of HTLV-I, by RIP. However, it is also true that about 20% of the PA-positive, IF-negative samples tested were found to be negative even by RIP. Thus, it is important to study whether this discrepancy is caused by a non-specific reaction, especially one due to impurity of the antigen preparations.

ELISA also has problems. In this method, the assessment is usually based on the degree of absorbance of negative control sera. Thus, results on samples that have an absorbance of about the cut-off value are not always reproducible, because the absorbance of the negative control sera themselves also differs somewhat from test to test.

The IF test is reliable and specific. But this procedure is not so suitable as PA or ELISA for screening large numbers of serum samples, such as those in blood centers. It sometimes provides false-negative results because it is less sensitive than RIP or PA. Moreover, it requires special expertise for reading of the results, and results for samples with low IF titers in particular may be misjudged. Furthermore, this procedure is not applicable to serum samples from patients with autoimmune diseases such as systemic lupus erythematosus, because these give various fluorescences that are not specific for HTLV-I. It is noteworthy that these sera also give a nonspecific reaction by the ELISA procedure.

Some serum samples appear positive only by PA or ELISA. However, no sera giving positive reactions in both PA and ELISA but a negative reaction in the IF test have been found. These facts suggest that the three assay systems may each show nonspecific reactions for different reasons. Thus, positive results must be confirmed by different procedures such as IF, WB, or RIP. When a large number of sera from blood donors are examined, the samples may be screened initially by a qualitative PA test, and then the positive sera may be examined by quantitative PA and ELISA.

CONTROL AND PREVENTION

The basic strategy to control any infectious disease is to prevent infection with the causative agent. Seroepidemiological studies showed that there are many ATLA-positive adults among healthy residents in ATL-endemic areas. Furthermore, results strongly suggested that the familial clustering of anti-ATLA-positive individuals results from perinatal infection from mother to child or transmission from husband to wife by sexual contact. The main cause of HTLV-I infection from mother to child is suspected to be breast feeding. Attempts to block this route of infection have already began; namely, babies of seropositive mothers are fed formula milk instead of mothers' milk. This type of approach for prevention of HTLV-I infection seems important. It is also possible to block virus transmission from carrier husbands to their wives by conventional ways that prevent introduction of infected semen.

It has been proven that IF antibody positive but not negative donors are definitely HTLV-I carriers (Gotoh et al. 1982; Hinuma et al. 1982b). Hence the spread of HTLV-I infection by blood transfusion can be prevented by screening donor blood for anti-HTLV-I (anti-ATLA) and discarding infected blood. In Japan there are thought to be about 1 million healthy carriers (Hinuma, 1985). Blood transfusion may be a main cause of increase in the number of infected persons, directly as well as indirectly through natural infection from seroconverted individuals to uninfected ones. For prevention of this, more than 8 million blood donors have been screened for anti-ATLA (HTLV-I) every year in Japan since 1987. Consequently the blood screening has lead to remarkable decrease of HTLV-I infection in recipients of transfused blood.

Several viral vaccines are very effective. The viral structural proteins, their precursors, and the genome structures encoding these polypeptides have been studied extensively. Viral glycoproteins encoded by the env gene of HTLV-I are present not only on the viral surface, but also on membranes of infected cells. Thus, they will be good targets for the immune surveillance system of the host. To develop a vaccine against ATL, we constructed recombinant vaccinia viruses containing the env gene of HTLV-I in the vaccinia virus hemagglutinin (HA) gene, a new site where foreign genes can be inserted. A single inoculation of the recombinant virus induced antibodies to the env proteins of HTLV-I in rabbits and had a protective effect against HTLV-I infection (Shida et al. 1987). Efficacy of the vaccine is now being examined in the experiments with cynamolgus monkeys. These steps have nearly been completed, and HTLV-I vaccination will become a reality in the near future. Possibly, totally different problems will prove more difficult. For example, the question may arise before vaccination of who should be immunized?

Since multiplication of HTLV-I *in vivo* is poor, continuous expression of the viral genome in ATL cells is probably no longer required for maintenance of the transformed state. This conclusion suggests that antiviral drugs, such as inhibitors of reverse transcriptase, might not be therapeutically effective against ATL. Therefore, drugs that interfere with development of ATL cells should be more effective. In this regard, it is interesting that 2'-deoxy-conformycin has been reported to be effective in some ATL patients who were resistant to current antileukemic agents (Daenen et al. 1984; Yamaguchi et al. 1986). These data appear to encourage and warrant further extensive studies.

REFERENCES

Asou N, Kumagai T, Uekihara S, Ishii M, Sato M, Sakai K, Nishimura H, Yamaguchi K, Takatsuki K (1986) HTLV-1 seroprevalence in patients with malignancy. Cancer 58:903-907.

Blattner WA, Blayney DW, Robert-Guroff M, Sarangadharan MG, Kalyanaraman PS, Saarin PS, Jaffe ES, Gallo RC (1983) Epidemiology of human T-cell leukemia/lymphoma virus. J Infect Dis 147:406-416.

Catovsky D, Greaves MF, Rose M, Galton DAG, Goolden SWG, McClusky DR, White JM, Lampert I, Bourikas G, Ireland R, Brownell AI, Bridges WA, Blattner WA, Gallo RC (1982) Adult T-cell lymphoma-leukemia in blacks from the West Indies. Lancet 1:639-643.

Daenen S, Rojer JW, Smit JW, Hais MR, and Nieweg HO (1984) Successful chemotherapy with deoxycoformycin in adult T-cell lymphoma-leukaemia. Br J Haematol 58:723-727.

Gallo RC, Wong-Staal F, Mongtagnier L, Haseltine WA, Yoshida M (1988) HIV/HTLV gene nomenclature Nature 333:504.

Gessain A, Barin F, Vernant JC, Gunt O, Maurs L, Calender A, de The G (1985) Antibodies to HTLV-I in patients with tropical spastic paraparesis. Lancet 2:407-410.

Gotoh Y, Sugamura K, Hinuma Y (1982) Healthy carriers of a human retrovirus, adult T-cell leukemia virus (ATLV): demonstration by clonal nature of ATLV-carrying T-cells from peripheral blood. Proc Natl Acad Sci USA 79:4780-4782.

Hayami M (1986) Simian T-cell leukemia virus, STLV. Cancer Rev 1:35-63.

Hayami M, Ishikawa K, Komuro Y, Kawamoto K, Nozawa K, Yamamoto K, Ishida T, Hinuma Y (1983) ATLV antibodies in cynamolgus monkeys in the wild. Lancet 2:620.

Hinuma Y (1985) Natural history of the retrovirus associated with a human leukemia. BioEssays 3:205-209.

Hinuma Y, Nagata K, Hanaoka M, Mitsuoka M, Nakai M, Matsumoto T, Kinoshita K, Shirakawa S, Miyoshi I (1981) Adult T-cell leukemia: antigen in an ATL cell line and detection of antibodies to the antigen in human sera. Proc Natl Acad Sci USA 78:6476-6480.

Hinuma Y, Gotoh Y, Sugamura K, Nagata K, Goto M, Nakai N, Kamada T, Matsumoto T, Kinoshita K (1982a) A retrovirus associated with human adult T-cell leukemia: *in vitro* activation. Gann 73:341-344.

Hinuma Y, Komoda H, Chosa T, Kondo T, Kohakura M, Takenaka M, Kikuchi M, Ichimaru M, Yunoki K, Sato I, Matsuo R, Takiuchi Y, Uchino H, Hanaoka M (1982b) Antibodies to adult T-cell leukemia virus-associated antigen (ATLA) in sera from patients with ATL and controls in Japan: a nationwide sero epidemiologic study. Int J Cancer 29:631-635.

Hinuma Y, Chosa T, Komoda H, Mori I, Suzuki M, Tajima, K, Pan IH, Lee M (1983) Sporadic retrovirus (ATLV)-seropositive individuals outside Japan. Lancet 1:824-825.

Hoshino H, Esumi H, Miwa M, Shimoyama M, Minato K, Tobinai K, Hirose N, Watanabe S, Imada N, Kinoshita K, Kamihara S, Ichimaru M, Sugimura T (1983) Establishment and characterization of ten cell lines derived from patients with adult T-cell leukemia. Proc Natl Acad Sci USA 80:6061-6065.

Hunsmann G, Hinuma Y (1987) Human adult T-cell leukemia virus and its association with disease. Adv Viral Oncol 5:147-172.

Ikeda M, Fujino R, Matsui T, Yoshida T, Komoda H, Imai J. ((1984) A new agglutination test for serum antibodies to adult T-cell leukemia virus. Gann 75:845-848.

Kalyanaraman VS, Sarangadharan MG, Poiesz B, Ruscetti FW, Gallo RC (1981) Immunological properties of a type-C retrovirus isolated from cultured human T-lymphoma cells and comparison to other mammalian retroviruses. J Virol 38:906-915.

Kalyanaraman VS, Sarangadharan MG, Robert-Guroff M, Miyoshi I, Blayney D, Golde D, Gallo RC (1982) A new subtype of human T-cell leukemia virus (HTLV-III) associated with a T-cell variant of hairy cell leukemia. Science 218:572-573.

Katsuki T, Yamaguchi K, Matsuda Y, Hinuma Y (1986) Impairment of T-cell control of Epstein-Barr virus-infected B-cells in patient with adult T-cell leukemia. AIDS Res 2:suppl S125-130.

Katsuki T, Katsuki K, Imai J, Hinuma Y (1987) Immune suppression in healthy carriers of adult T-cell leukemia retrovirus (HTLV-I): Impairment of T-cell control of Epstein-Barr virus-infected B-cells. Jpn J Cancer Res (Gann) 78:639-642.

Kobayashi S, Yoshida T, Hiroshige Y, Matsui T, Yamamoto N (1988) Comparative studies of commercially available particle agglutination assay and enzyme-linked immunosorbent assay for screening of human T-cell leukemia virus type I antibodies in blood donors. J Clin Microbiol 26:308-312.

Koyanagi Y, Hinuma Y, Schneider J, Chosa T, Hunsmann J, Kobayashi N, Hatanaka M, Yamamoto N (1984) Expression of HTLV-specific polypeptides in various human T-cell lines. Med Microbiol Immunol 173:127-140.

Kuo T, Chan HL, Su IJ, Eimoto T, Maeda Y, Kikuchi M, Chen MJ, Kuan YZ, Chen WJ, Sun CF, Shih LY, Chen JS, Takeshita M (1985) Serological survey of antibodies to the adult T-cell leukemia virus-associated antigen (HTLV-A) in Taiwan. Int J Cancer 36:345-348.

Maeda Y, Furukawa M, Takehara Y, Yoshimura K, Miyamoto K, Matsuura T, Morishima Y, Tajima K, Okochi K, Hinuma Y (1984) Prevalence of possible adult T-cell leukemia virus-carriers among volunteer blood donors in Japan: a nationwide study. Int J Cancer 33:717-721.

Miyoshi I, Kubonishi I, Sumida M, Yoshimoto, S, Hiraki S, Tsubota T, Kobashi H, Lai M, Tanaka T, Kimura I, Miyamoto K, Sato J (1979) Characteristics of a leukemia T-cell line derived from adult T-cell leukemia. Jpn J Clin Oncol 9:485-494.

Miyoshi I, Kubonishi I, Yoshimoto S, Akagi T, Ohtsuki Y, Shiraishi Y, Nagata K, Hinuma Y (1981) Type C virus particles in a cord T-cell line derived by co-cultivating normal cord leukocytes and human co-cultivating normal cord leukocytes and human leukemic T-cells. Nature 294:770-771.

Miyoshi I, Taguchi H, Fujishita M, Niya K, Kitagawa T, Ohtuski Y, Akagi T

(1982) Asymptomatic type-C virus carriers in the family of an adult T-cell leukemia patient. Gann 73:332-333.

Nakano S, Ando Y, Ichijo M, Moriyama I, Saito S, Sugamura K, Hinuma, Y (1984) Search for possible routes for vertical and horizontal transmission of adult T-cell leukemia virus. Gann 75:1044-1045.

Nakano S, Ando Y, Saito K, Moriyama I, Ichijo M, Toyama T, Sugamura K, Imai J, Hinuma, Y (1986) Primary infection of Japanese infants with adult T-cell leukaemia-associated retrovirus (ATLV): evidence for viral transmission from mothers to children. J Infec Dis 12:205-212.

Okochi K, Sato H, Hinuma, Y (1984) A retrospective study on transmission of adult T-cell leukemia virus by blood transfusion: seroconversion in recipients. Vox Sang 46:245-253.

Osame M, Usuku K, Izumo S, Iigichi N, Amitani H, Igara A, Matsumoto M, Tara M (1986) HTLV-I associated myelopathy, a new clinical entity. Lancet 1:1031-1032

Poiesz BJ, Ruscetti FW, Gazdar AF, Bunn PA, Minna JD, Gallo RC (1980) Detection and isolation of type-C retrovirus particles from fresh and cultured lymphocytes of a patient with cutaneous T-cell lymphoma. Proc Natl Acad Sci USA 77:7415-7419.

Poiesz BJ, Ruscetti FW, Reitz MS, Kalyanaraman VS, Gallo RC (1981) Isolation of a new type C retrovirus (HTLV) in primary uncultured cells of a patients with Sezary T-cell leukemia. Nature 294:268-271.

Popovic M, Lange-Wautzin G, Sarin PS, Mann D, Gallo RC (1983) Transformation of human umbilical cord blood T-cells by human T-cell leukemia/lymphome virus. Proc Natl Acad Sci USA 80:5402-5405.

Reitz MS, Poiesz BJ, Rescetti FW, Gallo RC (1981) Characterization and distribution of nucleic acid sequences of a novel type-C retrovirus isolated from neoplastic human T-lymphocytes. Proc Natl Acad Sci USA 78:1887-1891.

Reitz MS, Popovic M, Haynes BF, Clark SC, Gallo RC (1983) Relatedness by nucleic acid hybridization of new isolates of human T-cell leukemia-lymphoma virus (HTLV) and demonstration of provirus in uncultured leukemic blood cells. Virology 126:668-672.

Sato H, Okochi K (1986) Transmission of human T-cell leukemia virus (HTLV-I) by blood transfusion: demonstration of proviral DNA in recipients' blood lymphocytes. Int J Cancer 37:395-400.

Saxinger C, Gallo RC (1983) Application of the indirect enzyme-linked immunosorbent assay microtest to the detection and surveillance of human T cell leukemia-lymphoma virus. Lab Invest 49:371-377.

Schneider J, Yamamoto N, Hinuma Y, Hunsmann G (1984a) Sera from adult T-cell leukemia virus. Virology 132:1-11.

Schneider J, Yamamoto N, Hinuma Y, Hunsmann G (1984b) Precursor polypeptides of adult T-cell leukemia virus: detection with antisera against isolated polypeptides gp68, p24, and p19. J Gen Virol 65:2249-22a58.

Seiki M, Hattori S, Hirayama Y, Yoshida M (1983) Human adult T-cell leukemia virus: complete nucleotide sequence of the provirus genome integrated in leukemia cell DNA. Proc Natl Acad Sci USA 80:3618-3622.

Seiki M, Hikikoshi A, Taniguchi T, Yoshida M (1985) Expression of the pX gene of HTLV-I: General splicing mechanism in the HTLV family. Science 228:1532-1534.

Shida, H, Tochikura T, Sato T, Konno T, Hirayoshi K, Ito Y, Hatanaka M, Hinuma Y, Sugimoto M, Takahashi F, Maruyama T, Miki K, Suzuki K, Morita M, Sashiyama H, Yoshimura N, Hayami M (1987) Effect of the recombinant vaccine viruses that express HTLV-I envelope gene on HTLV-I infection. EMBO J 6:3379-3384.

Shimoyama M, Minato K, Tobinai K, Horikoshi N, Ibuka T, Deura K, Nagatani T, Ozaki Y, Inada N, Komoda H, Hinuma Y (1982) Anti-ATLA (antibody to the adult T-cell leukemia cell-associated antigen)-positive hematologic malignancies in the Kanto district. Jpn J Clin Oncol 12:109-116.

Shimoyama M, Kagami Y, Shimotohno K, Miwa M, Minato K, Tobinai K, Suemasu K, Sugimura T (1986) Adult T-cell leukemia/lymphoma not associated with human T-cell leukemia virus type I. Proc Natl Acad Sci USA 83:4524-4528.

Sugamura K, Fujii M, Kannagi M, Sakitani M, Takeuchi M, Hinuma Y (1984) Cell surface phenotype and expression of viral antigens of various human cell lines carrying human T-cell leukemia virus. Int J Cancer 34:221-228.

Taguchi H, Sawada T, Fujishita M, Morimoto T, Niiya K, Miyoshi I (1983) Enzyme-linked immunosorbent assay of antibodies to adult T-cell leukemia-associated antigens. Gann 74:185-187.

Tajima K, Hinuma Y. (1985) Epidemiological features of adult T-cell leukemia virus. In Mathe G, Reizenstein P (ed.), Pathological aspects of cancer epidemiology. Elmsford, N.Y.: Pergamon Press Inc., pp.75-87.

Tajima K, Tominaga S, Suchi T, Kawagoe T, Komoda H, Hinuma Y, Oda T, Fujita K (1982) Epidemiological analysis of the distribution of antibody to adult T-cell leukemia-virus-associated antigen: possible horizontal transmission of adult T-cell leukemia virus. Gann 73:893-901.

Takatsuki K (1989) Adult T-cell leukemia: An overview. In Roman GC, Vernant J-C, Osame M (ed.), HTLV-I and nervous system, New York, N.Y.: Alan R. Liss Inc., pp.57-63.

Takatsuki K, Uchiyama T, Sagawa K, Yodoi J (1977) Adult T-cell leukemia in Japan. In Seno S, Takaku F, Irino S (ed.), Excerpta Medica, Amsterdam:pp. 73-77.

Uchiyama T, Yodoi J, Sagawa K, Takatsuki K, Uchino H (1977) Adult T-cell leukemia: clinical and hematologic features of 16 cases. Blood 50:481-492.

Watanabe T, Seiki M, Yoshida M (1984) HTLV type I (U.S. isolate) and ATLV (Japanese isolate) are the same species of human retrovirus. Virology 133:238-241.

Yamaguchi K, Seiki M, Yoshida M, Nishimura H, Kawano F, Takatsuki K (1984) The detection of human T-cell leukemia virus proviral DNA and its application for classification and diagnosis of T cell malignancy. Blood 63:1235-1240.

Yamaguchi K, Lee SY, Oda T, Takatsuki K (1986) Clinical consequence of 2'-deoxyconformycin treatment in patients with refractory adult T-cell leukemia. Leukemia Res 10:989-993.

Yamamoto N, and Hinuma Y (1982) Antigens in an adult T-cell leukemia virus-producer cell line: reactivity with human serum antibodies. Int J Cancer 30:289-293.

Yamamoto N, Hinuma Y (1985) Viral aetiology of adult T-cell leukemia. J Gen Virol 66:1641-1660.

Yamamoto N, Matsumoto T, Koyanagi K, Tanaka Y, Hinuma Y (1982a) Unique cell lines harboring both Epstein-Barr virus and other T-cell leukemia virus established from leukemia patients. Nature 299:267-269.

Yamamoto N, Okada M, Koyanagi M, Hinuma Y (1982b) Transformation of human leukocytes by cocultivation with an adult T-cell leukemia virus producer cell line. Science 217:737-739.

Yoshida M, Miyoshi I, Hinuma Y (1982) Isolation and characterization of retrovirus from cell lines of human adult T-cell leukemia and its implication
in the disease. Proc Natl Acad Sci USA 79:2031-2035.

Yoshida M, Seiki K, Yamaguchi K, Takatsuki K (1984) Monoclonal integration
of human T-cell leukemia provirus in all primary tumors of T-cell
leukemia suggests causative role of human T-cell leukemia virus in the
disease. Proc Natl Acad Sci USA 81:2534-2537.

DISCUSSION

Bone D (DuPont Company, Wilmington, DE):

Is HAM a fatal disease? Do patients with HAM end up dying of HAM?

Hinuma Y:

I don't think so. It has a very long course. I don't know of any cases of
death by HAM itself.

Bone D:

In the United States, we are screening blood now for antibodies to HTLV-I,
but recent data, and in fact, information we heard yesterday suggests that
HTLV-II infection may be a problem even among normal blood donors. Are
there data from Japan which rule out the presence of HTLV-II infection?

Hinuma Y:

Unfortunately, so far, I don't know of any data on HTLV-II in Japan. My
own laboratory just started to screen by PCR.

Back A (San Francisco Health Department, San Francisco, CA):

Between the ELISA and the particle agglutination test for screening of
blood samples, which tests do you prefer and why?

Hinuma Y:

In Japan the the 74 Blood Centers use the agglutination test for screening.
It is an easy test and in Japan, the Blood Donor Centers are used to this technology because they use it for other agents. The problem is that there is a positive
rate almost twice as high than with ELISA. However, the Fujirebio Company
has now changed the antigen and the preliminary data shows results similar to
the ELISA.

Back A:

Which test is most commonly used for confirmation of screening tests?
Do you use Western blot?

Hinuma Y:

Some blood centers use Western blot, some blood centers use immunofluorescence. But not every blood center does them. Only large centers do it.

Heilman C (DuPont Company, Wilmington, DE):

Can you say how much of the envelope gene you have in your vaccine construct. Does it have all of the gp 46 and p21E? The envelope gene in your recombinant that you would like to use for a vaccine. Can you say how much of the envelope gene it has? Does it have the entire envelope?

Hinuma Y:

The entire envelope.

Lennette E:

I have a question on epidemiology. I'm not sure that I've got this straight. Maybe my demography and geography is amiss. In your slide you showed the annual incidence in Kyushu as one in 10,000? All other areas is one in 2,000. But, from your manuscript you read one in 10,000. Which is correct?

Hinuma Y:

One in 1,000.

An earthquake took place at this point in the discussion: 5:04 pm, October 17, 1989. End of Symposium.

Note Added in proof: Crude annual incidence rate of ATL among 10^5 male virus carriers aged over 30 was 145.3 and that for females was 55.2. The whole life span (0-79) risks of ATL for males and females are 6.9% and 3.0% of carriers, respectively (Kondo et al., Int J Cancer 43, 1061, 1989).

HUMAN HERPESVIRUS 6: BASIC BIOLOGY AND CLINICAL ASSOCIATIONS

John A. Stewart

Viral Exanthems and Herpesviruses Branch
Division of Viral Diseases
Centers for Disease Control
1600 Clifton Road, Atlanta, Georgia, USA

HISTORY

The recognition of a new herpesvirus was announced to the world by Salahuddin et al. (1986) from the National Cancer Institute, NIH. These investigators were involved in studies of the mechanisms regulating human hematopoiesis, especially defects leading to malignancy. In these studies, mononuclear cells from patients with various lymphoproliferative disorders were established in culture, and occasionally short-lived, large, refractile cells, containing intranuclear and intracytoplasmic inclusion bodies were observed. Electron microscopic (EM) examination demonstrated herpes-type virus particles in these cultures. Unenveloped icosahedral nucleocapsids developed in the nucleus and enveloped virions were seen in cytoplasmic vesicles, cisternae and at the cell surface (Figure 1). The original six isolates, from four leukemia and lymphoma patients and two lymphadenopathy patients, were similar antigenically, but differed markedly from the other human herpesviruses. Because B-cell surface markers were detected on infected cells, the virus was called human B-lymphotropic virus (HBLV). Infection with the virus seemed rare, as only 4 of 220 serum specimens from normal donors were weakly positive by immunofluorescence antibody (IFA) testing.

Similar isolates were obtained during attempts to isolate the human immunodeficiency virus type 1 (HIV-1) from AIDS patients in Uganda, Gambia, and at the Centers for Disease Control (CDC). However, unlike the originally reported B cell specificity of HBLV, the Ugandan isolates infected and propagated in T lymphocytes (Downing et al. 1987). The Gambian isolate (from an HIV-2-associated AIDS patient) also had a wider cell tropism, and prolonged cultivation was possible in J JHAN cells, a T-cell line (Tedder et al. 1987). Antigenic and molecular characterization of a CDC isolate, Z29, indicated that it was distinct from the five known human herpesviruses (Lopez et al. 1988), and DNA hybridization studies showed that it was closely related to the original NIH isolate, HBLV.

Since that time viruses with similar properties have been isolated from children with roseola, AIDS patients, patients with a variety of lymphoid abnormalities, and healthy adults. The virus is clearly distinct from the other

Medical Virology 9
Edited by L.M. de la Maza and E.M. Peterson
Plenum Press. New York. 1990

163

human herpesviruses and does not appear to represent a recent infection of man by one of the simian herpesviruses. In accordance with the provisional classification of the International Committee on the Taxonomy of Viruses for the herpesviruses this new virus should be classified in the family Herpesviridae, and it is called human herpesvirus 6 (HHV-6), a nomenclature independent of its cell tropism.

HHV-6 GROWTH CHARACTERISTICS AND CELL TROPISM

General Features

HHV-6 is differentiated from the other human herpesviruses by a number of characteristics. HHV-6 causes lytic infection of T-lymphocytes, whereas the other lymphotropic herpesvirus, Epstein-Barr virus (EBV), is usually non-lytic, infects B cells and some epithelial cells and may transform lymphocytes. Herpes simplex 1 and 2 (HSV-1) (HSV-2), cytomegalovirus (CMV), and varicella-zoster virus (VZV) infect a variety of cell types in culture and each have characteristic cytopathic effects.

The growth of HHV-6 in cell culture has been monitored by a number of procedures. Although the original observation of large refractile cells in lymphocyte cultures led to discovery of the virus, monitoring such cultures for a cytopathic effect is not reliable since some strains of HHV-6 do not cause such effects. EM was initially used to confirm the presence of virus but cannot be readily applied to monitor virus growth in large number of cultures. IFA and anticomplement immunofluorescence (ACIF) assays have been the most extensively used methods to routinely monitor virus growth. However, the number and intensity of fluorescent cells is not easy to quantify, so that the amount of viral DNA produced in rigorous growth studies has been quantified by using a slot blot procedure and a densitometer (Black et al. 1989).

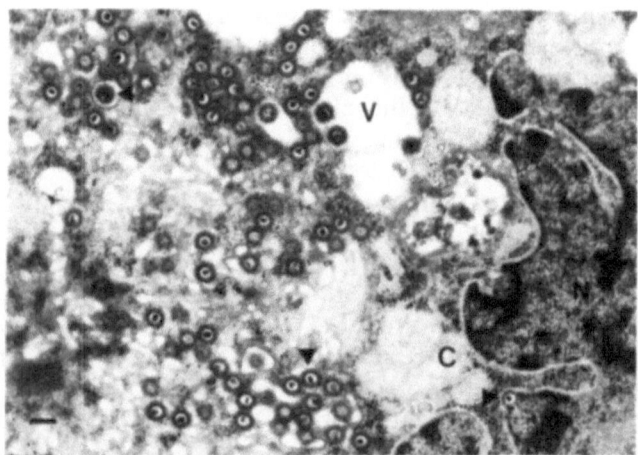

Figure 1. Thin section electron micrograph of HHV-6 virions in an infected lymphocyte. n=nucleus, c=cytoplasm, v=Golgi vesicle, lower arrow indicates nucleocapsid in nucleus, central arrow indicates nucleocapsid coated with tegument, and upper arrow indicates enveloped virion in Golgi apparatus, marker bar= 200 nm. (Provided by C. Goldsmith and J. Black, CDC, Atlanta, GA)

TABLE 1. Infection of Cord Blood Lymphocyte Subpopulations with HHV-6 Related Virus

Antigen	Level of infection in given subpopulation[a,b]			
	CD4+CD8+	CD4+CD8-	CD4-CD8+	CD3+CD4+
HHV-6	84	85	1.3	78
CD3	83 ,85	47 ,46	87, 95	63, 67
CD4	88, 99	99, 99	33, 97	95, 99
CD8	84, 96	20, 7	97, 99	47, 58

[a] First number is percent of positive cells for given antigen.
[b] Second number is percentage of surface marker-positive cells among HHV-6 infected cells.

From Takahashi et al. 1989.

HHV-6 Cell Tropism

HHV-6-infected cells described in the initial NIH report expressed B-cell surface markers and the virus was called HBLV (Salahuddin et al. 1986). However, subsequent *in vitro* studies indicated that most infected cells are of T-cell lineage. Infected cells express a variety of T-cell antigens, CD2, CD4, CD5, CD7 and are negative for B1 and B4. Twenty to 40% of infected cells express both the CD4 and CD8 antigens (Lusso et al. 1987). The NIH prototype strain HHV6(GS) has been shown to infect several permanent human cell lines, including megakaryocytes, glioblastoma cells, some B cells lines and T cells at various stages of maturation (Ablashi et al. 1987). Similarly, other HHV-6 strains are able to infect and proliferate in T-cells (Agut et al. 1988; Downing et al. 1987; Lopez et al. 1988; Tedder et al. 1987;). The current nomenclature, HHV-6, is independent of cell tropism (Ablashi et al. 1987; Lopez et al. 1988; Lusso et al. 1987).

The first detailed report of HHV-6 *in vivo* cellular tropism studied peripheral mononuclear cells from six exanthem subitum (roseola) patients (Takahashi et al. 1989). HHV-6 was isolated from CD4+CD8- and CD3+CD4+ mature T lymphocytes but could not be isolated from CD4-CD8+, CD4-CD8-, CD3- or B cells in the peripheral blood of roseola patients. *In vitro* studies with cord blood lymphocytes (Table 1) indicated that HHV-6(HST) strain predominantly infected CD4+CD8+, CD4+CD8- and CD3+CD4+ cells of mature phenotypes, rarely infected CD4-CD8+ cells, and did not infect B cells. Any generalization about HHV-6 tropism must be tempered by the realization that heterogeneity exists between strains and that a given strain's tropism may change during passage in culture. However, HHV-6 appears to preferentially infect CD4+ T-cells both *in vivo* and *in vitro*, and growth has been reported in megakaryocytes, glial cells, and some EBV-infected B-cell lines. These findings suggest that HHV-6 may behave similarly to HIV-1 and HTLV in the pathogenesis of CD4 T-cell disorders.

Growth Characteristics HHV-6 (Z29) Strain

In contrast to the NIH HHV-6(GS) strain that grows readily in several established lymphoid cell lines (Ablashi et al. 1987), the CDC strain Z29 does not grow efficiently in any established cell line and must be routinely propagated in primary cord blood lymphocytes (CBL) cultures (Lopez et al. 1988). Several aspects of viral growth have been studied in detail including: 1) susceptibility of various cell lines, 2) the effects on virus growth of agents that modulate cell growth or virus-cell interactions, 3) the influence of maternal antibody upon virus growth in CBL, 4) the time course of viral replication (Black et al. 1989).

The susceptibility of various cell lines was tested by cocultivation with HHV-6-infected CBL. The T-cell lines Molt-3, Molt-4, HT, and HuT 78 supported a low level of infection for about two weeks with no more than 2% of the cells positive by ACIF. The only significant growth (25% ACIF-positive cells at 7 days) was found in MT-4, an HTLV-I transformed T-cell line. None of the glial, fibroblast, megakaryocyte and B-cell lines or other cell culture systems tested supported the growth of strain Z29. Different CBL preparations vary greatly in susceptibility to HHV-6 infection (the percentage of ACIF positive cells ranges from 8-35%) so tests must be performed in triplicate. Optimal virus growth occurs in phytohemagglutinin-stimulated CBL cultured in media containing 32 units/ml interleukin-2 (IL2) and 0.01 mg/ml hydrocortisone. The level of maternal antibody in the plasma of the cord blood cells does not correlate with their ability to support virus growth. The growth cycle in CBL is approximately 5 days, virus capsids appear at about day 3, and mature virions at day 5. In the CBL system, HHV-6 appears to be one of the slower replicating herpesviruses.

In Vitro Sensitivity to Antiviral Drugs

While as yet no HHV-6 associated diseases are known that warrant antiviral therapy, the sensitivity of HHV-6 to several drugs has been evaluated. Cultures of a T-cell line, HSB-2, infected with HHV-6 were treated with phosphonoformic acid (PFA), acyclovir (ACV), and gancyclovir (GCV) (Steicher et al. 1988). PFA at a concentration of 20 ug/ml (66uM) significantly inhibited viral activity and showed minimum toxicity for HSB-2 cells. ACV (50 ug/ml [200uM]) and GCV (80 ug/ml [320uM]) produced minimum viral inhibition and had significant toxic effects on uninfected cells. In contrast, Agut et al. (1989) found that HHV-6 was readily inhibited by both PFA and GCV (Table 2). In addition, HHV-6 was inhibited by high concentrations of ACV but zidovudine (ZVD) had no effect on virus growth. Two additional studies have noted greater susceptibility of HHV6 to ACV (Kikuta et al. 1989; Russler et al. 1989). While further studies of antiviral activity which compare other HHV-6 strains grown in various cell lines are certainly needed, it is clear that HHV-6 is sensitive to GCV and PFA and may respond to ACV.

PHYSICAL PROPERTIES OF HHV-6

Electron Microscopic Examination of HHV-6

Biberfeld et al. (1987) reported that, during the nonlytic phase of infection, the nuclei of infected cells exhibited some clumping of chromatin but no

TABLE 2. Inhibitory Concentrations of Antiviral Drugs Against HHV-6

Drug	IC_{50} (μM) determined by IFA	IC_{90} (μM) determined by IFA
GVC	1.1	4.5
PFA	8.7	59
ACV	27	70
ZVD	>8	>8

From: Agut et al. 1989

margination, typical of other herpesvirus infections. The cytoplasm often displayed a fairly large Golgi apparatus, vesicles of varying sizes, and distended cisternae of the rough endoplasmic reticulum filled with parallel arrays of helical structures 10-13 nm in diameter (these may be the cytoplasmic inclusion bodies seen by light microscopy). Nucleocapids in the cytoplasm were always coated with dense amorphous material identical to the tegument described for several herpesviruses. Negatively stained EM preparations of gradient-purified extracellular virus contain almost exclusively enveloped virions. The nucleocapsid has icosahedral symmetry with 162 capsomers (5 capsomers on an edge), a distinguishing characteristic of herpesviruses. The lipid membrane, which envelopes most nucleocapsids, is covered with surface spikes. It has been noted with interest that newly infected lymphocytes continue to replicate, at least for a time following infection; however, as virus is produced, most infected cells eventually lyse. This suggests that HHV-6 infection involves both lymphoproliferative and lympholytic pathogenic mechanisms.

EM studies of infected lymphocyte cultures show that as many as 71% of cells have evidence of HHV-6 infection (Lopez et al. 1988). Infected lymphocytes show high concentrations of typically enveloped herpesvirus particles at the cell surface, non-enveloped particles in the cytoplasm, and nucleocapsids within the nuclei of infected cells (Figure 1). Nucleocapsids average 106 nm in diameter, and estimates of the whole virion diameter range from 160 to 200 nm. Although virions have been found in lymphoblasts and monocytes, nucleocapsids are not found in the monocytes. This indicates that virus has been phagocytosed and is not replicating in those cells. Groups of mature virions are often found in vacuoles of the infected cells.

Human lymphocytes and MT4 cells infected with a HHV-6 isolate from a patient with roseola showed many features in common with those described above (Yoshida M. et al. 1989). The most striking characteristic of the ultrastructure of this virus is the distinct coating of intra- and extracellular nucleocapsids with a moderately electron dense tegument. Only CMV, of the human herpesviruses, acquires a substantial amount of tegument that is not as sharply defined as with HHV-6. Another major difference from CMV is that the formation of large dense bodies has not been observed. Interestingly, scanning EM showed hundreds of particles on the cell surface membrane.

The proteins of HHV-6. Identification, isolation and functional characterization of HHV-6 proteins are fundamental to an understanding of the biology of the virus and to the preparation of diagnostic reagents. With a genome of about 170 kilobases (Kb) HHV-6 should code for at least 70 proteins. Balachandran et al. (1989) have used polyclonal and monoclonal antibodies to identify at least 20 proteins and 6 glycoproteins specific for HHV-6-infected HSB-2 cells. Some of the proteins could be virus-induced host cell proteins. A human serum used in immune precipitation and Western immunoblot reactions recognized 30 HHV-6-specific proteins and 7 glycoproteins. Shiraki et al. (1989) conducted similar studies by using virus purified from HHV-6-infected CBL. Again, a large array of at least 29 polypeptides, ranging in size from 280 to 30 Kda, was obtained from the virion. A 180 Kda polypeptide was detected in both the whole virion and nucleocapsid fractions and may correspond to the major capsid protein.

The physical properties and immunogenicity of the HHV6(Z29) structural components have been similarly studied by M.Yamamoto (personal communication). Nucleocapsids and whole virions were purified from infected human CBL cultures by banding on a Nycodenz gradient and then on two successive sucrose gradients. Nucleocapsids were prepared by solubilizing membranes with NP-40 and deoxycholate before layering onto the sucrose gradients. EM examination of the banded material showed progressive clearing of cellular debris with successive centrifugation steps. The structural components of the virus were analyzed by polyacrylamide gel electrophoresis after solubilization in sodium dodecyl sulfate. More than 20 polypeptides, ranging in size from 30 to 220 Kda, were found in the autoradiogram of purified whole virions. A 100 Kda polypeptide was found to be strongly reactive on Western blot analysis with HHV-6 positive human serum and a murine monoclonal antibody. Human serum samples that contained antibodies directed against the other human herpesviruses and that had no HHV-6 antibody detected by ACIF showed no reaction with this polypeptide, which may thus serve as a sensitive and specific marker for HHV-6 infection.

Immunologic characterization of HHV-6 isolates. Serum samples from the 6 original HHV-6 positive patients reported by Salahuddin et al. (1986) had IFA IgG antibody titers between 40 and 80. Acetone-fixed infected cells showed staining patterns ranging from punctate nuclear staining to diffuse staining of the entire cell. Cells infected by HHV6 were used to evaluate immunological cross-reactivities with the other human herpesviruses, EBV, CMV, HSV-l, HSV-2, and VZV. Monoclonal and well-characterized human polyclonal antibodies to the human herpesviruses, and hyperimmune serum to rhesus CMV, African green monkey CMV and *Herpesvirus saimiri* did not react with HHV-6 infected cells. Serum samples with reactivity to both HHV-6 and another human herpesvirus (EBV, CMV, HSV, VZV) had the homologous reactivities removed by adsorption with cells infected by these viruses without significantly affecting the antibody titer to HHV-6. The converse was also shown (Salahuddin et al. 1986). A similar evaluation using the ACIF for HHV-6 to test human sera with antibodies to various herpesviruses (EBV, CMV, HSV-1, HSV-2, VZV) failed to show cross-reactivity (Lopez et al. 1988).

To distinguish HHV-6 from nonhuman herpesviruses (HV), tests for reactivity against HHV-6 were performed with polyclonal antibodies directed to deer HV, pseudorabies virus of swine, equine HV, bovine HV, bovine rhinotracheitis virus, feline HV, guinea pig HV, murine CMV, HV of owl monkeys, monkey B virus, and Marek's disease virus. These reagents failed to react with HHV-6 in IFA, dot immunoblot, and Western blot assays (Ablashi et al. 1988).

HHV-6 DNA. HHV-6 contains a double-stranded DNA genome of approximately 170 Kb, which is consistent with it's morphologic classification as a herpesvirus. In studies conducted by Josephs et al. (1986) a 9-Kb-pair molecular clone was derived from the DNA of HHV-6(GS). This clone, named pZVH14, specifically hybridized to the DNA and the cytoplasmic messenger RNA of HHV-6 infected cord blood cells. The probe did not hybridize to genomic DNA of the other human herpesviruses or Herpesvirus saimiri.

In studies conducted at CDC, DNA from HHV-6, HSV-1, HSV-2, VZV, CMV, and EBV were digested with BamHI and the fragments separated by agarose gel electrophoresis. Ethidium bromide stained gels clearly show that the HHV-6 fragment pattern is distinct from those of other human herpesviruses. No significant nucleic acid cross-hybridization is seen when whole HHV-6 DNA is used as a probe on Southern blots preparations containing restriction endonuclease digestions of the other human herpesvirus DNAs. These studies confirm the previously shown antigenic and biologic distinctiveness of the virus and warrant it's designation as a new human virus, HHV-6 (Lopez et al. 1988). Other studies using cloned subgenomic fragments as probes, however, revealed limited regions of homology between HHV-6 and CMV (Efstathiou et al. 1988), HHV-6 and Marek's disease virus (Kishi et al. 1988), and under moderate stringent conditions between HHV-6 and HSV-1 (P. Pellett personal communication)

Collaborative studies between the National Cancer Institute and CDC, have shown that DNA from HHV-6(Z29), hybridizes under stringent conditions with DNA from HHV-6(GS). This is shown both with probes made from whole HHV-6(Z29) DNA and from the pZVH14 fragment. The DNA restriction enzyme pattern of the two viruses is similar (Josephs et al. 1988). Further studies of HHV-6(Z29) by Linquester and Pellett (personal communication) found that the G+C content is 42% and that the genome is 161 to 167 Kb in length and consists of a 141 Kb unique segment bracketed by a pair of directly repeated sequences that vary from 10 to 13 Kb in length (depending on virus passage level.

HHV-6 isolates from other parts of the world have also been shown to hybridize with the pZVH14 DNA probe under stringent conditions (Downing et al. 1987; Tedder et al. 1987). Studies in progress, using a whole virus probe made from HHV-6(Z29), have shown strong DNA homology among 8 Japanese isolates, an African isolate, and an United States isolate (Pellett and Lindquester, 1990). Restriction enzyme comparison of the isolates showed a remarkable degree of homogeneity (greater than that seen among HSV-1 isolates). Most of the observed heterogeneity has been mapped to a small region near the termini of the genome and has been found to vary upon serial passage. Minor variability in the region of the viral genome probed by the Hind III fragment pZVH14 has also been noted in 5 isolates from Japanese roseola patients (Kikuta et al. 1989).

Patients with AIDS have been the source of many HHV-6 isolates. Since T cells and glial cells can be infected with HHV-6 (*in vitro*) and these cells may also be infected with HIV-1, any evidence of interaction between the viruses will be quite important. The early gene products of other viruses, HSV, CMV and adenovirus, have been found to trans-activate the long terminal repeat of HIV-1 *in vitro*. It has recently been demonstrated that HHV-6 can trans-activate the HIV-l promoter in human T-cell lines (Horvat et al. 1989; Lusso et al. 1989). With the possibility that this might also occur *in vivo*, the search for a role for HHV-6 in HIV-l pathogenesis continues.

LABORATORY DIAGNOSIS OF HHV-6 INFECTION

Detection of HHV-6 in Cell Culture

HHV-6 has been isolated from peripheral blood both by primary culture of the mononuclear cells and by their cocultivation with either cord blood or adult lymphocytes. A higher percentage of infected cells are usually obtained upon culture with CBL, so they have been preferred for HHV-6 isolation. For HHV-6 cocultivation studies at CDC, mononuclear cells are purified from fresh human cord blood on Ficoll gradients and stimulated to blast formation by culturing for 1-3 days in growth media (RPMI 1640 with 10% fetal calf serum and antibiotics) in the presence of 0.002% phytohemagglutinin (PHA) and 5% IL-2. After washing, the CBL are co-cultivated with an equal volume of mononuclear cells from patients. HHV-6 replication is evaluated by ACIF at 5-7 day intervals. To propagate an isolate, a 1:10 to 1:100 dilution of infected cells is subcultured with fresh CBL and monitored by ACIF for HHV-6 antigen.

Detection of HHV-6 Genome by Polymerase Chain Reaction (PCR)

The PCR is an extremely sensitive technique for exponentially amplifying minute amounts of DNA from a predefined genomic segment. Buchbinder et al. (1988) used HHV-6 primer sequences (derived from DNA of the pZVH14 clone) to detect HHV-6 in the peripheral blood from 52 of 63 AIDS patients and in 20 of 23 tissue samples obtained from patients with lymphoproliferative disorders. As yet no comparative studies have been reported of normal individuals from the general population. So the role of HHV-6 in any particular pathologic condition in these patients cannot yet be established.

Serologic Tests for HHV-6

Immunofluorescence assays. The IFA is one of the easiest serologic tests to perform soon after discovery of a new virus. The IFA uses virus infected cells, convalescent serum from a patient from whom virus has been isolated, and anti-human IgG animal serum conjugated with fluorescein as detector. Specificity is evaluated by comparing fluorescence of infected versus uninfected cells, but cross-reactivity with closely related viruses is often difficult to exclude. Sensitivity is also difficult to evaluate since virus isolation may be achieved from only a small proportion of those who were infected in the past and yet maintain antibody titers. The initial IFA procedure described by Salahuddin et al. (1986), used infected cord blood CBL as antigen and detected antibody (at a 1:40 dilution) in all six patients from whom the virus was isolated. In contrast only 4/220 sera from normal donors were positive.

A more sensitive test was reported by Tedder et al. (1987) who used infected J Jhan cells as the antigen source and adsorbed the serum samples with acetone-treated uninfected J Jahn cells to decrease nonspecific staining before testing. They achieved intense IFA staining using the serum from the Gambian patient from whom the isolate was obtained. Using a conservative end point, 12 of 66 (18%) sera from the United Kingdom were strongly positive, with another 16 (24%) minimally positive. Because of the possibility that low levels of cross-reactivity with the other human herpesviruses might exist, the authors reported only the strongly positive results as positive.

Early difficulties in producing large amounts of virus for biologic and immunologic studies were overcome with the discovery that HHV-6(GS) would grow well in several T-cell lines. With the selection of infected HSB-2 cells as the HHV-6 antigen source, the IFA became easier to perform and possibly more reproducible. The seroprevalence rate in normal blood donors in the U.S.A. and Canada was shown to be 26% with this procedure while a 52% prevalence rate was found in sera collected from West Africa (Ablashi et al. 1988). Some of the problems of interpretation of the IFA are pointed out by the report of Krueger et al. (1988) who noted an antibody prevalence to HHV-6 of 26% in a normal population if strict criteria (a titer of 40) for positivity were applied. But if a titer of 10 was considered positive, antibody was noted in 63% of the population. The first authors to report an almost universal prevalence of antibody, detected by IFA, were Knowles and Gardner (1988). They regarded as positive sera diluted 1/10 that gave diffuse fluorescence only with the distinctly enlarged infected CBL, and they found that over 90% of children were positive at age five and 98% were positive at 17 years of age.

ACIF test. At CDC an ACIF assay was adopted that greatly reduced nonspecific fluorescence and gave a stronger signal with positive cells (Lopez et al.1988). In this test, heat inactivated sera are added to acetone fixed HHV-6 (Z29) infected CBL spotted onto microscope slides. After incubation and wash steps, a human serum containing complement is added and after further incubation complement fixation is detected by a fluorescein labeled goat antihuman C3. Cytoplasmic fluorescence is granular, whereas nuclear fluorescence is solid and bright. Granular cytoplasmic fluorescence predominated during the early stages of infection. In subsequent studies the ACIF assay has been used to detect HHV-6-infected cells in culture as well as to determine human antibody responses to the virus. The ACIF assay has also been effectively used to detect the antibody response in roseola patients (Yamanishi et al. 1988) and in seroprevalence studies (R Ashley et al. personal communication; Okuno et al. 1989b).

The specificity of the ACIF for HHV-6 was shown in the following ways (Lopez et al. 1988). 1) The percentage of infected cells by ACIF correlates with the proportion of infected cells found by EM. 2) No cross-reactivity was found with the other human herpesviruses using high-titered well characterized patient serum samples. 3) Cross adsorption of dually reactive sera was done with CMV, or HSV, or VZV infected cells before testing for residual antibody. Antibody to the homologous virus was effectively adsorbed while no change in response to HHV-6 was noted. 4) Ten of 12 serum pairs that showed seroconversion to either CMV or EBV failed to boost in HHV-6 reactivity, while the other two pairs showed a four-fold increase in HHV-6 titer.

Other serologic procedures. The other tests that have been employed to detect HHV-6 antibodies include Western immunoblots, immunoprecipitation, neutralization and enzyme immunoassay (EIA). The Western blot and immunoprecipatation assays (Balachandran et al. 1989, Shiraki et al. 1989), are key procedures in the detailed analysis of the individual polypeptide and proteins involved in the immune response to HHV-6. The neutralization test has only been evaluated in one study to date (Asano et al. 1989b), but the rapid appearance of antibody after day 4 of illness with roseola was correlated with a rapid decrease in the rate of virus isolation.

An EIA, developed by Saxinger et al. (1988), uses a soluble viral lysate harvested from HHV-6 infected HSB-2 cells adsorbed onto microtitration plates as antigen. Diluted serum specimens are added and bound antibody detected by alkaline phosphatase reagents. The specificity of the reaction was determined by pre-incubation of sera with soluble HHV-6, HSB-2 control, CMV, EBV, HSV, and VZV antigens. The other herpesvirus antigens showed no competitive cross-reactivity with IgG antibodies to HHV-6 antigens while the HHV-6 absorbing antigens reduced binding to HHV-6 by more than 90%. The EIA serum results were normally distributed over a broad range of values with no evidence of bimodality. This precluded the designation of a "cut-off" level between negative and positive sera. Sera showing HHV-6-specific absorption of at least 50% were regarded as positive.

CLINICAL ASSOCIATIONS

Roseola Epidemiology and Clinical Features

A major question regarding HHV-6 concerns its role in human disease. The most convincing data on this subject are those from Japan that strongly implicate HHV-6 as the causative agent of roseola or exanthem subitum, a mild childhood disease. It is interesting to compare what was known or suspected about roseola with what has been learned through recent studies using HHV-6. Roseola was thought to be an infectious disease based on the nature of reported outbreaks in hospital settings and on studies of transmission in human and nonhuman primates. In two hospital outbreaks the incidence of infection in infants of a susceptible age varied from 35-60% and the incubation period varied from 5-15 days. The disease was reproduced by Kempe et al. (1950) in a 6-month-old infant 9 days following the intravenous injection of 1 cc of serum collected from a roseola patient on day 3 of fever. Serum (given subcutaneously) and throat washings (given intranasally) from the index case also produced a febrile, leukopenic illness in rhesus monkeys. Serum taken from the monkeys on day 2 of fever and given subcutaneously and intranasally produced an identical disease in other monkeys. The human transmission experiments were confirmed by Hellstrom and Vahlquist (1951), who were able to reproduce the disease in 3/14 infants injected intramuscularly with heparinized blood taken from an index case on the first day of rash.

Roseola epidemiology fits very closely with what is being learned about HHV-6 infection. The age distribution of this disease is unique for exanthems: 82% of the patients are younger than two years of age and 93% younger than three. The peak incidence occurs in infants 6-18 months of age, and the rare occurrence in infants less than 6 months of age suggests that maternal antibody is protective. In a private pediatric practice, Breese (1941) found that 30% of children develop roseola in the first two years of life. However, the overall incidence of disease or infection must be higher than this, since the syndrome is defined by a rash which is typically sparse and may last for only a few hours. The high rate of disease in young children suggests that infection is spread by oral or respiratory secretions. The transmission pattern in hospital outbreaks and in family settings further suggests that adults (parents or nurses) may serve as asymptomatic carriers for the agent. In most cases symptomatic infants have only been exposed to well family members and secondary cases are infrequent.

The clinical picture of roseola has been well described in the early pediatric literature (Krugman and Katz, 1981). With no other prodrome, the patient experiences a sudden rise in temperature to 102-105°F. In most cases the temperature remains elevated for 2-5 days and rapidly falls to normal. The other symptoms are listlessness, irritability and drowsiness. Despite the fever, patients often continue to eat and play and do not seem nearly as sick as their temperature might indicate. Convulsions, which may sometimes occur after the onset of fever, are usually the only major concern. Other physical findings that may occur before the rash appears are palprebral edema and suboccipital, postauricular and posterior cervical lymphadenopathy. The skin lesions are pale pink, vary from 1-5mm in diameter, and are surrounded by a clear area that separates them from other lesions. The lesions are macular or sometimes papular but never vesicular. They appear first on the neck, behind the ears, and on the back and may spread quickly to involve the scalp, chest, abdomen, and thighs. The face and distal extremities are usually spared. The illness may last from 2-7 days. Since roseola is usually regarded as a minor illness of childhood, what are the complications that have been described with it? Some clinicians believe that the incidence of febrile convulsions is greater than in other febrile illnesses. Two articles have reported both transient and severe neurologic complications following prolonged seizures in the febrile phase of illness (Berenberg et al. 1949; Burnstine and Paine 1959).

HHV-6 Association with Roseola

Yamanishi et al. (1988) were the first to report the isolation of HHV-6 from 4 children with exanthem subitum. Primary cultures of peripheral blood lymphocytes, obtained two to three days after the onset of fever and before the onset of rash showed balloon-like syncytia in 3 patients, and antigen was detected from a fourth patient after cocultivation of mononuclear cells with CBL. Specific ACIF staining of fixed syncytial cells was not observed with the acute phase serum, but occurred with serum obtained 13-18 days after onset. EM ex-

TABLE 3.   Virus Isolation and Detection of Serum Antibody

Patient	Age (mo)	Days after Onset	Antibody Titer New agent	HHV-6 Z29	Virus Isolation
1	6	2	<10	<10	+
		18	320	320	
2	6	3	<10	<10	+
		13	40	40	
3	6	3	<10	<10	+
		14	40	40	
4	6	3	<10	<10	+
		14	20	20	

From: Yamanishi et al. (1988)

amination of the infected cells showed many viral particles that morphologically resembled herpesvirus particles. All four serum pairs had a significant rise in ACIF antibody, titers ranged from <10 to 320 against both one of the new roseola agents and HHV-6(Z29) (Table 3). Serum samples from 10 other patients were tested against HHV-6 and all 7 samples, collected at least 14 days after onset of illness, were positive.

In a follow-up study, Takahashi et al. (1988) tested serum samples, serially collected between birth and 18 months of age, from 12 infants being followed because their mothers were seropositive for HTLV-1. All the maternal serum and cord blood samples had HHV-6 antibody as measured by an ACIF assay with the Jurkat T-cell line infected with the MA strain. The antibody titers in the infants decreased over time and 11 of the 12 seroconverted to HHV-6 after subsidence of maternally derived antibody. Seven of these 11 (64%) infants had roseola within 9 months of birth and the time of seroconversion was consistent with onset of rash. Three more of the infants had a febrile episode between 3 and 7 months of age, but no rash was observed. This study suggests a higher apparent rate of infection with roseola than the 30% rate described by Breese (1941).

The findings of Yaminishi et al. (1988), which etiologically associated HHV-6 with roseola, were further supported by Ueda et al. (1989) in serologic studies. An eight-fold or greater increase in IFA antibody titer was observed in serum pairs from 6 infants (aged 5-21 months) with roseola. The acute-phase serum sample (obtained prior to onset of rash) was negative in 5 of the 6 patients and had a titer of 20 in the remaining 5 month old. Seroconversion or a four-fold increase in antibody titer was also noted in paired serum specimens from 14 other infants (drawn for other viral studies), that spanned their time of illness with roseola. A more extensive description of the role of HHV-6 in roseola was reported this year (Asano et al. 1989b). Thirty-nine virus strains were isolated from mononuclear cell samples collected between days 0-7 of disease from 38 of the 43 children examined with roseola. The rate of virus isolation from mononuclear cells was 100% (26/26) on days 0-2 (before appearance of rash), 82% (9/11) on day 3, 20% (2/10) on day 4, 17% (2/12) on days 5-7, and zero (0/37) on day 8 or later. The rate of virus isolation from plasma samples was 28% (10/36) on days 0-4, and zero (0/48) for samples collected after day 4 of illness. Neutralizing antibody to HHV-6 was first detected on day 3 of disease (2/11), and was positive in all 37 samples collected more than 7 days after onset. Thus,clearance of the virus from the blood was associated with the induction of specific immunity, neutralizing antibody, to the virus.

These studies clearly indicate that HHV-6 is the causative agent of roseola. They also indicate, as do the seroepidemiologic studies of several populations, that HHV-6 infection is very common and usually occurs early in life. Infants are protected from infection by maternal antibodies and infection is most frequent between 6 and 18 months of age. The route of infection in infants is unknown, but oral or respiratory secretions from other family members are a likely source. While breast milk has been implicated in the transmission of CMV, the study of Takahashi et al. (1988) found that 11 of 12 HHV-6-infected infants with roseola had not received breast milk.

Two other recent studies fit with some of the hypotheses generated from clinical epidemiologic observations of roseola. One was the report by Suga et al. (1989) of "atypical roseola" in five patients (4 to 7 months of age) who had fever but no subsequent rash and yielded HHV-6 isolates from peripheral blood mononuclear cells during the acute stage of illness. In addition, paired serum samples from two patients showed seroconversion by IFA and neutralization tests against the HHV-6 strain FG1 obtained from a patient with roseola. The second was the report of roseola without fever by Asano et al. (1989a). They isolated HHV-6 from the blood of two infants (5 and 7 months of age) who had the sudden onset of a skin rash but failed to develop fever at any point. A clear-cut rise in HHV-6 antibody titers was noted in both children.

Association of HHV-6 With Illness in Adults

Among the many studies reported to date, only a handful appear to show a definite association between HHV-6 infection and disease in adults. A number of studies (using IFA) have reported a higher seroprevalence to HHV-6 in patients with sarcoidosis, lymphomas, lymphoid hyperplasia, and immune suppression/deficiency compared to low rates in "normal population" groups. It now appears that many of the IFAs were insensitive and the actual seropositive rate in the normal population exceeds 90 to 95%. Thus, the studies showing an increased seroprevalence rate in certain diseases may simply indicate that higher mean antibody titers to HHV6 are present in these patients as compared to controls. However, this is true for a number of viruses in patients with connective tissue disease, rheumatoid arthritis, leukemia, lymphoid hyperplasia and even chronic fatigue syndrome.

One of most convincing reports described three patients with serologic evidence of HHV-6 infection that had mild, afebrile illnesses with nonspecific symptoms (Niederman et al. 1988). In each case, the characteristic clinical feature was the presence of bilateral, enlarged, non-tender, cervical nodes early in the illness which persisted for up to 3 months. Fever and chills were not noted but one patient had a mild sore throat. Two patients had transiently raised serum liver enzymes with SGOT levels of 90 and 244 U/l without hepato- or splenomegaly. Leucopenia was an early finding and differential WBC counts showed 37-56% mononuclear cells, of which 3-14% were atypical lymphocytes. IgG antibody titers to HHV-6 were 160 to 2,560 during early illness with no rise in titer found. IgM responses were found at low titers (10) in 2 patients, but were not present after 1 and 3 months respectively. None of the patients had antibodies to CMV or toxoplasma and only one of them had positive but stable titers to EBV. Of particular interest was one previously EBV negative patient who was noted to have classic infectious mononucleosis with typical hematological and EBV serologic responses 15 months after his HHV-6-associated illness. At that time he had a sharp increase in his HHV-6 IgM titer to 160 and the IgG titer remained elevated at 2,560. In view of the well described adsorption studies that show no cross-reactivity between HHV-6 and the other herpesviruses, it was believed that this dual rise in antibody titer could have been due to polyclonal B-cell activation or reactivation of HHV-6 infection by primary EBV infection. Another clear-cut case of HHV-6 infection was noted in a 21-year-old patient following liver transplantation from a HHV-6 seropositive donor (Ward et al. 1989). The patient developed fever, grand mal seizures and hepatitis, HHV-6 was isolated from the blood and an IgM and IgG seroconversion to HHV-6 occurred.

While primary infection with HHV-6 in adults appears to be rare because of the high rate of infection in childhood, it may have some of the features of a mild "mono"-like illness. It would appear to be a mild illness in healthy adults but may well be more serious in the transplanted or immunosuppressed patient. Careful assessment of suspected cases is needed with simultaneous studies of antibodies to HHV-6, CMV, and EBV and cultures for HHV-6.

175

TABLE 4. Prevalence of HHV-6 Antibody in Infants and Older Children

Age (mo)	Infants proportion positive		Age (yr)	Older children proportion positive	
<2	9/22	41%	1+2	25/42	60%
2+3	7/23	30%	3+4	14/23	61%
4+5	1/14	7%	5+6	14/22	64%
6+7	10/22	45%	7+8	11/25	44%
8+9	7/15	47%	9+10	19/31	61%
10+11	10/16	63%	11+12	25/41	61%
			13+14	25/45	56%
			>14	15/23	65%

Antibody measured by indirect IFA at a 1:50 dilution using the J Jhan T-lymphocyte line infected with HHV-6(AJ).

From: Briggs et al. 1988.

SEROEPIDEMIOLOGIC STUDIES

As antibody detection procedures have evolved, so have the estimates of HHV-6 seroprevalence in normal populations. From an initial estimate of 2% positive in the USA (Salahuddin et al. 1986), to 18% in Britain (Tedder et al. 1987), to 26% in healthy USA, Canadian and European donors (Ablashi et al. 1988) current studies estimate seroprevalence rates between 52% (Briggs et al. 1988) and more than 80% (Linde et al. 1988).

A highly sensitive IFA was developed by using HHV-6 infected HSB-2 cells. Up to 95% of these cells could be infected, and a marked cytopathic effect with many giant cells was noted (Brown et al. 1988). This assay detected virtually ubiquitous infection in children 2 to 4 years of age (46/49, 94% positive) with consistent declines in prevalence to 60-70% in adults older than 30. The geometric mean titers (GMT) peaked in young children 2 to 3 years of age (GMT-272), probably reflecting the effect of recent primary infections. In older age groups, there were consistent declines in GMT that fell to a level of 16-25 in groups older than 16 years of age. A somewhat higher seroprevalence rate of 88% was noted in a study of Swedish adults (Linde et al. 1988) that also employed an IFA with infected HSB-2 cells as antigen.

The high rate of seropositivity noted in more recent studies by IFA is similar to that Saxinger et al. (1988) reported earlier by using an EIA with purified disrupted virions as antigen. Serum samples showing significant reactivity were found in a high percentage of blood donors in Minneapolis (81%) and Kansas City (88%) and in 97% of a random USA population survey of persons 6 to 74 years of age.

Seroprevalence of HHV-6 in Infancy and Childhood

Briggs et al. (1988) studied 252 London children (from St. Bartholomew's Hospital) 1-17 years of age and found a constant prevalence of antibody throughout this age span, with 59% (148/252) positive overall. Their study of 112 infant samples from a London Hospital (Table 4) showed a decline in prevalence over the first five months of life, and from a nadir of 7% rose to 63% by 11 months of age. Saxinger et al. (1988) plotted the HHV-6 IgG response by EIA versus age and showed that the quantitative levels at birth are at the adult level or slightly higher. Antibody levels decrease gradually over time to reach a minimum between 3 to 6 months of age and rise again to reach adult levels between 2 to 3 years of age. In a hospital-based prevalence study, frequent IgM and IgG seroconversions by EIA were apparent among infants 6 to 12 months of age, indicating that infection with HHV-6 occurs after passively acquired maternally antibody declines.

The prevalence of HHV-6 antibody as detected by IFA was determined in 119 Japanese infants without a history of roseola (Ueda et al. 1989). The percent positive was 89% age O-1m., 30% age 2-3m., 26% age 4-5m., and 75% age 1O-11m. They concluded that the acquisition of HHV-6 antibody during late infancy by many infants without a history of roseola indicates that subclinical infection with HHV-6 is common. Additional antibody studies of 182 Japanese children (by ACIF) (Okuno et al. 1989b) showed similar results with the lowest rate of 5% positive noted at 5 months of age and the number of children with antibody increasing to 83% at 12 months of age. These studies provide further evidence that HHV-6 is a ubiquitous virus acquired early in life.

Antibody Studies in Pregnancy

Antibody prevalence to HHV-6 was determined in pregnant and control women of similar ages in Thailand by using HHV6(Z29) infected CBL as the IFA antigen (Balachandra et al. 1989). The positive rate was 41% in the pregnant women and 42% and 45% in control groups. The antibody titers remained unchanged between the first and third trimester of pregnancy. Thus, no evidence of active HHV-6 infection during pregnancy was noted in contrast to what occurs with human CMV. In agreement with the studies in infancy cited above, they also noted that the number of children with maternal antibody gradually fell over the first months of life before the percentage positive began to increase at 6 months of age.

Seroprevalence of HHV-6 in HIV-1 Risk Groups

Since many HHV-6 isolates have been obtained from patients infected with HIV-1 it was natural to consider that infection with HHV-6 might be a co-factor in acquiring HIV-1 infection or in decreasing the interval between infection with HIV and the onset of clinical disease. No significant differences in either HHV-6 antibody prevalence (43% to 47%) or titer between 3 groups of homosexual men were found (Fox et al. 1988). One group of thirty patients had antibody to HIV-1 and their serum specimens contained p24 antigen, another 30 men were HIV-1 antibody positive but antigen negative, and the third group of 30 men were negative for both HIV-1 antibody and antigen. The antibody profile to HHV-6 was very similar to that found earlier in male blood donors of similar age.

Another study in the Atlanta area showed no significant differences in the rate of HHV-6 seropositivity or in the distribution of HHV-6 ACIF titers between HIV-1 seropositive men with the HIV-1-associated lymphadenopathy syndrome (LAS) (66% positive) when compared to HIV-1 seronegative homosexual or bisexual men (64% positive). Although HHV-6 titers and the percentage positive in the LAS group gradually fell over the four year follow-up period, the HHV-6 prevalence (44%) or titer distribution did not differ between men with LAS who progressed to AIDS compared to those who had not (Spira et al. 1990).

In contrast with these results a much lower prevalence of IFA antibody to HHV-6 was noted in several groups (Huemer et al. 1989). Only 8% of normal blood donors, 16% of HIV positive drug addicts, and 6% of HIV negative drug addicts were positive. The results were not significantly different between groups. The reason for these low results is not apparent but their test appears to be less sensitive.

Latency and Reactivation of HHV-6

Since latency is a feature of the five other human herpesviruses, evidence of HHV-6 reactivation was looked for in several ways. Serum samples banked from individuals over a 20 year period were found to have persistent antibody but with slowly declining titers. This may indicate that reactivation of latent virus occurs relatively infrequently. Clear-cut evidence for reactivation or re-infection was found in a normal HIV-1 negative lymphocyte donor. The donor's lymphocytes were successfully used for the cultivation of HIV-1 for months before an obvious inhibiting agent was detected in a number of samples over a 6 months period. Herpesvirus particles were found in the cells and HHV-6 was identified from the cultures. Antibody titers to HHV-6 gradually increased by four-fold over the virus-detection period (Table 5) (P Feorino personal communication). Serological evidence of reactivation of HHV-6 was also evaluated in cardiac or renal-transplant recipients in comparison with CMV. Six patients were found in which a significant rise in HHV-6 IgG titer occurred simultaneously with serologically proven CMV infection (Irving et al. 1988). All 6 patients had a clinical illness compatible with active CMV infection at the time of antibody increase. One of the patients became IgM-positive to both CMV and HHV-6, while three others were IgM-positive to CMV alone. It was not possible to tell if this represented dual infection acquired from the donor or if infection with one virus reactivated the other latent virus.

Evaluation of reactivation is complicated by a number of reports of apparent concurrent infection with HHV-6 and CMV. To define the specificity of serologic tests, HHV-6 antibody titers were assayed in paired serum specimens from patients who showed a 4-fold or greater rise in antibody titer to HSV, VZV, or CMV and from whom the respective herpesviruses were isolated (Morris et al. 1988). A HHV-6 antibody response was detected in none of three patients infected with HSV, in two of five patients infected with VZV, and in nine of ten patients infected with CMV. The patient with no increase in HHV-6 antibody showed the highest increase in CMV titer. HHV-6 antibody was assayed in an additional 8 serum pairs from renal transplant recipients (seronegative to CMV using a sensitive assay) from whom CMV was not isolated during the 6 months after transplantation. Five of the 8 pairs showed a significant HHV-6 antibody response. The latter findings exclude major cross-

TABLE 5. Reactivation of Reinfection with HHV-6

Date	Inhibition of HIV Reverse Transcriptase	HHV-6 ACIF Titer
3-85	neg	nd[a]
8-85	neg	nd
10-85	neg	nd
1-86	pos	160
2-86	pos	nd
3-86	pos	320
6-86	pos	640
7-86	neg	nd
10-86	neg	>640
2-87	neg	>640

[a] nd = not done

reactivity between CMV and HHV-6 and indicate that active HHV-6 infection frequently occurs in renal transplant recipients. An additional series of 17 renal transplant recipients was studied in an attempt to define episodes of disease that could be associated with HHV-6 infection (Morris et a1. 1989). Of the four patients who seroconverted to HHV-6, two remained well and two developed self-limiting febrile illnesses without exanthem, leukopenia or thrombocytopenia. One of the four also acquired a primary CMV infection. A second group of 10 patients had HHV-6 antibody initially and showed at least four-fold increases in antibody after transplantation. Eight of them remained well, and the two patients who developed nonspecific febrile illnesses also had primary CMV infection. The last group of three patients had no serologic evidence of HHV-6 infection. Since three of the four patients with minor illnesses had concurrent primary CMV infection (a well recognized cause of fever in transplant recipients), no obvious disease associated with HHV-6 could be defined in these patients.

More evidence of active disease caused by HHV-6 in renal transplant recipients was found in a study of 21 patients (Okuno et al. 1989a). All of the kidney donors and recipients were HHV-6 antibody positive. Eight patients showed a 4- to 64-fold increase in antibody titer after surgery and all 8 experienced a severe rejection episode and required the use of additional immunosuppressive therapy. In two cases virus was isolated from peripheral blood leukocytes about two weeks following the time of rejection and one week before the antibody rise occurred. One patient showed a decrease in urinary output, fever and an increase of serum creatinine while the other patient was asymptomatic. The rejected kidneys of 9 other patients were examined histologically with the use of monoclonal antibody. HHV-6 antigens were detected in the epithelial cells of the medullary tubules and in lymphocytes and histiocytes that had infiltrated into the interstitial tissue in 5 of the 9 kidneys, but no antigen was found in the glomeruli. The isolation of virus from leucocytes, detection of antigen in rejected kidneys, and significant rises in antibody titers indicate that HHV-6 was latent and reactivated in either the donor kidney or from other recipient tissues. The timing of rejection, anti OKT3 antibody use, virus isolation, and antibody response in these patients suggest that HHV-6 reactivation was due to increased immunosuppressive drugs used to treat rejection instead of rejection being caused by virus reactivation.

SUMMARY

1. Human herpesvirus 6 is genetically and serologically distinct from the other known human herpesviruses.
2. Infection is commonly acquired early in life but the proportion of seropositive persons appears to decrease after 40 years of age as titers fall below detectable levels in some assays.
3. Laboratory and epidemiologic data are consistent with an etiological association between HHV-6 infection and roseola.
4. Clinical-epidemiological studies suggest that HHV-6 infection is commonly transmitted from asymptomatic adults to children, but the suspected oral route of infection has not been established.
5. HHV-6 is a lymphotropic virus which appears to undergo latency and may be reactivated in immunocompromised persons. A few mild illnesses and one fairly severe hepatitis infection have been associated with reactivation of reinfection with the virus at this time.

REFERENCES

Ablashi DV, Salahuddin SZ, Josephs SF, Imam F, Lusso P, Gallo RC, Hung C, Lemp J, Markham PD (1987) HBLV (or HHV-6) in human cell lines. Nature 329:207.

Ablashi DV, Joesphs SF, Buchbinder A, Hellman K, Nakamura S, Llana T, Lusso P, Kaplan M, Dahlberg J, Memon S, Imam F, Ablashi KL, Markham PD, Kramarsky B, Krueger GRF, Biberfeld Wong-Staal F, Salahuddin SZ, Gallo RC (1988) Human B-lymphotropic virus (human herpesvirus-6). J Virol Methods 21:29-48.

Agut H, Guetard D, Collandre H, Drauguet C, Montagnier L, Miclea J-M, Baurmann H, Gessain A (1988) Concomitant infection by human herpesvirus 6, HTLV-l, and HIV-2. Lancet 1:712.

Agut H, Collandre H, Aubin J-T, Guetard D, Favier V, Ingrand D, Montaignier L, Huraux J-M (1989) *In vitro* sensitivity of human herpesvirus-6 to antiviral drugs. Res Virol 140:219-228.

Asano Y, Suga S, Yoshikawa T, Urisu A, Yazaki T (1989a) Human herpesvirus type 6 infection (exanthem subitum) without fever. J Pediatr 115:2264-265.

Asano Y, Yoshikawa T, Suga S, Yazaki T, Hata T, Nagai T, Kajita Y, Ozaki T, Yoshida S (1989b) Viremia and neutralizing antibody response in infants with exanthem subitum. J Pediatr 114:535-539.

Balachandra K, Ayuthaya PI, Auwanit W, Jayavasu C, Okuno T, Yamanishi K, Takahashi M (1989) Prevalence of antibody to human herpesvirus 6 in women and children. Microbiol Immunol 33:515-518.

Balachandran N, Amelse RE, Zhou WW, Chang CK (1989) Identification of proteins specific for human herpesvirus 6 infected human T cells. J Virol 63:2835-2840.

Berenberg W, Wright, S Janeway CA (1949) Roseola infantum (exanthem subitum). N Engl J Med 241:253-259.

Biberfeld P, Kramarsky B, Salahuddin SZ, Gallo RC (1987) Ultrastructural characterization of a new human B lymphotropic DNA virus (human herpesvirus 6) isolated from patients with lymphoproliferative disease. J Natl Cancer Inst 79:933-941.

Biberfeld P, Petren AL, Eklund A, Lindemalm C, Barkhem T, Ekman M, Ablashi D, Salahuddin Z (1988) Human herpesvirus 6 (HHV-6, HBLV) in sarcoidosis and lymphoproliferative disorders. J Virol Methods 21:49-59.

Black JB, Sanderlin KC, Goldsmith CS, Gary HE, Lopez C, Pellett PE (1989) Growth properties of human herpesvirus 6 strain Z29. J Virol Methods 26:133-145.

Breese B (1941) Roseola infantum (exanthema subitum). NY State Med. 41:1854.

Briggs M, Fox J, Tedder RS (1988) Age prevalence of antibody to human herpesvirus 6. Lancet 1:1058-1059.

Brown NA, Sumaya CV, Liu C-R, Ench Y, Kovacs A, Coronesi M, Kaplan MH (1988) Fall in human herpesvirus 6 seropositivity with age. Lancet 2:396.

Buchbinder A, Josephs SF, Ablashi D, Salahuddin SZ, Klotman, ME, Manak, M, Kruger, GRF, Wong-Staal F, Gallo R (1988) Polymerase chain reaction amplification and in situ hybridization for the detection of human B-lymphotropic virus. J. Virol Methods 21:191-197.

Burnstine RC, Paine RS (1959) Residual encephalopathy following roseola infantum. Am J Dis Child 98:144-152.

Downing RG, Sewankambo N, Serwadda D, Honess R, Crawford D, Jarrett R, Griffin BE (1987) Isolation of human lymphotropic herpesviruses from Uganda. Lancet 2:390.

Efstathiou S, Gompels UA, Craxton MA, Honess RW, Ward K (1988) DNA homology between a novel human herpesvirus (HHV6) and human cytomegalovirus. Lancet 1:63-64.

Fox J, Briggs M, Tedder RS (1988) Antibody to human herpesvirus 6 in HIV-1 positive and negative homosexual men. Lancet 2:396-397.

Hellstrom B, Vahlquist B (1951) Experimental inoculation of roseola infantum. Acta Paediatrica 40:189-197.

Horvat RT, Wood C, Balachandran N (1989) Transactivation of human immunodeficiency virus promoter by human herpesvirus 6. J Virol 63:970-973.

Huemer HP, Larcher C, Wachter H, Dierich MP (1989) Prevalence of antibodies to human herpesvirus 6 in human immunodeficiency virus l-seropositive and -negative intravenous drug addicts. J Infect Dis 160:549-550.

Irving WL, Cunningham AL, Keogh A, Chapman JR (1988) Antibody to both human herpesvirus 6 and cytomegalovirus. Lancet 2:630-631.

Josephs SF, Salahuddin SZ, Ablashi DV, Schachter F, Wong-Staal F, Gallo RC (1986) Genomic analysis of the human B lymphotropic virus (HBLV). Science 234:60-603.

Josephs SF, Ablashi DV, Salahuddin SZ, Kramarsky B, Franza RB, Pellett P, Buchbinder A, Memon S, Wong-Staal F, Gallo RC (1988) Molecular studies of HHV-6. J Virol Methods 21:179-190.

Kempe CH, Shaw EB, Lackson JR, Silver HK (1950) Studies on the etiology of exanthem subitum (roseola infantum). J Pediatr 37:561-568.

Kikuta H, Lu H, Matsumoto S (1989) Susceptibility of human herpesvirus 6 to acyclovir. Lancet 2:861.

Kishi M, Harada H, Takahashi M, Tanaka A, Hayashi M, Nonoyama M, Josephs SF, Buchbinder A, Schachter F, Ablashi DV, Wong-Staal F, Salahuddin SZ, Gallo RC (1988) A repeat sequence, GGGTTA, is shared by DNA of human herpesvirus 6 and Marek's disease virus. J Virol 62:4824-4827.

Knowles WA and Gardner SD (1988) High prevalence of antibody to human herpesvirus 6 and seroconversion associated with rash in two infants. Lancet 2:912-913.

Krueger GRF, Koch B, Ramon A, Ablashi DV, Salahuddin SZ, Josephs SF, Streicher HZ, Gallo RC, Habermann U (1988) Antibody prevalence to HBLV (human herpesvirus-6, HHV-6) and suggestive pathogenicity in the general population and in patients with immune deficiency syndromes. J Virol Methods 21:125-131.

Krugman S and Katz SL, (eds) (1981) Exanthema subitum (roseola infantum) In:Infectious diseases of children (Seventh Ed.) St Louis: C V Mosby Company, 59-62.

Linde A, Dahl H, Wahren B, Fridell E, Salahuddin Z, Biberfeld P (1988) IgG antibodies to human herpesvirus-6 in children and adults both in primary Epstein-Barr virus and cytomegalovirus infections. J Virol Methods 21:117-123.

Lopez C, Pellett P, Stewart J, Goldsmith C, Sanderlin K, Black J, Warfield D, Feorino P (1988) Characteristics of human herpesvirus-6. J Infect Dis 157:1271-1273.

Lusso P, Salahuddin SZ, Ablashi DV, Gallo RC, di Marzo Veronese F, Markham PD (1987) Diverse tropism of human B-lymphotropic virus (human herpesvirus 6). Lancet 2:743-744.

Lusso P, Markham PD, Tschachler E, Veronese FdM, Salahuddin SZ, Ablashi DV, Pahwa S, Krohn K, Gallo RC (1988) In vitro cellular tropism of Human B-lymphotropic virus (human herpesvirus-6). J Exp Med 167:1659-1670.

Lusso P, Ensoli B, Markham PD, Ablashi DV, Salahuddin SZ, Tschachler E, Wong-Staal F, Gallo RC (1989) Productive dual infection of human CD4+ T lymphocytes by HIV-l and HHV6. Nature 337:370-373.

Morris DJ, Littler E, Jordan D, Arrand JR (1988) Antibody responses to human herpesvirus 6 and other herpesviruses. Lancet 2:1425-1426.

Morris DJ, Littler E, Arrand JR, Jordan D, Mallick NP, Johnson RWG (1989) Human herpesvirus 6 infection in renal-transplant recipients. Lancet 1:1560-1561.

Niederman JC, Liu C-R, Kaplan MH, Brown NA (1988) Clinical and serological features of human herpesvirus 6 infection in three adults. Lancet 2:817-819.

Okuno T, Higashi K, Shiraki K, Yamanishi K, Takahashi M, Kokado Y, Ishibashi M, Takahara S, Sonoda T, Tanaka K, Baba K, Yabuuchi H, Kurata T (1989a) Human herpesvirus 6 (HHV-6) infection in renal transplantation. Transplantation (in press).

Okuno T, Takahashi K, Balachandra K, Shiraki K, Yamanishi K, Takahashi M, Baba K (1989b) Seroepidemiology of human herpesvirus 6 infection in normal children and adults. J Clin Microbiol 27:651-653.

Pellett PE, Lindquester GJ, Feorino P, Lopez C (1990) Genomic heterogeneity of human herpesvirus 6 isolates. In: Immunobiology and Prophylaxis of Human Herpesvirus Infection (eds) Lopez C, Mori R, Roizman B, Whitley R. New York: Plenum Press (in press).

Pietroboni GR, Harnett GB, Farr TJ, Bucens MR (1988) Human herpes virus type 6 (HHV-6) and its *in vitro* effect on human immunodeficiency virus (HIV). J Clin Pathol 41:1310-1312.

Russler SK, Tapper MA, Carrigan DR (1989) Susceptibility of human herpesvirus 6 to acyclovir and gancyclovir. Lancet 2:382.

Salahuddin SZ, Ablashi DV, Markham PD, Josephs SF, Sturzenegger S, Kaplan M, Halligan G, Biberfeld P, Wong-Staal F, Kramarsky B, Gallo RC (1986) Isolation of a new virus, HBLV, in patients with lymphoproliferative disorders. Science 234:596-601.

Saxinger C, Polesky H, Eby N, Grufferman S, Murphy R, Tegtmeier G, Parekh V, Memon S, Hung C (1988) Antibody reactivity with HBLV (HHV-6) in U.S. populations. J Virol Methods 21:199-208.

Shiraki K, Okuno T, Yamanishi K, Takahashi M (1989) Virion and nonstructural polypeptides of human herpesvirus-6. Virus Research 13:173-178.

Spira TJ, Bozeman LH, Sanderlin KC, Warfield D, Feorino PM, Holman R, Kaplan JE, Fishbein DB, Lopez C (1990) Lack of correlation between human herpesvirus 6 infection and the course of human immunodeficiency virus infection. J Infect Dis 161: 567-570.

Streicher HZ, Hung CL, Ablashi DV, Hellman K, Saxinger C, Fullen J, Salahuddin SZ (1988) *In vitro* inhibition of human herpesvirus-6 by phosphonoformate. J Virol Methods 21:301-304.

Suga S, Yoshikawa T, Asano Y, Yazaki T, Hirata S (1989) Human herpesvirus-6 infection (exanthem subitum) without rash. Pediatrics 83:1003-1006.

Takahashi K, Sonoda S, Kawakami K, Miyata K, Oki T, Nagata T (1988) Human herpesvirus 6 and exanthem subitum. Lancet 1:1463.

Takahashi K, Sonoda S, Higashi K, Kondo T, Takahashi H, Takahashi M, Yamanishi K (1989) Predominant CD4 T-lymphocyte tropism of human herpesvirus 6-related virus. J Virol 63:3161-3163.

Tedder RS, Briggs M, Cameron CH, Honess R, Robertson D, Whittle H (1987) A novel lymphotropic herpesvirus. Lancet 2:390-392.

Ueda K, Kusuhara K, Hirose M, Okada K, Miyazaki C, Tokugawa K, Nakayama M, Yamanishi K (1989) Exanthem subitum and antibody to human herpesvirus-6. J Infect Dis 159:750-752.

Ward KN, Gray JJ, Efstathiou S (1989) Brief report: Primary human herpesvirus 6 infection in a patient following liver transplantation from a seropositive donor. J Med Virol 28:69-72.

Yamanishi K, Okuno T, Shiraki K, Takahashi M, Kondo T, Asano Y, Kurata T (1988) Identification of human herpesvirus-6 as a causal agent for exanthem subitum. Lancet 1:1065-1067.

Yoshida M, Uno F, Bai ZL, Yamada M, Nii S, Sata T, Kurata T, Yamanishi K, Takahashi M (1989) Electron microscopic study of a herpes-type virus isolated from an infant with exanthem subitum. Microbiol Immunol 33:147-154.

Yoshida T, Yoshiyama H, Suzuki E, Harada S, Yanagi K, Yamamoto N (1989) Immune response of exanthema subitum to human herpesvirus type 6 (HHV-6) polypeptides. J Infect Dis 160:901-902.

DISCUSSION

Chang R (University of California, Davis, CA):

Dr. Stewart, I think a lot of the earlier seroepidemiological studies on HHV-6 are very confusing. I am glad that you have showed the data indicating that if we use a more sensitive technique we can demonstrate close to 100% seropositivity. Recently, I did a survey on students at UC Davis and found that 96% of our college freshman were already positive. I feel very uncomfortable about it because other people are talking about 30% and 40% seropositivity. So I am glad to see your data. Maybe you should stress this point more because a lot of the earlier studies are very misleading.

Stewart J:

Exactly.

Al-Nakib W (Kuwait University, Kuwait, Kuwait):

John, has reactivation been associated with disease in renal transplant patients?

Stewart J:

We have looked at three illnesses in which there was a rise in titer to HHV-6, but in each one of them there was an associated rise of CMV titer, and CMV was isolated from the blood. So, it would seem more likely to assume that CMV was involved. Whether there might be a co-factor-role involved there will still have to be determined. There has also been one report of apparent primary infection after liver transplantation with a febrile illness and hepatitis observed in that patient. That is probably one of the best examples of no confounding factors in a HHV-6 infection in an adult patient.

Wright J (Gull Laboratories, Salt Lake City, UT):

What would be the significance of a serological test for this virus when there is such a high prevalence in the population at such an early age?

Stewart J:

I think that is a very important question. I saw one manufacturer that was advertising a test a couple of months after the announcement of the virus isolation. I get called by either doctors, or more often, by patients saying "I have human herpesvirus-6, how can it be treated?" It takes me ten minutes to try to explain to them that my own antibody titers are higher than their's and that I'm healthy and that no, this does not mean that herpes-6 is the cause of their chronic fatigue syndrome. Unfortunately, that is one of the slides that I left out in my over-preparation and I didn't have time to show. Yes, titers to herpesvirus-6 may be somewhat elevated with chronic fatigue syndrome, but so are CMV, EB and in our hands, even measles titers. We have now looked at over 40 patients. In studies done with both the Lake Tahoe group and Dr. Nelson Gantz, there does not appear to be any help in trying to diagnose chronic fatigue syndrome by testing for HHV-6. My attitude is, there is nothing to be gained by testing the adult patient for HHV-6 unless you have an acute illness, a mono-type illness in someone who is both EB and CMV negative. I think you are going to find some of those patients with it. But, in someone who has been ill for six or more months, there is very little point in looking for antibody to her-pes-6.

Lennette E (California Public Health Foundation, Berkeley, CA):

As your data showed, the virus is very widely spread through the popula-tion. It has been theorized that the virus is probably involved in oncological processes, and as you said, even chronic fatigue syndrome. It has also been at-tributed as the cause of a number of other syndromes. So, we have to start sort-ing some of these things out. How do you feel about that? There has to be more than roseola involved.

Stewart J:

One interesting thing about roseola is that there is a high instance of seizures in the febrile stage which seems to be more than what you would ex-pect with just a high fever. There appears to be, in some cases, an encephalopa-thy with transient neurologic problems, paralysis and in some cases, persistence of symptoms. It suggests, along with finding that the virus would grow in some glioblastoma cell lines, that there may be some neurotropism. We have had one patient so far, with "encephalitis" in whom we were able to show very high titers of HHV-6, that subsequently declined, while all the other herpes-viruses were negative. We were pleased to see that this child, though comatose for a period of a week or so, recovered completely and went back to normal school function. I think that in the neurologic and lymphoproliferative disor-ders, it is going to take careful sorting out to determine what is going on.

Epstein M (Oxford University, Oxford, England):

I am interested to hear you say that, because as I recall, in the primary in-fection in the liver transplant recipient, the viremia coincided with fits.

Stewart J:

I think that is another correlation.

Lennette E:

The virus may play another role, which is yet undetermined. There was a recent report of a young child who had a dual infection with HSV, and EBV, with perforation of the gut. So, the question arises, which virus is responsible for this occurrence, or was it just synergistic reaction between the two.

Lennette E:

There was one study by A. Buchbinder in which, this is in volunteers, they found seropositivity in about 80-85% of the people tested. With PCR, it's only about 50-55%, which is essentially, no correlation.

Stewart J:

Because so many of the HHV-6 isolates have come from AIDS patients or HIV positive patients, many think that there is a very common infection there. However, I think that in his lab at CDC, Paul Feorino has only found something like seven positive out of 2,000 cultures. Whenever this virus is reactivated, it does not seem be reactivated as commonly in AIDS patients as CMV. We have been looking at another project with CMV and are isolating CMV from over 55% of semen cultures. While this is a different system, and we'll be looking at a number of body secretions by PCR, HHV-6 doesn't show the degree of reactivation that HSV and CMV have.

Hendry M (California Department of Health Services, Berkeley, CA):

Have you or anyone else looked at HHV-6 in association with Kawasaki disease?

Stewart J:

A group from Tufts looked at Kawasaki disease and did not find any association.

STRAINS OF RESPIRATORY SYNCYTIAL VIRUS: IMPLICATIONS FOR VACCINE DEVELOPMENT

Larry J. Anderson

Department of Health and Human Services
Division of Viral Diseases
Center for Infectious Diseases
Centers for Disease Control
Atlanta, Georgia, 30330 USA

INTRODUCTION

Within a few years after its discovery in 1957, respiratory syncytial virus (RSV) was shown to be the single most important pathogen of acute lower respiratory tract illness among infants and young children worldwide. In the 1960's and 1970's multiple attempts to develop a vaccine failed. The first vaccine was formalin-inactivated, alum precipitated virus grown in primary monkey kidney cells that was evaluated in four field trials (Chin et al. 1969; Fulginiti et al. 1969; Kapikian et al. 1969; Kim et al. 1969). The vaccine induced a good serologic response to RSV but failed to protect the vaccinee from infection and, vaccinees <2 yrs old experienced more severe RSV disease than those not vaccinated. This increase in serious disease is illustrated in Table 1 for one of the trials. Live virus vaccines also failed (Belshe et al. 1982; McKay et al. 1988; Tyeryar, 1983; Wright et al. 1976, 1982). Some vaccine strains reverted to wild phenotype virus during clinical trials and others failed to induce an adequate antibody response. These failures have led researchers to conclude that a better understanding of the virus and the virus host interaction is needed to increase the chances of successfully developing an RSV vaccine.

One feature of the virus that could be important in vaccine development is antigenic differences among isolates. Differences among strains of RSV were first demonstrated by cross neutralization studies in the early 1960's (Coates et al. 1963); but were not felt to be clinically significant (Beem, 1967). The advent of monoclonal antibodies (MAbs) made it possible to more easily and precisely characterize antigenic differences between strains (Anderson et al. 1983), and to identify two groups of RSV strains (Anderson et al. 1985; Mufson et al. 1985). Subsequent studies have confirmed the presence of two major groups and demonstrated antigenic differences within the two groups (Akerlind et al. 1988; Finger et al. 1987; Hendry et al. 1986, 1989; Morgan et al. 1987; Mufson et al. 1988; Storch and Park, 1987; Tsutsumi et al. 1988).

These antigenic differences may be important in RSV disease in a fashion similar to antigenic drift and shift being important in influenza virus disease. Immunity induced by one strain may not provide an adequate level of protection against another strain. It is also possible that some strains may be

TABLE 1.   RSV Vaccine Trial, Washington, D.C., 1966-1967[a]

Age (mo)	RSV Vaccine[b]			No Vaccine		
	Infected	Pneum	Hospital	Infected	Pneum	Hospital
6-23	67% (15)	60%	33%	48% (48)	8%	0%
≥24	36% (22)	18%	0%	31% (61)	11%	2%

[a] Subjects were exposed to RSV nine months after the vaccine trial was initiated. Pre-outbreak sera demonstrated neutralizing antibody >1:100 in all vaccinated and 44% of unvaccinated. Neutralization titer was not associated with risk of pneumonia but with duration of viral shedding.

[b] Percent of total patients studied (total number studied). Pneum: RSV pneumonia; hospital: hospitalized for RSV disease.

From: Kapikian et al. 1969.

more virulent than others. Such differences may contribute both to an individuals susceptibility to disease and a communities susceptibility to outbreaks. In this review, I will discuss the features of RSV disease that suggest strain differences might be important, the types of differences present between RSV strains, the available information on the importance of strain differences in protective immunity, and the epidemiology of different strains.

STRAIN DIFFERENCES AND RSV DISEASE

Three features of RSV disease could be explained in part by variation in RSV strains. First, the peak age for serious RSV disease, pneumonia and bronchiolitis, is 2 to 6 months, a time when maternal neutralizing antibodies are present in the infants serum (Figure 1) (Parrott et al. 1973). Several studies have demonstrated that the presence of high titers of serum antibodies correlate with but do not ensure protection from either infection or disease (Glezen et al. 1981, 1986; Ogilvie et al. 1981). Studies of passively administered antibody in cotton rats showed that increasing amounts of neutralizing antibody provided increasing levels of protection against lower respiratory tract infection (Prince et al. 1985). Thus, it is possible that even relatively small differences in a child's neutralizing titer against different strains could correlate with significant differences in that child's susceptibility to severe disease when infected by the respective strains. For example, maternally acquired antibody may protect the infant against severe disease with infection by some strains but not with infection by others.

Second, persons can be reinfected and suffer severe disease throughout life, even yearly, as illustrated by studies of day-care center attendees (Henderson et al. 1979), RSV infected hospital staff (Hall et al. 1978), and an outbreak in a nursing home (Mathur et al. 1980). In children attending a day-care center, the percent of lower respiratory tract disease decreased with but did

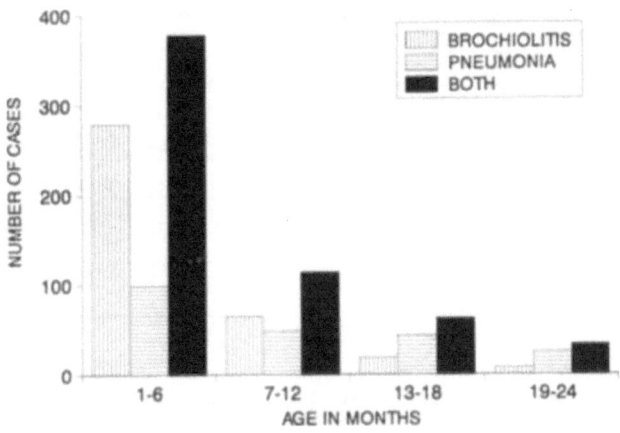

Figure 1. Hospitalized children with RSV pneumonia or bronchiolitis. Children <2 years of age hospitalized with pneumonia or bronchiolitis from January, 1960 through June, 1970 in Washington, D.C. and with RSV infection confirmed by isolation or 4-fold rise in RSV antibodies.

Adapted from Parrott et al. 1973.

not disappear with second and third infections (Table 2). It is possible that naturally acquired immunity provides better protection against the homologous than heterologous strains; thus, disease with second or third infections might be more common when caused by a different strain than that of the first infection.

Third, the severity of community outbreaks of RSV vary from year-to-year (Centers for Disease Control, 1986). It is possible that shifts from one year to the next in the outbreak strain, or strains, could result in individuals in the community having low levels of protective immunity and lead to a more severe community outbreak.

ANTIGENIC DIFFERENCES AMONG STRAINS

Our present understanding of differences between RSV strains is based primarily on their reactivity against panels of MAb. MAbs identify differences at the level of epitopes (the site on a protein where an individual MAb reacts) and antigenic sites (a contiguous region in the 3-dimensional structure of a protein that includes one or several epitopes) (Yewdell & Gerhard, 1981). Epitopes and antigenic sites can be differentiated from each other by the methods enumerated in Table 3. Although investigators have found epitope differences on all the structural proteins of RSV (Table 4) except the 1A and L proteins (Anderson et al. 1985; Gimenez et al. 1984; Mufson et al. 1985; Orvell et al. 1987; Routledge et al. 1987), we have focused on differences on the two surface glycoproteins, F and G. F and G are felt to be most important for inducing a protective immune response (Olmsted et al. 1986; Routledge et al. 1988; Stott et al. 1987; Walsh et al. 1987; Wathen et al. 1989).

TABLE 2. Effect of Age and Repeat Infections on RSV Illness

	Primary Infection			Repeat Infection		
	2-6 mo	6-12 mo	1-3 yr	1st	2nd	3rd
Number[a]	14	28	19	61	44	18
Bronchiolitis[b]	57%	39%	21%	38%	18%	11%
Pneumonia[b]	14%	11%	11%	12%	5%	0%

[a] Number of patients in group.
[b] Percent of total with RSV associated illness.

From: Henderson et al. 1979.

Immunofluorescence, enzyme immunoassay (EIA), immunoprecipitation, and Western blot have been used to study RSV strains. We have found EIA to be the simplest and most reproducible way to characterize strains and consequently have chosen it for our studies. Table 5 illustrates the reaction pattern of selected MAbs against different strains of RSV. In studies of over 1,000 RSV isolates, we have consistently found two groups of strains, group A that gives a positive reaction against the MAb at epitope Flb and a negative

TABLE 3. Definitions of Epitopes and Antigenic Sites

An epitope is the site on a protein where a monospecific antibody (e.g. a monoclonal antibody) reacts. Epitopes can be distinguished by:

1) Different patterns in blocking antibody assays.

2) Different reaction patterns in the same test against different virus strains, or

3) Different reaction patterns in different tests (e.g. ELISA and neutralization) against the same virus strain.

An antigenic site is an epitope or group of epitopes that represent a contiguous area on the 3-dimensional structure of a protein. Antigenic sites can be distinguished by:

1) MAbs that do not block each other in antibody blocking assays, or

2) Variant selection and sequence studies that identify distinct sites on the genome that encode for distinct areas on the 3-dimensional structure of the protein.

From: Yewdell and Gerhard, 1981.

TABLE 4.   The Proteins in RSV

Protein	Molecular weight by PAGE[a]	Sequence[b]	Location in Virion	Function
1C	11K - 18K	15,567	non-structural	unknown
1B	11K - 18K	14,674	non-structural	unknown
N	40K - 45K	42,600	nucleocapsid	structural protein of nucleocapsid
P	31K - 37K	27,150	nucleocapsid	part of the poly-merase complex
M	27K - 29K	28,717	envelope	unknown
1A	9.5K	7,536	envelope	unknown
G	79K - 92K	32,587	envelope	attachment to cell
F	66K - 70K	63,453	envelope	fusion, neutraliza-tion
22K	22K - 25K	22,156	envelope	unknown
L	160K - 200K	unknown	nucleocapsid	? polymerase

[a] Polyacrylamide gel electrophoresis.
[b] Calculated from the amino acid sequence deduced from the gene sequence.

reaction against the MAb at Flc and group B that gives a negative reaction against the MAb at epitope Flb and a positive reaction against the MAb at Flc. We use the above noted MAb reaction pattern to define the two groups of strains and differences in reaction patterns to other MAbs, primarily G protein MAbs, to identify different strains within the two groups.

EPITOPE DIFFERENCES ON RSV F AND G PROTEINS

To better understand the importance of differences at these epitopes and antigenic sites, we and others are studying 1) the function of the epitopes; 2) the ability of the two major groups to induce cross neutralizing antibodies or cross protection in animals; 3) the level of cross neutralizing antibodies induced after infection in humans; 4) the pattern of RSV groups found in natural reinfections in humans; and 5) the ability of changes in outbreak strains of RSV to explain differences in the severity of community outbreaks.

The function of epitopes has been investigated by studying the ability of the corresponding MAb to neutralize or inhibit fusion of the virus. We have identified three antigenic sites (Fl, F2, and F3) on the F protein; two, F2 and F3, are associated with strong neutralizing and anti-fusion activity (Table 6) (Anderson et al. 1988a). Binding activity against these two sites is conserved

TABLE 5. Examples and Interpretation of EIA Reaction Patterns

				MAb Reacted Against the Isolates[a]					
Isolate	Yr.	Loc.	Sbgp.	130-6d	143-5a	130-5f	130-9g	92-11c	102-10b
2584	84/85	MN	A1	1.387	1.309			0.961	
Long	56	MD	A2	0.818		0.651		1.185	
MV78	86	WV	A3	0.912				1.276	
82-776	82	MA	A4	0.835	0.499	1.179		0.989	
A2	62	Aust	A5	0.602		0.665	0.826	0.989	
9320	77	MA	B1		0.360	1.199			0.741
1122	84/85	MN	B2			1.212			0.755
18537	62	DC	B3			0.174			0.913

[a] Entry = specific absorbance signal. MAb 130-6d reacts at epitope G12, 143-5a at G5a, 130-5f at G4, 130-9g at G3b, 92-11c at f1b and 102-10b at F1c. Yr = year; Loc = location; and Sbgp = Subgroup. A = positive reaction against 92-11c and 131-2a and a negative reaction against MAb 102-10b. B = negative reaction against 92-11c and a positive reaction against MAbs 102-10b and 131-2 *Subgroups* (A1, A2, etc.) are based on differences in reaction patterns against MAbs 143-5a, 130-5f, and 130-9g.

among different strains of RSV but in studies by Beeler and Coelingh (1989) neutralizing activity at those two sites is not always conserved among different strains. The third antigenic site, Fl, includes epitopes Fla, Flb, Flc, and Fld and has some neutralizing activity. In our studies (Anderson et al. 1988a), the MAbs at site Fl were less effective at neutralizing RSV than those at F2 or F3 but all had some neutralizing activity either by themselves (partial neutralization) or when mixed with other MAbs (enhanced neutralization). Partial neutralization was defined as a significant decrease in infectious virus without complete neutralization of the virus as illustrated in Figure 2. Enhanced neutralization was defined as 1) an increase in the efficacy of neutralization from partial or no with the individual MAbs to complete neutralization with a mixture of the MAbs or 2) an increase in neutralizing titer with a mixture of two MAbs compared to the individual MAbs. Enhanced neutralization is illustrated in Figure 3. Half of the epitopes at site Fl were present on all strains by EIA (Anderson et al. 1988a) but some present on all strains by EIA were not present by neutralization (Beeler and Coelingh, 1989). An example of variability in neutralization activity against different strains is given in Figure 4. This MAb, at Fld, neutralizes strain 18537, a group B strain, but not strain A2 that was the immunizing strain for this MAb. Figures 5 schematically depicts the function of the different F protein epitopes and Figure 6 their variability. Note that the

TABLE 6. Neutralization of RSV F Protein MAbs

Mab	Epitope	Neut[a]	Fusion Inhibit	Partial Neut[b]	Enhance[c] Neut	Epitope Variability
131-2a	Fla	<20	<20	+/-	+	All
92-11c	Flb	<20	<20	+/-	+	Gp1
102-10b	Flc	<20	<20	<20	+	Gp2
130-7e	Fld	+/-[d]	<20	<20	+	All
133-1h	F2	2,560	50		+	All
143-6c	F3	20,480	>500		+	All

[a] Neutralization: complete neutralization in an EIA infectivity assay. Titer is taken as the highest dilution of the MAb that gave specific absorbance <0.100. The reciprocal neutralizing titer listed in the table is the median titer against the A2 stain. MAbs at epitopes Fla, Flb, F2 and F3 were tested against the Long, A2, and 82-776 strains of RSV. The MAb at Flc was tested against the 18537 and 8/60 strains of RSV.

[b] Partial neutralization: neutralization in an EIA neutralization assay in which no dilution gave complete neutralization. All MAbs were tested against RSV A2 except MAb 102-10b (epitope Flc) that was tested against RSV 18537; <20 = negative results in all tests; +/- = positive result in some but <50% of 4 to 8 tests; Blanks are either not done or not applicable.

[c] Entries are + = enhancement of neutralization from none or partial to complete neutralization.

[d] 130-7e neutralized strain 18537 to a median titer of 1:640, but did not neutralize strains A2, Long, 8/60, or 82-776.

majority of antigenic sites/epitopes are broadly reactive (dark shading) and all have some neutralizing activity (dark or intermediate shading). The conservation of most F protein epitopes is consistent with sequence studies of the F protein gene that show a high degree of homology at the genomic and deduced amino acid sequence levels between the F genes of the two groups of RSV strains (Table 7). The function and variability studies of F protein epitopes suggest that this protein should induce high levels of broadly reactive immunity.

None of the G protein MAbs neutralized RSV but some gave partial or enhanced neutralization (Figures 2 and 3, and Table 8). The reaction patterns of G protein MAbs against strains of RSV suggest a great deal of epitope variability on this protein (Table 9); most of the variable epitopes gave some neutralizing activity (Table 8). Note that multiple epitope differences are present between the strains within groups and not just between groups. The results in Table 9 show that group A strains can be as different from each other (at the G protein epitopes studied) as they are from group B strains. The function of the epitopes and antigenic sites on G is schematically depicted in Figure 5 and their variability in Figure 6. Note that only one of the epitopes is broadly reactive (dark shading in Figure 6) and none give complete neutralization or fusion inhibition (dark shading in Figure 5). The extensive antigenic variability of G correlates with a low level of genomic and deduced amino acid sequence homology

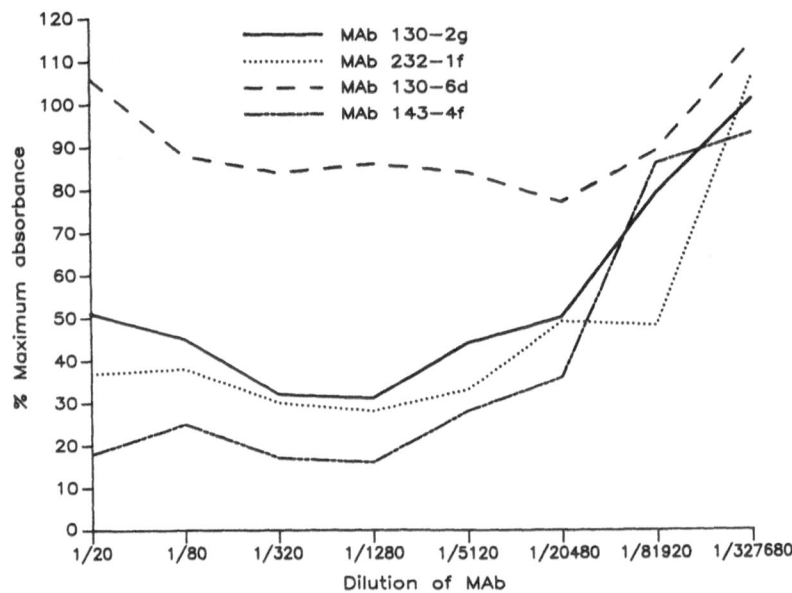

Figure 2. Neutralization of RSV by MAbs 130-6d (epitope G12), 130-2g (epitope G2), 232-1f (epitope G3a), and 143-4f (epitope G6) in an EIA neutralization assay. Absorbance readings from the EIA are expressed as the percent of maximum specific absorbance (maximum specific absorbance = mean absorbance for virus infected control cells - mean absorbance for uninfected control cells); a higher absorbance signal indicates greater replication. Note that MAb 130-2g gives <50% maximum absorbance (indicative of partial neutralization) to a titer of 1:20,480, MAb 232-1f to a titer of 1:81,920, and MAb 143-4f to a titer of 1:20,480. MAb 130-6d shows no neutralization activity at the dilutions tested.

between the G genes of the two groups of RSV strains (Table 7). The function and variability studies of the G protein suggest that it will induce lower levels of protection than the F protein and this protection will tend to be group specific.

STRAIN DIFFERENCES AND PROTECTIVE IMMUNITY

Studies in the early 1960s demonstrated 3- to 32-fold differences in cross neutralization between strains later characterized as group A or group B (Coates and Chanock, 1962; Coates et al. 1963; Doggett and Taylor-Robinson, 1965). Thus, immunity induced by a virus from one group provides some immunity against viruses in the other group but at a lower level than against viruses in the same group. Studies with the cloned expressed proteins or affinity purified proteins have provided an opportunity to look at protection induced by individual proteins (Olmsted et al. 1986; Routledge et al. 1988; Stott et al. 1987; Walsh et al. 1987; Wathen et al. 1989). These studies have demonstrated that, as expected from the function and variability studies, the highest level of neutralizing antibodies and the greatest degree of protection is induced by the F protein; these antibodies and this protection tends to be good for viruses from

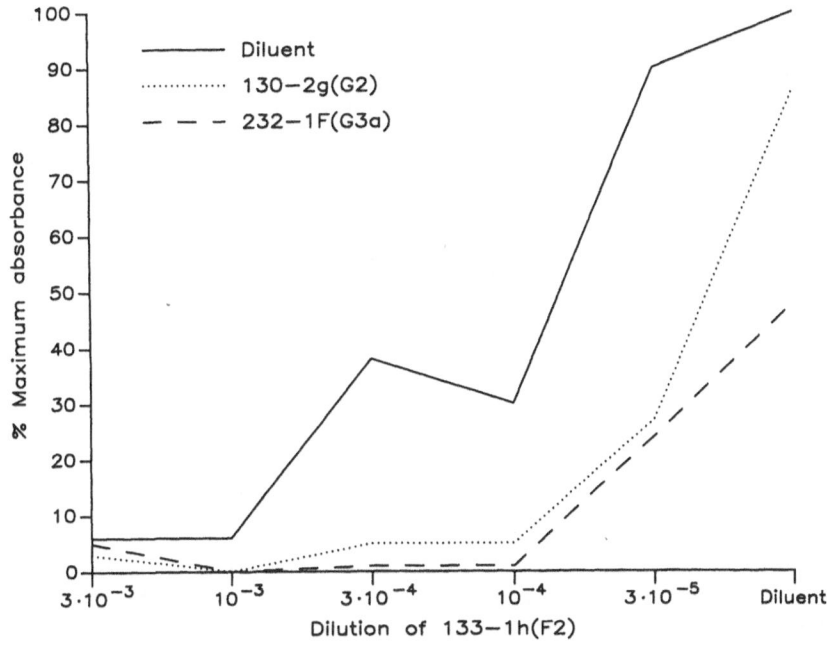

Figure 3. Enhanced neutralization between MAb 133-1h (epitope F2) and MAbs 130-2g (epitope G2) or 232-1f (epitope G3a). Neutralization was determined in an EIA neutralization assay. Absorbance readings from the EIA are expressed as percent maximum specific absorbance (maximum specific absorbance = mean absorbance for virus infected control cells - mean absorbance uninfected control cells), a higher absorbance signal indicates greater replication. MAbs 130-2g and 232-1f did not neutralize RSV by themselves. The "diluent" dilution for their respective curves gives the values for the MAbs by themselves. When mixed with MAb 133-1h, they shifted the curve of 133-1h (the diluent curve gives the values for 133-1h by itself) to the right and increased neutralization titer between 10- and 30-fold.

TABLE 7.   Sequence Divergence Between Group A and B RSV Strains

	Protein				
	1B	1C	N	F	G
Nucleotide homology	78%	78%	86%	79%	68%
Amino acid homology	93%	87%	96%	89%	53%

Percent homology between nucleotide sequences of RSV A2 compared with RSV 18537 and corresponding deduced amino acid sequences.

From: Johnson and Collins, 1989.

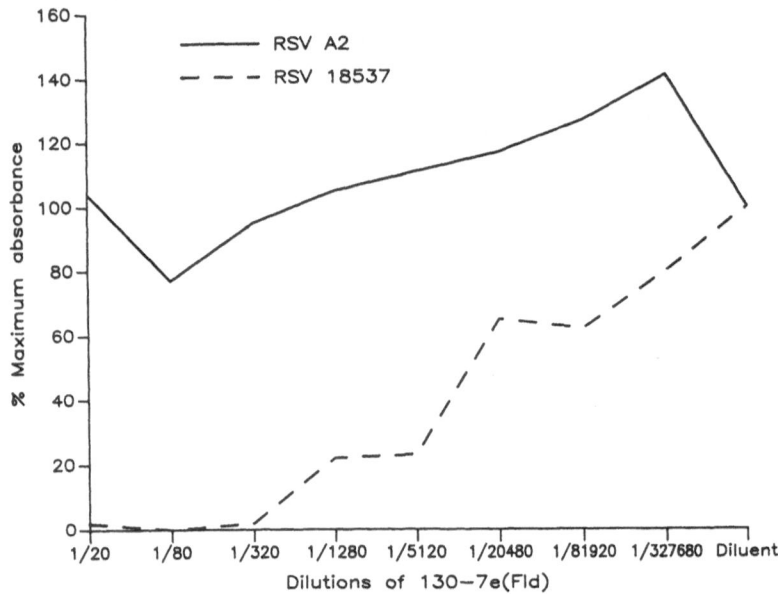

Figure 4. Neutralization of RSV by MAb 130-7e (epitope F1d) in an EIA neutralization assay. Absorbance readings from the EIA are expressed as the percent maximum specific absorbance (maximum specific absorbance = mean absorbance for virus infected control cells - mean absorbance for uninfected control cells); a higher absorbance signal indicates greater replication. Note that 130-7e neutralizes RSV 18537 to a titer of 1:320 but has no neutralizing activity against RSV A2.

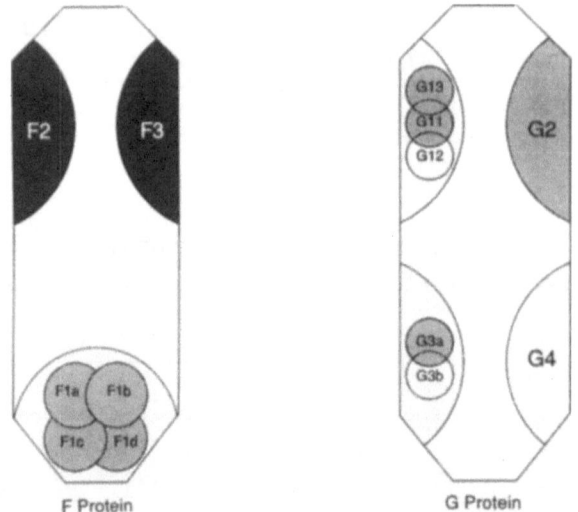

Figure 5. Epitopes on RSV surface glycoproteins. Schematic depiction of the function of epitopes and antigenic sites on the RSV F and G proteins. Antigenic sites are represented by the larger circles. The MAbs reacting at antigenic sites with the darkest shading inhibit fusion and completely neutralize RSV; MAbs reacting at antigenic sites/epitopes with intermediate shading do not inhibit fusion and partially neutralize, completely neutralize some but not all strains, or enhance neutralization of RSV; and MAbs reacting at antigenic sites/epitopes with the lightest shading do not inhibit fusion or neutralize RSV.

TABLE 8. Neutralization of RSV G Protein MAbs

Mab	Epitope	Neut[a]	Partial Neut[b]	Enhance Neut[c]	Epitope Variability
63-10f	G11	<20	10,240	+	Gp1
130-6d	G12	<20	<20	+	Gp1
131-2g	G13	<20	<20	0	All
130-2g	G2	<20	20,480	+	+/- Gp1
232-1f	G3a	<20	10,240	+	+/- Gp1
130-9g	G3b	<20	<20	0	+/- Gp1
130-5f	G4	<20	<20	0	+/- Gp1 & 2
143-5a	G5a	<20	+/-	+	+/- Gp1 & 2
142-12g	G5b	<20	>100	+	+/- Gp1
143-4f	G6	<20	10,240	+	+/- Gp1 & 2

[a] Neutralization: Complete neutralization in an EIA infectivity assay.

[b] Partial neutralization = neutralization in an EIA neutralization assay in which no dilution gave complete neutralization (specific absorbance <0.100) but some dilutions gave a specific absorbance <50% of maximum specific absorbance. All MAbs were tested against RSV A2 except MAbs 143-5a (epitope G5a), 142-12g (epitope G5b), and 143-4f (epitope G6) that were tested against RSV 82-776. + = positive result in >60% of 3 to 8 tests (in some tests partial neutralization was not seen); 0 = negative results in all tests; +/- = positive result in some but <50% of 4 to 8 tests.

[c] Entries are + = enhancement of neutralization from none or partial to complete neutralization. 0 = no enhancement.

TABLE 9. G Epitope Differences Between RSV Long and Other Strains[a]

Virus	Yr	Loc	Gp	G11	G12	G2	G3a	G3b	G4	G5a	G5b	G6	Total[b]
Long	56	MD	A	+	+	0	+	0	+	0	0	0	
A2	61	Aust	A			+		+					2
82-776	82	MA	A			0				+	+	+	4
2584	84	MN	A			0		0	+			+	4
18537	62	DC	B	0	0	0			+/-				4
1122	85	MN	B	0	0	0							3
9320	77	MA	B	0	0	0				+/-			4

[a] Blank is no difference between the isolate and RSV Long in its reaction with the respective MAb by EIA. + = positive reaction; 0 = negative reaction; +/- = low positive reaction; Yr = year; Loc = location; and Gp = group.

[b] Total number of epitope differences between the isolate and RSV Long.

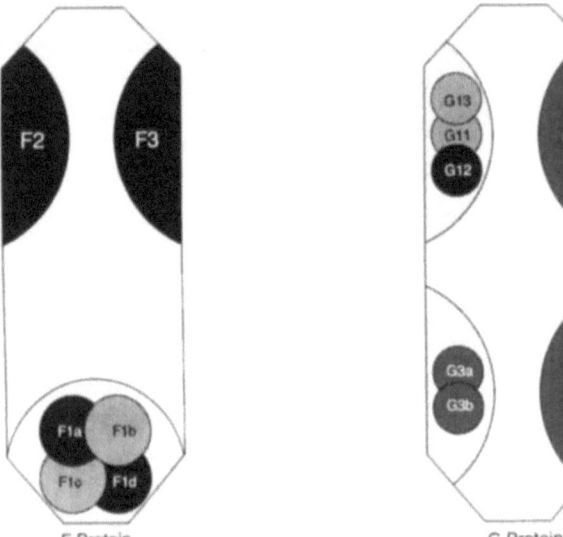

F Protein

G Protein

Figure 6. Epitopes on RSV surface glycoproteins. Schematic depiction of the variability of epitopes and antigenic sites on the RSV F and G proteins. Antigenic sites are represented by the larger circles and epitopes within antigenic sites by the smaller circles. The MAbs reacting at antigenic sites with the darkest shading react against all strains by EIA; MAbs reacting at antigenic sites/epitopes with intermediate shading react with some but not all isolates in either or both groups; MAbs reacting at antigenic sites/epitopes with the lightest shading react specifically with either group A or group B isolates.

the same as from the other group of strains. The G protein also induces neutralizing antibodies and protective immunity but these antibodies and this immunity tends to be group specific. Data from a study by Johnson et al. (1987) illustrates the differences in the immunity induced by the two proteins (Table 10). Note that most animals immunized with vaccinia-A2-F had no RSV in their lungs after challenge while most immunized with vaccinia-A2-G had RSV in their lungs. Note also that the reduction in titer of RSV in the lungs after challenge with the homologous versus the heterologous virus was not significantly different for animals immunized with vaccinia-A2-F but was for animals immunized with vaccinia-A2-G. This group also estimated, from binding antibody studies, that the F protein of the two groups has 30 to 60% antigenic relatedness and the G protein 5% antigenic relatedness and, from neutralizing antibody studies, the two groups have an overall antigenic relatedness of 25%.

The antibody response to RSV infections in humans parallels that in animals. In one study of primary infection in infants, the antibody response by EIA to group A and group B F proteins was similar for the infecting and heterologous group (Hendry et al. 1988). The antibody response by EIA to the group A and group B G proteins however was better to the infecting than the heterologous group (Table 11). The neutralizing titer also tended to be higher

TABLE 10. Protective Efficacy of Vaccinia Virus-RSV F and G Protein

Immunogen titer	No. of Animals	Challenge Virus[a]	No. + RSV in lungs	Titer RSV in lungs[b]	Reduction in RSV
Vaccinia-Bgal	8	A2	8 (100%)	6.6	
	8	18537	8 (100%)	4.8	
Vaccinia-A2-F	9	A2	0 (0%)[c]	2.0[c]	$\geq$4.6
	10	18537	1 (10%)[c]	2.1[c]	$\geq$2.7
Vaccinia-A2-G	11	A2	9 (82%)	3.6[c]	3.0[d]
	10	18537	7 (78%)	2.9[c]	1.9[d]

[a] Dose was $10^{5.4}$ PFU of the indicated RSV strain intranasally at 21 days after immunization.

[b] Geometric mean PFU/g of tissue. Animals with no virus detected (<2.0 PFU/g) were assigned a value of 2.0 PFU/g.

[c] Significant reduction in virus titer or number positive compared with Vaccinia-Bgal group at P<0.001.

[d] Mean reduction in PFU/g greater for RSV A2 vs. RSV 18537 at P>0.05.

From: Johnson et al. 1987.

TABLE 11. Strain Specific Antibody Response to RSV Infection

Infection (No.)[a]	Assay	Antigen Gp1	Gp2
Gp 1 (24)	EIA F	62%	71%
Gp 2 (19)		67%	58%
Gp 1 (24)	EIA G	67%[b]	21%
Gp 2 (19)		32%	58%
Gp 1 (24)	Neut	62%	54%
Gp 2 (19)		42%	68%

[a] Primary infection in infants. Mean age = 6 mo (0.5 to 23 months). Entries = % with > 4-fold rise.

[b] Significant at P<.05.

From: Hendry et al. 1988.

to the infecting than the heterologous group but this difference was not significant. There is limited information about the pattern of strains isolated from sequential infections in humans. The one published study demonstrated a significant increase in likelihood of having a group B rather than a group A infection if the first infection were group A; suggesting that strain differences, at least at the group level, may play some role in susceptibility to reinfection (Mufson et al. 1987). Several studies have suggested that group A strains may be more virulent than group B strains (Mufson et al. 1988; Taylor et al. 1989), but this observation needs to be confirmed by additional study.

Although the role of strain differences in protection or disease associated with the cellular immune response to RSV is yet to be determined, the available data suggest it could play a role. As measured by the cytotoxic T lymphocyte (CTL) response, the F and N proteins are important in cellular immunity, the G protein is of little importance in cellular immunity, and the cellular immune response includes both broadly reactive and group or strain specific components (Bangham and Askonas, 1986; Bangham et al. 1985, 1986; Cannon and Bangham, 1989; Openshaw et al. 1988; Pemberton et al. 1987).

EPIDEMIOLOGY OF RSV STRAINS

The temporal and geographic patterns of RSV isolations provide the foundation for determining the role that shifts in outbreak strains may have on the severity of associated outbreaks. Since the initial studies with RSV strains, it became apparent that group A strains were more common than group B strains, that both groups could circulate simultaneously in the same community, and that the two groups could circulate with different temporal and geographic patterns (Akerlind and Norrby, 1986; Anderson et al. 1985; Finger et al. 1987; Hendry et al. 1986, 1989; Morgan et al. 1987; Mufson et al. 1985, 1988; Storch and Park, 1987; Tsutsumi et al. 1988). It is also evident that there are multiple antigenically distinct strains within each major group and that during outbreaks multiple distinct strains co-circulate. A recent study of isolates from a number of laboratories in the United States and Canada illustrates this phenomena (Anderson et al. 1988b). In one year of this study, laboratories in Chicago, Denver, and St. Louis found predominantly group A isolates while a laboratory in Washington, D.C. found predominantly group B isolates (Table 12). Differences in strains circulating in different communities became even more apparent when differences within the two groups were studied (Table 13). Note that each of the three laboratories summarized in this table had a unique pattern of RSV strains. The presently available data suggest that community outbreaks of RSV are not linked by a national outbreak strain or strains; RSV outbreaks apparently are community, possibly regional, but not national phenomena.

Variability in strains circulating in a community has made it possible to study patterns of nosocomial transmission of RSV and infer efficacy of infection control efforts (Finger et al. 1987). Studies of the impact of antigenic shifts in outbreak strains on the severity of RSV outbreaks are in progress.

CONCLUSIONS

Antigenic and genetic differences among strains of RSV isolates have been clearly identified. Based on studies with monoclonal antibodies, strains of RSV can be placed into two major groups, A and B. There are also multiple antigenically distinct strains within these two groups. Most of the antigenic differences between isolates occurs on the G protein but 5 of the 7 other the structural proteins have also been shown to have differences. Sequence studies of

TABLE 12. Group 1 and 2 Isolates for the 84/85 RSV Season

Laboratory	No. Tested	% Total Group A	% Total Group B	% Total Group ?[a]
Denver, CO	20	85	15	0
Chicago, IL	19	84	16	0
St. Louis, MO	20	65	35	0
Washington, DC	20	40	55	5
TOTAL	79	68	30	1

[a] Isolates that did not grow well enough to be subgrouped.

TABLE 13. Subgroups of RSV Isolates for Selected Laboratories

Laboratory	Year	Group A 1a	1c	1d	1f	Group B 2	2a	2b	All Isolates ?[a]	Total
St. Louis	84/85	2	11				7		0	20
D.C.	84/85	6		2			6	4	2	20
Chicago	84/85	6	7		3	3			0	19

[a] Isolates that did not grow well enough to be subgrouped.

RSV genes show differences consistent with those found by MAb studies. Animal studies of the immune response to individual proteins of RSV have shown that the F and G proteins are the most important for inducing protective immunity, the F protein induces a high level of protective immunity against all strains, the G protein induces a lower level of protective immunity that tends to be group specific, and virus from one group induces better protection against viruses from the same than viruses from the other group. Limited studies of different strains in human infection and disease are consistent with these animal studies; thus, strain differences are likely to be a factor in the clinical and epidemiologic features of RSV disease and may be important for vaccine development. How important they are, however, is yet to be determined.

REFERENCES

Akerlind B, Norrby E (1986) Occurrence of respiratory syncytial virus subtypes A and B strains in Sweden. J Med Virol 19:241-247.

Akerlind B, Norrby E, Orvell C, Mufson MA (1988) Respiratory syncytial virus: Heterogeneity of subgroup B strains. J Gen Virol 69:2145-2154.

Anderson LJ, Hierholzer JC, Tsou C, McIntosh K (1983) Characterization of respiratory syncytial virus strains using monoclonal antibodies. In, Program and abstracts of the Twenty-third Interscience Conference on Antimicrobial Agents and Chemotherapy, New Orleans, Louisiana: American Society of Microbioloby, Abstract #926.

Anderson LJ, Hierholzer JC, Tsou C, Hendry RM, Fernie BF, Stone Y, McIntosh K (1985) Antigenic characterization of respiratory syncytial virus strains with monoclonal antibodies. J Infect Dis 151:626-633.

Anderson LJ, Bingham P, Hierholzer JC (1988a) Neutralization of respiratory syncytial virus by individual and mixtures of F and G protein monoclonal antibodies. J Virol 62:4232-4238.

Anderson LJ, Hendry RM, Pierik LT, McIntosh K (1988b) Multicenter study of strains of respiratory syncytial virus. In, Program and abstracts of the Twenty-eighth Interscience Conference on Antimicrobial Agents and Chemotherapy. Washington,D.C.: American Society for Microbiology, Abstract #211.

Bangham CRM, Askonas BA (1986) Murine cytotoxic T cells specific to respiratory syncytial virus recognize different antigenic subtypes of the virus. J Gen Virol 67:623-629.

Bangham CRM, Cannon MJ, Karzon DT, Askonas BA (1985) Cytotoxic T-cell response to respiratory syncytial virus in mice. J Virol 56:55-59.

Bangham CRM, Openshaw PJM, Ball LA, King AMQ, Wertz GW, Askonas BA (1986) Human and murine cytotoxic T cells specific to respiratory syncytial virus recognize the viral nucleoprotein (N), but not the major glycoprotein (G), expressed by vaccinia virus recombinants. J Immunol 137:3973-3977.

Beeler JA, Coelingh KVW (1989) Neutralization of epitopes of the F glycoprotein of respiratory syncytial virus: Effect of mutation upon fusion function. J Virol 63:2941-2950.

Beem M (1967) Repeated infections with respiratory syncytial virus. J Immunol 98:1115-1122.

Belshe RB, Van Voris LP, Mufson MA (1982) Parenteral administration of live respiratory syncytial virus vaccine: Results of a field trial. J Infect Dis 145:311-319.

Cannon MJ, Bangham CRM (1989) Recognition of respiratory syncytial virus fusion protein by mouse cytotoxic T cell clones and a human cytotoxic T cell line. J Gen Virol 70:79-87.

Centers for Disease Control (1986) Respiratory syncytial virus Oklahoma. MMWR 35:162-164.

Chin J, Magoffin RL, Shearer LA, Schieble JH, Lennette EH (1969) Field evaluation of a respiratory syncytial virus vaccine and a trivalent parainfluenza virus vaccine in a pediatric population. Am J Epidemiol 89:449-463.

Coates HV, Chanock RM (1962) Experimental infection with respiratory syncytial virus in several species of animals. Am J Hyg 76:302-312.

Coates HV, Kendrick L, Chanock RM (1963) Antigenic differences between two strains of respiratory syncytial virus. Proc Soc Exp Biol Med 112:958-964.

Doggett JE, Taylor-Robinson D (1965) Serological studies with respiratory syncytial virus. Arch ges Virusforsch 15:601-608.

Finger F, Anderson LJ, Dicker RC, Harrison B, Doan R, Downing A, Corey L

(1987) Epidemic infections caused by respiratory syncytial virus in institutionalized young adults. J Infect Dis 155:1335-1339.

Fulginiti VA, Eller JJ, Sieber OF, Joyner JW, Minamitani M, Meiklejohn G (1969) Respiratory virus immunization: I. A field of two inactivated respiratory virus vaccines; an aqueous trivalent parainfluenza virus vaccine and an alum-precipitated respiratory syncytial virus vaccine. Am J Epidemiol 89:435-448.

Gimenez HB, Cash P, Melvin WT (1984) Monoclonal antibodies to human respiratory syncytial virus and their use in comparison of different virus isolates. J Gen Virol 65:963-971.

Glezen WP, Paredes A, Allison JE, Taber LH, Frank AL (1981) Risk of respiratory syncytial virus infection for infants from low-income families in relationship to age, sex, ethnic group, and maternal antibody level. J Pediatr 98:708-715.

Glezen WP, Taber LH, Frank AL, Kasel JA (1986) Risk of primary infection and reinfection with respiratory syncytial virus. Am J Dis Child 140:543-546.

Hall WJ, Hall CB, Speers DM (1978) Respiratory syncytial virus infection in adults: Clinical, virologic, and serial pulmonary function studies. Ann Intern Med 88:203-205.

Henderson FW, Collier AM, Clyde WA, Denny FW (1979) Respiratory-syncytial-virus infections, reinfections and immunity: a prospective longitudinal study in young children N Engl J Med 300:530-534.

Hendry RM, Talis AL, Godfrey E, Anderson LJ, Fernie BF, McIntosh K (1986) Concurrent circulation of antigenically distinct strains of respiratory syncytial virus during community outbreaks. J Infect Dis 153:291-297.

Hendry RM, Burns JC, Walsh EE, Graham BS, Wright PF, Hemming VG, Ridriquez WJ, Kim HW, Prince GA, McIntosh K, Chanock RM, Murphy BR (1988) Strain-specific serum antibody responses in infants undergoing primary infection with respiratory syncytial virus. J Infect Dis 157:640-647.

Hendry RM, Pierik LT, McIntosh K (1989) Prevalence of respiratory syncytial virus subgroups over six consecutive outbreaks: 1981-1987. J Infect Dis 160:185-190.

Johnson PR, Jr., Olmsted RA, Prince GA, Murphy BR, Alling DW, Walsh EE, Collins PL (1987) Antigenic relatedness between glycoproteins of human respiratory syncytial virus subgroups A and B: Evaluation of the contributions of F and G glycoproteins to immunity. J Virol 61:3163-3166.

Johnson PR, Collins PL (1989) The IB (NS1) and N proteins of human respiratory syncytial virus (RSV) of antigenic subgroups A and B: Sequence conservation and divergence within RSV genomic RNA. J Gen Virol 70:1539-1547.

Kapikian AZ, Mitchell RH, Chanock RM, Shvedoff RA, Stewart CE (1969) An epidemiologic study of altered clinical reactivity to respiratory syncytial (RS) virus infection in children previously vaccinated with an inactivated RS virus vaccine. Am J Epidemiol 89:405-421.

Kim HW, Canchola JG, Brandt CD, Pyles G, Chanock RM, Jensen K, Parrott RH (1969) Respiratory syncytial virus disease in infants despite prior administration of antigenic inactivated vaccine. Am J Epidemiol 89:422-434.

Mathur U, Bentley DW, Hall CB (1980) Concurrent respiratory syncytial virus and influenza A infections in the institutionalized elderly and chronically ill. Ann Intern Med 93:49-52.

McKay E, Higgins P, Tyrrell D, Pringle C (1988) Immunogenicity and pathogenicity of temperature-sensitive modified respiratory syncytial virus in adult volunteers. J Med Virol 25:411-421.

Morgan LA, Routledge EG, Willcocks MM, Samson ACR, Scott R, Toms GL (1987) Strain variation of respiratory syncytial virus. J Gen Virol 68:2781-2788.

Mufson MA, Orvell C, Rafnar B, Norrby E (1985) Two distinct subtypes of human respiratory syncytial virus. J Gen Virol 66:2111-2124.

Mufson MA, Belshe RB, Orvell C, Norrby E (1987) Subgroup characteristics of respiratory syncytial virus strains recovered from children with two consecutive infections. J Clin Microbiol 25:1535-1539.

Mufson MA, Belshe RB, Orvell C, Norrby E (1988) Respiratory syncytial virus epidemics: Variable dominance of subgroups A and B strains among children, 1981-1986. J Infect Dis 157:143-148.

Ogilvie MM, Vathenen AS, Radford M, Codd J, Key S (1981) Maternal antibody and respiratory syncytial virus infection in infancy. J Med Virol 7:263-271.

Olmsted RA, Elango N, Prince GA, Murphy BR, Johnson PR, Moss B, Chanock RM, Collins PL (1986) Expression of the F glycoprotein of respiratory syncytial virus by a recombinant vaccinia virus: Comparison of the individual contributions of the F and G glycoproteins to host immunity. Proc Natl Acad Sci USA 83:7462-7466.

Openshaw PJM, Pemberton RM, Ball LA, Wertz GW, Askonas BA (1988) Helper T cell recognition of respiratory syncytial virus in mice. J Gen Virol 69:305-312.

Orvell C, Norrby E, Mufson MA (1987) Preparation and characterization of monoclonal antibodies directed against five structural components of human respiratory syncytial virus subgroup B. J Gen Virol 68:3125-3135.

Parrott RH, Kim HW, Arrobio JO, Hodes DS, Murphy BR, Brandt CD, Camargo E, Chanock RM (1973) Epidemiology of respiratory syncytial virus infection in Washington, D.C. II. Infection and disease with respect to age, immunologic status, race and sex. Am J Epidemiol 98:289-300.

Pemberton RM, Cannon MJ, Openshaw PJM, Ball LA, Wertz GW, Askonas BA (1987) Cytotoxic T cell specificity for respiratory syncytial virus proteins: Fusion protein is an important target antigen. J Gen Virol 68:2177-2182.

Prince GA, Horswood RL, Chanock RM (1985) Quantitative aspects of passive immunity to respiratory syncytial virus infection in infant cotton rats. J Virol 55:517-520.

Routledge EG, Willcocks MM, Morgan L, Samson ACR, Scott R, Toms GL (1987) Heterogeneity of the respiratory syncytial virus 22K protein revealed by western blotting with monoclonal antibodies. J Gen Virol 68:1209-1215.

Routledge EG, Willcocks MM, Samson ACR, Morgan L, Scott R, Anderson JJ, Toms GL (1988) The purification of four respiratory syncytial virus proteins and their evaluation as protective agents against experimental infection in BALB/c mice. J Gen Virol 69:293-303.

Storch GA, Park CS (1987) Monoclonal antibodies demonstrate heterogeneity in the G glycoprotein of prototype strains and clinical isolates of respiratory syncytial virus. J Med Virol 22:345-356.

Stott EJ, Taylor G, Ball LA, Anderson K, Young KK-Y, King AMQ, Wertz GW (1987) Immune and histopathological responses in animals vaccinated with recombinant vaccinia viruses that express individual genes of human respiratory syncytial virus. J Virol 61:3855-3861.

Taylor CE, Morrow S, Scott M, Young B, Toms GL (1989) Comparative virulence of respiratory syncytial virus subgroups A and B. Lancet 1:777-778.

Tsutsumi H, Onuma M, Suga K, Honjo T, Chiba Y, Chiba S, Ogra PL (1988) Occurrence of respiratory syncytial virus subgroup A and B strains in Japan, 1980 to 1987. J Clin Microbiol 26:1171-1174.

Tyeryar FJ (1983) Report of a workshop on respiratory syncytial virus and parainfluenza viruses. J Infect Dis 148:588-598.

Walsh EE, Hall CB, Briselli M, Brandriss MW, Schlesinger JJ (1987) Immunization with glycoprotein subunits of respiratory syncytial virus to protect cotton rats against viral infection. J Infect Dis 155:1198-1204.

Wathen MW, Brideau RJ, Thomsen DR (1989) Immunization of cotton rats with the human respiratory syncytial virus F glycoprotein produced using a baculovirus vector. J Infect Dis 159:255-264.

Wright PF, Shinozaki T, Fleet W, Sell SH, Thompson J, Karzon DT (1976) Evaluation of a live, attenuated respiratory syncytial virus vaccine in infants. J Pediatr 88:931-936.

Wright PF, Belshe RB, Kim HW, Van Voris LP, Chanock RM (1982) Administration of a highly attenuated, live respiratory syncytial virus vaccine to adults and children. Infect Immun 37:397-400.

Yewdell JW, Gerhard W (1981) Antigenic characterization of viruses by monoclonal antibodies. Annu Rev Microbiol 35:185-206.

PLANS FOR HUMAN TRIALS OF A VACCINE AGAINST EPSTEIN-BARR VIRUS INFECTION

M.A. Epstein

Nuffield Department of Clinical Medicine
University of Oxford
John Radcliffe Hospital
Oxford, OX3 9DU
United Kingdom

INTRODUCTION

Evidence linking Epstein-Barr (EB) virus to two human cancers, endemic Burkitt's lymphoma (BL) (Burkitt, 1963) and undifferentiated nasopharyngeal carcinoma (NPC) (Shanmugaratnam, 1971), has accumulated steadily over the years (Epstein and Morgan, 1983; Lenoir and Bornkamm, 1987; Whittle et al. 1984). Already in 1976 enough was known for it to be evident that EB virus was an essential link in a complicated chain of events leading inexorably to the malignant tumors, and it therefore seemed reasonable to suppose that if infection by the virus could be prevented, the incidence of these tumors in populations at risk would be decreased (Epstein, 1976). In view of the compelling analogy afforded by the effect of avoiding cigarette smoking on the incidence of bronchogenic carcinoma (Doll and Peto, 1976), it seemed unethical not, at least, to explore the possibility of vaccine intervention to prevent EB virus infection in the context of BL and NPC.

This concept has sometimes been criticized on the grounds that the mechanisms whereby EB virus might bring about malignant change were not known, nor even that the virus was in fact naturally carcinogenic. However, the mechanisms whereby cigarette smoking induces lung cancer are likewise not understood yet no-one would argue that cigarette smoking should not be discouraged until they are; and as regards EB virus and the causation of cancer in man, the only possible definitive proof in the human situation would come from the demonstration that a vaccine program against the virus reduced the incidence of the associated malignant diseases. Accordingly, a sustained effort has been made over the last dozen years to investigate the possibility of elaborating an anti-viral vaccine to prevent infection with EB virus.

RATIONALE FOR AN EB VIRUS VACCINE

Before the work was undertaken, a number of questions required consideration. In the first place, would the effort to develop such a vaccine be worthwhile? Although BL is the most common cancer of children in endemic zones, more common there than all other children's cancers added together (Burkitt, 1963), in World cancer terms it is not numerically significant, and where it is endemic there are far more pressing medical and public health problems to be resolved. NPC on the other hand, is the most common cancer of men and the second most common cancer of women of Southern Chinese origin and is thus the major cancer problem for huge populations in Southern China and wherever Southern Chinese have spread throughout South East Asia and indeed, the world (Shanmugaratnam, 1971). In addition, there is an intermediate high incidence zone right across North Africa and down through the Sudan as far as the Kenya highlands (Cammoun et al. 1974; Clifford, 1970). Amongst the Inuit, NPC is also the most common cancer (Lanier et al. 1980). Thus, attempts at vaccine intervention would certainly be worthwhile.

The second question which had to be asked concerned possible precedents for this approach. Marek's disease of chickens was recognized as a curiosity many decades ago (Marek, 1907), but after World War II when huge commercial chicken flocks were established for battery egg-laying and the raising of broilers, infection with the herpesvirus of Marek's disease led to a steady 25% loss of birds from the malignant lymphomas induced by the agent (Payne et al. 1976). This important economic problem was solved with the introduction of vaccines against Marek's herpesvirus (Churchill et al. 1969; Okazaki et al. 1970) which virtually eliminated the disease and its lymphomas. This represented the first example of antiviral vaccine intervention leading to a reduction in the incidence of a naturally occurring cancer.

Although the vaccine in current use is an apathogenic live viral vaccine, experimental vaccines based on membranes from cells infected with the herpesvirus of Marek's disease or even virus-determined antigens purified from such membranes (Kaaden and Dietzchold, 1974; Lesnick and Ross, 1975), proved highly effective. For this reason, and also because antibodies directed against the EB virus-determined membrane antigen (MA) were known to be virus-neutralizing (de Schryver et al. 1974), MA was chosen from the start as an appropriate vaccine immunogen.

ELABORATION AND VALIDATION OF A PROTOTYPE SUBUNIT VACCINE

Work on the nature of MA from several laboratories showed that it consisted of two high molecular weight glycoprotein molecules of 270 and 340Kd (MA gp270 and gp340) and further studies indicated that the former was a precursor of the latter (reviewed in Epstein, 1984). Efforts were therefore made to develop a vaccine against EB virus based on MA gp340.

For this program there were certain essential requirements. First, supplies of the only animal known to respond to EB virus infection with lesions had to be assured. The animal in question, the cottontop tamarin (*Saguinus oedipus oedipus*) (Miller et al. 1977), was placed on the endangered species list in the late 1970's and it was necessary therefore to establish a breeding colony. For this, the diet, bio-energetics, husbandry, and handling procedures for breeding had to be worked out (Kirkwood et al. 1983, 1985) and it was also necessary to establish a challenge dose of EB virus which would ensure the induction of tumors in 100% of normal animals of this outbred population (Epstein et al. 1985). At the same time, a sensitive test for the gp340 antigen was essential to optimize yields of antigen and a radio-iommunoassay (RIA) was therefore established (North et al. 1982); with it an efficient MW-based preparative procedure was introduced (Morgan et al. 1983). The purified gp340 was incorporated into artificial liposomes to increase its immunogenicity. Finally, the antibodies induced were monitored by a specific ELISA (Randle and Epstein, 1984) and standard virus neutralization tests (Moss and Pope, 1972; de Schryver et al. 1974). As soon as these requirements had been fulfilled, the prototype MA gp340-based vaccine was validated in cottontop tamarins by demonstrating its ability to protect vaccinated animals against the 100% carcinogenic dose of challenge EB virus (Epstein et al. 1985).

ASSESSMENT OF VACCINES FOR USE IN MAN

Over the last few years work has concentrated on the development of a gp340-based vaccine suitable for human use. The gene coding for gp340 has been known for some time and has been sequenced (Biggin et al. 1984; Hummel et al. 1984), but although it has been cloned and expressed in bacteria (Beizel et al. 1985), yeast (Schultz et al. 1987) and mammalian cells systems (Conway et al. 1988; Whang et al. 1987) it has not proved possible to purify the product for use as a vaccine. Accordingly, other approaches have been followed up. Recombinant vaccinia viruses capable of expressing gp340 in tissue culture have been made (Mackett and Arrand, 1985) and have been tested in tamarins. The recombinant made from the rather virulent WR strain of vaccinia caused massive lesions when scarified on the skin (4-5 cms in diameter with a crop of satellite secondary pustules one week later) and the considerable amount of virus replication responsible for these lesions gave sufficient expression of the gp340 gene to ensure protection in some animals (Morgan et al. 1988b). However, the WR strain of vaccinia virus is quite unsuitable for use in man and when the experiments were repeated with the Wyeth New York Board of Public Health vaccine strain of vaccinia, the minimal virus replication of this attenuated strain gave insufficient expression of the gp340 gene to induce immunity (Morgan et al. 1988b).

A recombinant varicella virus has also been made using similar genetic engineering techniques to insert the gp340 gene (Lowe et al. 1987), starting from the outset with the Oka vaccine strain of varicella (Takahashi et al. 1974). This strain has been employed extensively to vaccinate young children and no adverse reactions have been reported. It has been shown recently that the common marmoset (*Calithrix jacchus*) can be infected experimentally with Oka varicella (Provost et al. 1987) and although this species does not respond with disease manifestations when infected with EB virus it is closely related to the cottontop tamarin which does. It may therefore be possible to investigate the protective effects of the recombinant Oka varicella against tumor induction by EB virus using tamarins and this hope has been strengthened by current experiments indicating that the recombinant will grow in cottontop tamarin fibroblast cultures (Morgan et al. unpublished).

A FIRST GENERATION MA gp340 VACCINE FOR HUMAN TRIALS

In the meantime, more recent approaches have concentrated on new purification procedures for MA gp340 and the use of novel adjuvants. During a preliminary analysis of the structure of MA gp340 it was found that the carbohydrate moiety was heavily sialiated (Morgan et al. 1984) and it became clear that the molecule should therefore remain negatively charged at a relatively low pH. Making use of this characteristic a purification procedure was developed using anion exchange chromatography on an automated fast protein liquid chromatography (FPLC) system (David and Morgan, 1988). With this method the membrane fraction of EB virus-infected cells (expressing therefore MA gp340) is solubilized in a synthetic non-ionic detergent (MEGA-9) (Hildreth, 1982) and passed down an ion exchange matrix (Mono-Q) under conditions of low ionic strength at pH 5.0. The MA gp340 binds to the anion exchanger very efficiently and can be readily eluted with a salt gradient. Although virtual purity is achieved with this single step, a final gel filtration guarantees the homogeneity of the product (David and Morgan, 1988). The advantages of this procedure are that it isolates all MA gp340 molecules, avoids denaturing conditions, is both automated and reproducible and has the potential for easy scale-up. With FPLC-purified MA gp340 readily available, its efficiency as a vaccine has been investigated in conjunction with two very efficient new adjuvants.

FPLC-purified gp340 has been incorporated in Quil A-derived immunostimulating complexes (iscoms) (Morein et al. 1984, 1987), which have powerful adjuvant properties. This combined material was highly immunogenic on inoculation into cottontop tamarins inducing high titer virus-neutralizing antibodies and conferring protection against the standard 100% lymphomagenic dose of virus (Morgan et al. 1988a). Although there are preliminary observations suggesting that iscoms are not significantly toxic locally (Speijers et al. 1988), and they have already been used successfully in veterinary practice, a great deal more investigation will be required before they can be deemed suitable for use in Man.

Accordingly, another new adjuvant preparation has been tested with FPLC-purified gp340. This is the synthetic threonyl analogue of muramyl dipeptide (MDP) which has the advantage of retaining the highly efficient adjuvant properties of MDP whilst being entirely free from the well known side effects of that compound (Allison and Byars, 1986, 1987). Threonyl MDP emulsified in squalane together with a copolymer (L121) gives a suspension of microspheres on whose surface antigens are displayed and which activate complement. This adjuvant formulation is suitable for human use (Masterfile submitted to US Food and Drug Administration, 1988) and when tested with FPLC gp340 in tamarins, rapidly induced high titer virus-neutralizing antibodies after three injections and conferred protection against tumor induction by the standard dose of challenge virus (Morgan et al. 1989). With this adjuvant formulation doses of antigen as low as 5 mg have proved effective.

It is clear from the foregoing that MA gp340 can now be easily prepared in tractable amounts by the FPLC purification procedure and that the MDP formulation provides a safe, easy to use, highly effective adjuvant giving excellent results with extremely small doses of antigen. A first generation subunit vaccine against EB virus is thus now available for trial in man.

HUMAN TRAILS

The Phase I Trial

A Phase I human trial has been approved by two separate United Kingdom expert scientific committees and funds for such a trial have been allocated; it is proposed that FPLC gp340 immunogen will be prepared on contract in the commercial sector under conditions of Good Manufacturing Practice. The objectives of the trail are as follows:

1. To determine the immunogenicity in man of FPLC-purified EB virus MA gp340 administered with the MDP adjuvant formulation by intra-muscular injection.
2. To make a preliminary estimate of its tolerability including when given to EB virus seropositive individuals.

The protocol of the Phase I trial is planned in the following steps:

1. Serological screening for antibodies to EB virus of up to 100 young, informed, consenting human volunteers.
2. Selection of 12 seronegative individuals and 12 seropositive individuals with low levels of antibody to MA gp340.
3. Full clinical and laboratory screening of those chosen.
4. Full investigation of EB virus status of those chosen (serology, salivary virus shedding, establishment of B cell lines as subsequent targets for cell mediated immunity).
5. Administration of vaccine (three doses at two-weekly intervals).
6. Post immunization clinical monitoring and scrutiny of daily diary for possible symptoms.
7. Immunological assessment of antibodies and cell mediated responses to EB virus.
8. Analysis of possible local or systemic reactions to the vaccine, assessment in seronegative individuals or induced antibody and cellular reactivity to gp340, and in seropositive individuals of a boost in these measures of immunity.

Priority Use of the Vaccine Against Duncan Syndrome

As soon as the Phase I trial proves satisfactory, and even before further studies of the vaccine, it would be ethically acceptable to use it in young boys belonging to families known to suffer from Duncan or X-linked lymphoproliferative syndrome (Purtilo et al, 1986), since in this condition, primary EB virus infection of males is accompanied by life-threatening complications with a very high frequency of deaths.

Future Double Blind Trial

Should all go well in the Phase I trial, it will be necessary to extend human studies to a larger scale double-blind placebo trial. For this, a group of seronegative young adult volunteers will need to be investigated and will be randomly assigned to treatment with the vaccine preparation or an indistinguishable placebo. Thereafter, the rate of seroconversion from natural infection will be monitored with the two groups noting whether the vaccine prevents seroconversion and decreases the incidence of infectious mononucleosis.

Future Field Trials

If this trial likewise proves successful, it will be necessary to assess the performance of the vaccine in a field trial in some area where the incidence of endemic BL is high. Since such regions are restricted to Africa and New Guinea

(Burkitt, 1963) where primary EB virus infection occurs at a very early age (Henle and Henle, 1969; De Thé, 1979) it will be necessary to vaccinate infants in the first few months after birth. The logistics for this are no more difficult than those required for similar vaccine programs against hepatitis B virus infection in high incidence areas (Deinhardt and Jilg, 1986; Zuckerman, 1985), and indeed the EB virus project could readily be linked to an ongoing WHO hepatitis B program in the Gambia (International Agency for Research on Cancer, 1985). This small West African country has a number of special advantages for such work: 1) its geography is such that all of its 500 villages can readily be reached by road; 2) the population of some 700,000 is sufficient for the proposed vaccine studies; 3) the presence of a British Medical Research Council Tropical Medicine Research Unit over the last forty years has provided substantial baseline information on health, social make up and practices of the inhabitants; 4) for an EB virus vaccine trial, the existence of about 40 Burkitt's lymphomas per year is enough to monitor the ability of the vaccine to reduce the incidence of this cancer.

FINAL COMMENT

Once field studies have demonstrated that vaccination against EB virus infection can bring about a decrease in the incidence of BL in a high risk population, there will be an irresistible case for devoting funds and efforts to the development of more sophisticated vaccines against EB virus with the long-term aim of preventing primary infection in the huge populations for whom EB virus-associated NPC is the major cancer problem. Because this is a tumor of middle and later life (Shanmugaratnam, 1971) immunity to EB virus would probably need to be maintained for many decades and much study and effort will have to be devoted to the elaboration of methods to do this. Nevertheless, progress with a vaccine against EB virus has continued steadily and success at each step generates acceleration in the rate of advance.

REFERENCES

Allison AC, Byars NE (1986) An adjuvant formulation that selectively elicits the formation of antibodies of protective isotype and cell mediated immunity. J Immunol Methods 95:157-168.

Allison AC, Byars NE (1987) Vaccine technology: adjuvants for increased efficiency. Biotech 5:1041-1045.

Beisel C, Tanner J, Matsuo T, Thorley-Lawson D, Kezdy F, Keiff E (1985) Two major outer envelope glycoproteins of Epstein-Barr virus encoded by the same gene. J Virol 54:665-674.

Biggin M, Farrell PJ, Barrell BG (1984) Transcription and DNA sequence of Gam HIL fragment of B95-8 Epstein-Barr virus. EMBO J 3:1083-1090.

Burkitt D (1963) A lymphoma syndrome in tropical Africa. In Richter GW, Epstein MA (eds) International Review of Experimental Pathology. New York, London. Academic Press, Inc, pp. 67-138.

Cammoun M,. Hoerner GV, Mourali N (1974) Tumors of the nasopharynx in Tunisia: an anatomic and clinical study based on 143 cases. Cancer 33;184-192.

Churchill AE, Payne LN, Chubb RC (1969) Immunization against Marek's disease using a live attenuated virus. Nature 221:744-747.

Clifford P (1970) A review: on the epidemiology of nasopharyngeal carcinoma. Int J Cancer 5:287-309.

Conway M, Morgan A, Mackett M (1989) Expression of Epstein-Barr virus antigen gp340/220 in mouse fibroblasts using a bovine papilloma virus vector. J Gen Virol 70:729-734.

David EM, Morgan AJ (1988) Efficient purification of Epstein-Barr virus membrane antigen gp340 by fast protein liquid chromatography. J Immunol Methods 108:231-236.

Deinhardt F, Jilg W (1986) Vaccines against hepatitis. Ann Inst Pasteur, Virol 137E:79-95.

De Schryver A, Klein G, Heweston J, Rocchi G, Henle W, Henle G, Moss DJ, Pope JH (1974) Comparison of EBV neutralization tests based on abortive infection or transformation of lymphoid cells and their relation to membrane reactive antibodies (anti MA). Int J Cancer 13:353-362.

De-Thé G (1979) Demographic studies implicating the virus in the causation of Burkitt's lymphoma; prospects for nasopharyngeal carcinoma. In Epstein MA, Achong BG (eds), The Epstein-Barr Virus, Berlin, Heidelberg, New York: Springer, pp. 417-473.

Doll R, Peto R (1976) Mortality in relation to smoking: 20 years' observation on male British doctors. Brit Med J 2:1525-1536.

Epstein MA (1976) Epstein-Barr virus - is it time to develop a vaccine program? J Nat Cancer Inst 56:697-700.

Epstein MA (1984) A prototype vaccine to prevent Epstein-Barr (EB) virus-associated tumours. Proc Roy Soc B Lond 221:1-20.

Epstein MA, Morgan AJ (1983) Clinical consequences of Epstein-Barr virus infection and possible control by an antiviral vaccine. Clin Exp Immunol 53:257-271.

Epstein MA, Morgan AJ, Finerty S, Randle BJ, Kirkwood JK (1985) Protection of cottontop tamarins against Epstein-Barr virus-induced malignant lymphoma by a prototype subunit vaccine. Nature 318:287-289.

Henle W, Henle G (1969) The relation between the Epstein-Barr virus and infectious mononucleosis, Burkitt's lymphoma and cancer of the postnasal space. E African Med J 46:402-406.

Hildreth JEK (1982) N-D-gluco-N-methylalanamide compounds, a new class of non-ionic detergents for membrane biochemistry. Biochem J 207:363-366.

Hummel M, Thorley-Lawson DA, Kieff E (1984) An Epstein-Barr virus DNA fragment encodes messages fo the two major envelope glycoproteins ($gp^{350}/300$ and $gp^{220}/200$). J Virol 49:413-417.

International Agency for Research on Cancer (1985) An intervention study to evaluate the effectiveness of hepatitis B vaccine for the prevention of hepatocellular carcinoma in a high risk population. IARC Working Paper 3/6:1-46.

Kaaden OR, Dietzchold B (1974) Alterations of the immunological specificity of plasma membranes of cells infected with Marek's disease and turkey herpes viruses. J Gen Virol 25:1-10.

Kirkwood JK, Epstein MA, Terlecki AJ (1983) Factors influencing population growth of a colony of cotton-top tamarins. Lab Animals 17:45-41.

Kirkwood JK, Epstein MA, Terlecki AJ, Underwood SJ (1985) Rearing of a second generation of cotton-top tamarins (*Saguinus oedipus oediupus*) in captivity. Lab Animals 19:269-272.

Lanier A, Bender T, Talbot M, Wilmeth S, Tschopp C, Henle W, Henlw G, Ritter D, Terasaki P (1980) Nasopharyngeal carcinoma in Alaskan Eskimos,

Indians and Aleuts: a review of cases and study of Epstein-Barr virus, HLA and environmental risk factors. Cancer 46:2100-2106.

Lenoir GM, Bornkamm GW (1987) Burkitt's lymphoma, a human cancer model for the study of the multistep development of cancer; proposal for a new scenario. In Klein G (ed) Advances in Viral Oncology Series Vol 7. New York, Raven Press, pp. 173-106.

Lesnick F, Ross LJN (1975) Immunization against Marek's disease using Marek's disease virus-specific antigens free from infectious virus. Int J Cancer 16: 153-163.

Lowe RS, Keller PM, Keech BJ, Davison AJ, Whang Y, Morgan AJ, Kieff E, Ellis RW (1987) Varicella-zoster virus as a live vector for the expression of foreign genes. Proc Natl Acad Sci USA 84:3896-3900.

Mackett M, Arrand JR (1985) Recombinant vaccinia virus induces neutralizing antibodies in rabbits against Epstein-Barr virus membrane antigen gp340. EMBO J 3229-3234.

Marek J (1907) Multiple Nervenentzündung (polyneuritis) bei Hühern. Deutsch Tierärztl Wschr 15:412-421.

Miller G, Shope T, Coope D, Waters C, Pagano J, Bornkamm GW, Henle W (1977) Lymphoma in cotton-top marmosets after inoculation with Epstein-Barr virus: tumor incidence, histologic spectrum, antibody responses, demonstration of viral DNA, and characterization of viruses. J Exp Med 145:948-967.

Morein B, Sundquist B, Höglund S, Dalsgaard K, Osterhaus A (1984) Iscom, a novel structure for antigenic presentation of membrane proteins from enveloped viruses. Nature 308:457-460.

Morein B, Lövgren K, Höglund S, Sundquist B (1987) The iscom: an immuno-stimulating complex. Immunol Today 8:333-338.

Morgan AJ, North JR, Epstein MA (1983) Purification and properties of the gp340 component of Epstein-Barr (EB) virus membrane antigen (MA) in an immunogenic form. J Med Virol 25:189-195.

Morgan AJ, Smith AR, Barker RN, Epstein MA (1984) A structural investigation of the Epstein-Barr (EB) virus membrane antigen glycoprotein, gp340. J Gen Virol 65:397-404.

Morgan AJ, Finerty S, Lovgren K, Scullion FT, Morein B (1988a) Prevention of Epstein-Barr (EB) virus-induced lymphoma in cottontop tamarins by vaccination with the EB virus envelope glycoprotein gp340 incorporated into immune-stimulating complexes. J Gen Virol 69:2093-2096.

Morgan AJ, Mackett M, Finerty S, Arrand J, Scullion F, Epstein MA (1988b) Recombinant vaccinia virus expressing Epstein-Barr virus glycoprotein gp340 protects cottontop tamarins against EB virus-incuded lymphomas. J Med Virol 25:189-195.

Morgan AJ, Allison AC, Finerty S, Scullion FT, Byars NE, Epstein MA (1989) Validation of a first generation Epstein-Barr virus vaccine preparation suitable for human use. J Med Virol (in press).

Moss DJ, Pope JH (1972) Assay of the infectivity of Epstein-Barr virus by transformation of human leucocytes in vitro. J Gen Virol 17:233-236.

North JR, Morgan AJ, Thompson JL, Epstein MA (1982) Quantification of an EB virus-associated membrane antigen (MA) component. J Virol Methods 5:55-65.

Okazaki W, Purchase HG, Burmester BR (1970) Protection against Marek's disease by vaccination with a herpesvirus of turkeys. Avian Dis 14:413-429.

Payne LN, Frazier JA, Powell PC (1976) Pathogenesis of Marek's disease. In Richter GW, Epstein MA (eds) Internat Rev Exp Path, Vol. 16, New York, San Francisco, London; Academic Press, pp. 59-154.

Provost PJ, Keller PM, Banker FS, Keech BJ, Klein HJ, Lowe RS, Morton DH, Phelps AH, McAleer WJ, Ellis RW (1987) Successful infection of the common marmoset (*Callithrix jacchus*) with human varicella-zoster virus. J Virol 61:2951-2955.

Purtilo DT, Sakamoto K, Barnabei V, Seeley J, Beechtold T, Rogers G, Yetz J, Harada S (1986) Epstein-Barr virus-induced diseases in boys with the X-linked lymphoproliferative syndrome (XLP). Am J Med 73:49-56.

Randle BJ, Epstein M A (1984) A highly sensitive enzyme-linked immunosorbent assay to quantitate antibodies to Epstein-Barr virus membrane antigen gp340. J Virol Methods 9:201-208.

Schultz LD, Tanner J, Hofmann K, Emini E, Kieff E, Ellis RW (1987) Expression and analysis of EBV gp350 in yeast *Saccharomyces cerevisiae*. In Levine PH, Ablashi DV, Nonoyama M, Pearson GR, Glaser R (eds), Epstein-Barr Virus and Human Disease. Clifton, New Jersey, Humana Press, pp. 475-478.

Shanmugaratnam K (1971) Studies on the etiology of nasopharyngeal carcinoma. In Richter GW, Epstein MA (eds) Int Rev Exp Path, New York, London, Academic Press Inc, 10: pp. 361-413.

Speijers G, Danse L, Bewary E, Strik J, Vos J (1988) Local reactions of the saponin Quil A and a Quil A-containing iscom measles vaccine after intramuscular injection of rats: a comparison with the effects of DPT-polio vaccine. Fundamental and Appl Toxicol 10:425-430.

Takahaski M, Otsuka T, Okuno Y, Asano Y, Yazaki T, Isomura S (1974) Live varicella vaccine used to prevent the spread of varicella in children in hospital. Lancet 2:1288-1290.

Whang Y, Silberklang M, Morgan A, Munshi S, Lenny AB, Ellis RW, Kieff E (1987) Expression of Epstein-Barr virus gp350-220 gene in rodent and primate cells. J Virol 61:1796-1807.

Whittle HC, Brown J, Marsh K, Greenwood BM, Seidelin P, Tighe H, Wedderburn L (1984) T cell control of B cells infected with EB virus is lost during *P. falciparum* malaria. Nature 312:229-250.

Zuckerman AJ (1985) Prevention of hepatocellular carcinoma by immunization against hepatitis B. In Epstein MA, Richter GW (eds) Int Rev Exp Path. Orlando, San Diego, New York, London, Toronto, Montreal, Sydney, Tokyo, Academic Press Inc, 25:pp. 59-81.

ADVANCES IN THE TREATMENT OF HIV-1 INFECTIONS

Martin S. Hirsch

Department of Medicine
Massachusetts General Hospital
Harvard Medical School
Boston, Massachusetts 02114, USA

INTRODUCTION

As we move into the second decade of the global AIDS epidemic, the search for safe and effective therapies intensifies. Understanding of the human immunodeficiency virus type 1 (HIV-1) replicative cycle and pathogenetic mechanisms has increased at a rapid pace, and with this increased understanding have come dozens of new therapeutic agents for evaluation. New and promising drugs and combinations of drugs have outstripped our ability to study them in a deliberate and thorough fashion, raising the possibility of new and radical approaches to the conduct of clinical trials.

This review will discuss new and promising approaches to HIV-1 inhibition, as well as critical issues related to the clinical evaluation of new treatments. It will be a selective, rather than an encyclopedic review of our current state of therapy, emphasizing various points of antiviral attack in the virus replication cycle.

HIV-1 BINDING AND ENTRY

The initial steps in HIV-1 infection are attachment to and penetration of cells bearing appropriate target receptors. In most permissive cell types this is accomplished through attachment of virus envelope glycoprotein (gp120) to the external N-terminal domain of the cell membrane CD4 receptor molecule, followed by membrane fusion and internalization (Haseltine, 1989). Blockage of these reactions is a major target of attack, and both direct and indirect inhibitors have been developed.

The most promising inhibitors of attachment are recombinant soluble preparations of the receptor itself lacking the transmembrane and cytoplasmic domains (rsCD4). Several rsCD4 preparations are potent inhibitors of HIV-1 replication and HIV-1 induced cell fusion *in vitro* at concentrations that are not cytotoxic (Fisher et al. 1988; Hussey et al. 1988; Smith et al. 1987). Given the conserved nature of CD4 binding by diverse isolates of HIV-1 and HIV-2, rsCD4 has broad activity against all isolates. Moreover, rsCD4-based approaches may have benefits beyond direct antiviral effects. It is possible that free envelope gp120 can bind to CD4 receptors of uninfected cells, sensitizing them to immune destruction by antibody dependent cellular cytotoxicity or cytotoxic T lymphocytes. By blocking gp120 binding, rsCD4 could prevent such immune destruction.

Clinical trials of several rsCD4 preparations have been underway since 1988. In one Phase I trial 17 patients with AIDS received escalating dose regimens of rsCD4, administered intramuscularly or intravenously, for up to 28 consecutive days (Schooley et al. 1990). Doses up to 30 mg/day were well-tolerated. Serum levels of rsCD4 following intramuscular administration every 8 h were in the concentration ranges required to inhibit HIV-1 replication *in vitro*. Following repeated intramuscular administration, serum concentrations rose to steady-state inhibitory levels (50-300 ng/ml) for the 30 mg/day cohort, perhaps related to muscular deposition with slow release into the vascular compartment. Serum HIV-1 p24 antigen declined in patients receiving 30 mg of rsCD4 daily from a mean pretreatment value of 1,341 pg/ml to a mean posttreatment value of 789 pg/ml (p<0.03). No significant changes in immunologic function or CD4 cell numbers were observed.

Other Phase I studies of a different rsCD4 preparation showed similar pharmacokinetic profiles with minimal toxicity (Kahn et al. 1989; Yarchoan et al. 1988, 1989a). It is clear that the maximum tolerated dose of rsCD4 has not been reached, and higher dose regimen studies are planned to enhance antiviral and clinical effects. Of interest, limited studies of rsCD4 in rhesus monkeys infected with the simian immunodeficiency virus (SIV) suggest reduced amounts of recoverable virus and improved bone marrow cell growth following treatment (Watanabe et al. 1989).

Several approaches are under evaluation to improve upon current preparations of rsCD4. Truncated preparations may allow a more concentrated dose to be delivered to infected cells, increase organ penetration (particularly of the central nervous system), and reduce the risk of toxicity (Lifson et al. 1988; Traunecker et al. 1988). It has recently been reported that the MHC-binding and gp120 binding sites are separable, suggesting that design of HIV-1 inhibitors without immune toxicity is feasible (Lamarre et al. 1989). Second generation rsCD4 hybrids or complexes are also under study. Hybrid proteins between rsCD4 and immunoglobulin Fc domains ("immunoadhesins") have been developed which have prolonged serum half-lives in rabbits (Capon et al. 1989). Phase I studies of these immunoadhesins are underway in patients with AIDS. Hybrids between rsCD4 N-terminal amino acids and a segment of pseudomonas exotoxin A responsible for translocation and ADP-ribosylation have also been developed (Chaudhary et al. 1989), as have hybrids between rsCD4 and the deglycosylated plant toxin, ricin (Till et al. 1989). These CD4-toxin hybrids may have a selective killing effect on virus-infected cells and may be particularly useful against cells resistant to the cytopathic effects of HIV-1, such as certain macrophages. A concern with these larger molecules, however, is the possibility of increased immunogenicity resulting in the production of autoantibodies against CD4. Still another possible approach under development is insertion of a vascular implant containing genetically engineered cells capable of secreting recombinant sCD4; this approach is at an early stage of development.

Although CD4 is clearly the major HIV-1-receptor, recent studies indicate that there are other mechanisms by which certain cells may become infected (Clapham et al. 1989). Thus, it is unlikely that rsCD4 preparations as single agents will be sufficient for virus control. Combinations with other agents, e.g., zidovudine (AZT), will be discussed subsequently.

REVERSE TRANSCRIPTASE

The HIV-l pol gene encodes three proteins important to its own replication, a protease, a reverse transcriptase (RT) (which includes a DNA polymerase and a ribonuclease H) and an integrase. Ribosomal frame shifting is required to express the gag-pol fusion product of HIV-l, a necessary first step for production of the individual enzymes (Jacks et al. 1989). Inhibitors of this reaction are being sought. The viral DNA polymerase has been the major target for antivirals developed to date. The initial step in synthesis of HIV-l proviral DNA is binding of a tRNA primer to the RT, and inhibitors of this binding might inhibit subsequent events (Haseltine, 1989). Thereafter, a deoxynucleotide is added to the 5' end of the tRNA primer, followed by addition of nucleotides to the 3' end of the growing chain. This is where dideoxynucleoside analogues (e.g. AZT) appear to have their major activity. They become incorporated into the growing chain and terminate elongation since they do not contain the 3'hydroxyl group needed for subsequent addition of nucleotides. Once an initial portion of DNA synthesis occurs at the 5' end of the genome, the polymerase jumps to the 3' end of the second genome and the additional RNA template is degraded. The DNA anneals to complementary RNA at the 3' end of the second DNA strand and serves as a primer for a second elongation reaction, resulting in a DNA/RNA hybrid. Subsequently, the RNA is degraded and a second DNA strand is synthesized. The HIV-l ribonuclease H is probably involved in several of these reactions (e.g. digestion of RNA, removal of the tRNA precursor), and mutants defective in this enzyme have shown defective proviral synthesis. Thus, ribonuclease H is another possible target for anti-HIV-1 drugs.

ZIDOVUDINE (AZIDOTHYMIDINE, AZT)

AZT was synthesized in 1964, initially as a potential anti-cancer drug (Horwitz et al. 1964). Anti-retroviral activity was demonstrated against Friend murine leukemia virus in 1974 (Ostertag et al. 1974) and against HIV-l in 1985 (Mitsuya et al. 1985). AZT is converted by a cellular thymidine kinase to a monophosphate form (AZT-MP), by a cellular thymidylate kinase into a diphosphate and then by other cell enzymes to an active triphosphate form (AZT-TP). AZT-TP inhibits HIV-l reverse transcriptase approximately 100 times more effectively than it does cellular polymerase alpha. HIV-l replication is inhibited by at least two mechanisms, chain termination (as mentioned previously) and competitive inhibition of cellular nucleoside-5'-triphosphates by AZT-TP.

After oral dosing, AZT is absorbed rapidly. Peak serum concentrations occur at 30-90 minutes and the half life is approximately one hour (Yarchoan et al. 1989a). Average oral bioavailability is 65% and urinary recovery is about 90%. Drug levels are affected by hemodialysis, and dosing should be performed after dialysis sessions (Deray et al. 1988). AZT is metabolized primarily by glucuronidation, and drugs that inhibit this step increase its half-life (Kornhauser et al. 1989). Penetration of the blood-brain barrier is effective, resulting in antiviral CSF concentrations.

Several double-blind, placebo-controlled clinical trials have established the efficacy of AZT in a number of HIV-1 associated conditions. From February 1986 to the end of June 1986, 282 patients with AIDS or ARC were enrolled in a clinical trial at 12 different centers in the United States (Fischl et al. 1987). Patients with AIDS were within four months of their first attack of *Pneumocystis carinii* pneumonia (PCP), and patients with ARC had severe clinical signs (e.g., significant weight loss or oral thrush). One hundred forty-five patients received an oral dose of 250 mg of AZT every 4 h and 137 patients received placebo.

By September 1986, it was apparent that differences in survival had emerged, and the study was terminated after consultation with an independent data safety monitoring board. Nineteen patients (12 with AIDS, seven with ARC) in the placebo group had died, compared with only one who received AZT. Furthermore, there was a significant difference in opportunistic infections (OI) in patients who received placebo (45 OI) compared with AZT recipients (24 OI). Patients who received AZT generally gained weight, whereas placebo recipients lost weight. Karnofsky scores of functional capability also improved in AZT recipients but did not improve in the placebo groups. In addition, AZT recipients had improved cognition, as measured by a battery of neuropsychological tests (Schmitt et al. 1988). Individuals who received AZT generally showed an increase in CD4 cells, although this effect was lost after five months in patients with AIDS.

Of patients who received AZT, 29% developed skin test reactivity to at least one antigen, but only 9% of placebo recipients lost skin anergy. The exact mechanisms by which these effects occurred are not fully elucidated, but data from several centers suggest that HIV-l p24 antigenemia is reduced in AZT recipients (Jackson et al. 1988), a finding perhaps reflecting diminished virus replication *in vivo*.

The decreased mortality rates observed in patients treated with AZT were also seen during extended follow-up of the originally enrolled patients. Among those randomized to receive AZT, survival rates were 84.5% and 57.6% at 12 and 21 months after initiation of therapy (Fischl et al. 1989). Patients in this study who received little or no AZT had only 36% survival one year after diagnosis, a similar survival rate observed for patients with AIDS in previous historical natural history studies.

AZT toxicity was considerable in the initial study of patients with AIDS. (Richman et al. 1987). Macrocytic anemia occurred frequently in the AZT group and made transfusions necessary in 40 patients (11 in the placebo group). Another side effect that also made AZT dose reduction necessary was neutropenia in approximately 16%. Headache was the most-common adverse symptom observed, and nausea, myalgia, and insomnia were also reported more frequently in recipients of AZT. These adverse effects were reversible and generally tolerable for patients with AIDS or severe ARC.

As a result of these studies, AZT was licensed in the United States during 1987 for the "management of certain adult patients with symptomatic HIV-1 infection (AIDS and advanced ARC) who have a history of cytologically confirmed PCP or an absolute CD4 lymphocyte count of less than $200/mm^3$ in the peripheral blood before therapy is begun" (FDA package insert). Since more widespread availability of AZT in late 1986, thousands of patients have been treated for these indications. Data have been reviewed on 4,805 of such patients under a compassionate plea program administered by Burroughs-Wellcome and the National Institutes of Health (NIH) (Creagh-Kirk et al. 1988). Overall survival at 44 weeks after initiation of therapy was 73%. Positive associations were observed between survival and pretherapy hemoglobin levels, functional ability and the stage of disease as measured by time since diagnosis of PCP. In-patients with baseline hemoglobin levels of $\geq$120 g/l, Karnofsky scores >90, and PCP diagnosis within 90 days, 44 week survival was 88%.

Major adverse effects continue to be anemia and granulocytopenia. Patients with poor bone marrow reserve secondary to opportunistic infection or vitamin B12 deficiency have more toxicity than patients with adequate marrow reserve. Other adverse effects, including drug fever, rash (including Stevens Johnson syndrome), nail pigmentation, and myopathy occur rarely. Acute meningo-encephalitis on dose reduction of AZT has been reported (Helbert et al. 1988), but not yet confirmed.

Several studies begun in 1987 by the AIDS Clinical Trials Group or ACTG (a collaborative group of investigators sponsored by the NIH) have recently expanded our knowledge concerning the proper doses and appropriate indications for AZT. In one study conducted in patients with recent onset of AIDS, two different dose regimens were compared (Fischl et al. in preparation). Patients were randomized to receive either 1,500 mg AZT daily (250 mg every 4 hours) or a lower dose regimen (1,200 mg daily for one month followed by 600 mg daily, 100 mg every four hours). Over a nearly 2 year follow-up period, equivalent survival and equivalent CD4 cell changes in peripheral blood were observed at the two doses studied. However, hematologic toxicity, particularly neutropenia, was significantly reduced in the low dose group.

ACTG trials of AZT have also been conducted between 1987-89 in patients with early symptomatic HIV-1 infection (one or two symptoms of AIDS Related Complex). In this study (ACTG 016), 713 patients were randomized to receive 200 mg AZT every 4 h (1200 mg daily) or placebo (Fischl et al. in preparation). The study was terminated on August 2, 1989, at the suggestion of an independent data and safety monitoring board. At that time 45 study endpoints (primarily development of AIDS) had been observed in the stratum of patients who enrolled with 200-500 CD4 cells/mm^3; 34 of these endpoints occurred in placebo recipients and 11 in AZT recipients (p<0.0006). Events occurring in the stratum with 500-800 CD4 cells/mm^3 were too few for analysis. Although hematologic toxicity developed in 4% of AZT recipients, this was much less frequent than that observed in earlier trials conducted in patients with more advanced disease (30-40%).

In another ACTG study (019) over 3,200 subjects with asymptomatic HIV-1 infection were randomized to receive high dose AZT (300 mg, 5 times daily, total 1500 mg), low dose AZT (100 mg, 5 times daily, total 500 mg), or placebo (Volberding et al. in preparation). On August 16, 1989, a data and safety monitoring board recommended termination of one stratum of this trial, that with subjects who enrolled with $\leq$500 CD4 cells/mm^3 in peripheral blood. In this stratum, study endpoints (largely development of AIDS) were observed in 38 of 428 placebo recipients (8.9%), 17 of 453 recipients of low dose AZT (3.7%) and 19 of 457 recipients of high dose AZT (4.2%). Progression rates, as measured by endpoints per 100 patient years of treatment were 7.6, 3.6 and 4.2 (P = 0.003 and 0.05 for the low and high dose groups, respectively). Significant hematologic toxicity occurred in 11.8% of patients receiving high dose AZT, 2.9% receiving low dose AZT, and 2.1% receiving placebo. In the stratum with >500 CD4 cells/mm^3 too few episodes of progression were observed at the time of analysis for conclusions to be drawn. Evaluation of this stratum will continue until adequate endpoint numbers are reached for evaluation.

Based on the results of these ACTG studies, it has been recommended that HIV-1-positive subjects, with or without symptoms, who have $\leq$500 CD4 cells/

mm^3 in peripheral blood be treated with 100 mg AZT five times daily. No recommendations regarding treatment of individuals with >500 CD4 cells/ mm^3 can be made until more information becomes available from ongoing studies.

AZT trials have also been conducted in children with symptomatic HIV-1 infection and in adults with Kaposi's sarcoma or thrombocytopenia. Twenty-one children ranging from 14 months to 12 years received AZT at 4 dose levels by continuous infusion (Pizzo et al. 1988). Improvement in neurodevelopmental abnormalities occurred in most children. Increased appetite and weight, decreased lymphadenopathy and hepatosplenomegaly, decreased immunoglobulin levels, and increased CD4 cell numbers were also observed. Toxicities included transfusion-requiring anemia and dose-limiting neutropenia. Bolus injection was accompanied by rapid disappearance of AZT from peripheral blood, suggesting that continuous infusion may be more useful in long term management. Further studies in infants and older children are underway. Lane et al. (1989) have conducted a small placebo-controlled trial of various AZT regimens in patients with HIV-1 infection, Kaposi's sarcoma, CD4 counts >$200/mm^3$ and no systemic symptoms or history of OI. Although antiviral effects of AZT were observed (declines in serum HIV-1 antigen, clearance of HIV-1 from cerebrospinal fluid), no differences in tumor progression or CD4 cell numbers were observed. In a cross-over study of AZT in patients with thrombocytopenia, patients received 2 g/day for 2 weeks and 1 g/day for 6 weeks followed by placebo for 8 weeks, or vice versa (Swiss Group for Clinical Studies on AIDS, 1988). Platelet counts increased in all patients while on AZT, but not on placebo during the course of the study.

Prophylactic AZT is also under study in health care workers occupationally exposed to HIV-1 (Henderson and Gerberding, 1989). However, the rate of transmission in these circumstances is so low (~0.4%) that it will be difficult to demonstrate any protective effect for AZT, if it exists. Moreover, animal studies using analagous retroviruses suggest that AZT is only partially protective, at best, and must be administered within hours of HIV-1 exposure for any benefit to be observed.

A major problem in the long term use of AZT may be emerging, i.e., the development of drug-resistant HIV-1 with prolonged therapy. Larder et al. (1989) have evaluated the *in vitro* sensitivity of clinical isolates from patients with advanced HIV-1 infection before and during AZT therapy utilizing a plaque assay on CD4+ HeLa cells. Prior to therapy, isolates were uniformly sensitive to AZT concentrations between 0.01 μM and 0.05 μM (mean 0.03 μM). Most isolates from patients who had received AZT for six months or more were less sensitive, with some isolates demonstrating 100 fold increases in ID_{50} values. Such resistant isolates remained sensitive to several other RT inhibitors, but were cross-resistant with 3'-azido-2'-3'-deoxyuridine (AZdU). Clinical correlations could not be made in the small number of patients and isolates studied, although it is becoming clear that clinical benefits from AZT are maximal during the first year of therapy (Dournon et al. 1988; Creagh-Kirk et al. 1988; Fischl et al. 1989). The molecular mechanisms underlying AZT-resistance are unclear, although mutation in the RT gene is most likely. Resistance may occur more frequently in patients with advanced disease because of a higher rate of viral replication, with greater chances of mutagenesis and a decreased ability to clear viruses that may be partially defective. In any case, the discovery of AZT-resistance indicates the need for alternative or combination approaches to therapy of HIV-1 infection.

DIDEOXYCYTIDINE AND DIDEOXYINOSINE

Two other dideoxynucleoside analogs are in late stages of clinical evaluation. Both dideoxycytidine (ddC) and dideoxyinosine (ddI) were shown to have anti-HIV activity in 1986 (Mitsuya and Broder. 1986). ddI is closely related to dideoxyadenosine (ddA); ddA is converted to ddI by the enzyme adenosine deaminase, and ddI is converted again within cells to its active form ddA-triphosphate (ddA-TP). Both ddA-TP and ddC-TP are thought to act as chain terminators and inhibitors of RT. Unlike AZT-TP and ddC-TP, ddA-TP has a long half-life (>12 hours) in exposed cells. Because both ddC and ddI showed good therapeutic/toxic ratios *in vitro*, they have been brought rapidly into animal toxicology and phase 1 human trials.

Two phase I studies of ddC have been conducted in patients with AIDS or advanced ARC. In one, five dose regimens were administered intravenously for two weeks, then orally for ≥4 weeks, to 20 patients (Yarchoan et al. 1988). At doses of 0.03-0.09 mg/kg every 4 h, transient increases in CD4 cell numbers and falls in serum p24 antigen were observed in the majority of patients. Dose related toxic effects included skin rashes, fever, mouth sores, and a painful peripheral neuropathy occurring 6-14 weeks after onset of therapy. In the second study, conducted by the ACTG, 61 patients with AIDS or advanced ARC and ≥100 pg/ml of serum p24 antigen were enrolled (Merigan et al. 1989). Doses ranging from 0.005-0.06 mg/kg were administered orally every 4 h for 3-6 months. Similar beneficial effects on p24 antigen and CD4 cell numbers were observed, with similar toxicities, at doses of 0.01 mg/kg every 4 h or greater. In both studies, peripheral neuropathy was the limiting toxicity. Lower doses (0.03 mg/kg/day) or different regimens of administration may reduce the frequency and severity of the neuropathy (Hoffmann-LaRoche, unpublished observations). Because of the largely non-overlapping toxicities of ddC and AZT, trials of weekly or monthly alternating regimens are underway. Preliminary results suggest that some individuals can tolerate alternating weekly alternating ddC and AZT for at least 1.5 years without substantial toxicity from either drug.

ddI is under intense clinical investigation. Yarchoan et al. (1989b) evaluated eight dose regimens intravenously for 14 days, followed by twice the dose orally, in 26 patients with AIDS or advanced ARC for periods up to 42 weeks. Oral bioavailability averaged 35%, if ddI was given on an empty stomach after ingestion of antacids. CSF/plasma ratios averaged 0.19 1 h after completion of an intravenous infusion. At the higher dose regimens (≥1.6 mg/kg intravenously followed by ≥3.2 mg/kg orally at least every 12 h), increased CD4 cells (p< 0.0005), CD4/CD8 ratios (p<0.0l) and decreased p24 antigen (p<0.05) were seen. Delayed type cutaneous hypersensitivity and lymphocyte proliferative *in vitro* responses improved in some patients. At the higher doses, 14 patients reported increased energy and reduced fatigue. Overall weight gain averaged 1.6 kg by week 10. Toxicity included increase in serum uric acid (1-3 mg/dl), triglycerides, and amylase. Pancreatitis and dysesthesias of the feet were observed rarely. Headaches, restlessness, and insomnia were more common, but usually subsided by week 5. In subsequent studies, using higher doses of ddI (>10 mg/kg/day) neuropathies and pancreatitis have occurred more frequently. Several large ACTG trials of ddI are now underway in patients with advanced HIV-1-infection. One compares ddI with AZT in AIDS, another looks at ddI in AZT-intolerant patients, and a third evaluates ddI in patients who have been on long term AZT (a situation where AZT-resistance might be present.

A number of other dideoxynucleoside analogs are under investigation. Many have shown anti-HIV-1 activity in the laboratory and some are in early clinical trials.

PROTEASE (PROTEINASE) AND INTEGRASE

In addition to RT, HIV-1 codes for two other enzymes important in replication, a protease and an integrase. The protease gene appears to be located near the junction of gag and pol and may be essential for core protein processing and viral infectivity. Precursor gag and pol polyproteins are processed by the HIV-1-specific protease into mature virion components. The protease cleaves itself, as well as RT, from precursor to active forms, and cleaves the HIV-1 gag polyprotein into four polypeptides needed for viral nucleocapsid assembly and positioning. Inhibition of HIV-1-protease is, thus, an attractive target to limit virus maturation at possibly two or more sites of the replication cycle. Three-dimensional structural studies indicate that the HIV-1 protease is homologous to the family of microbial aspartic proteases, and reactive amino acid residues on the protein substrate have been predicted (Navia et al. 1989; Wlodawer et al. 1989). It appears that two 99-amino acid chains come together in a dimer to form the active enzyme. Enzyme activity could be blocked by preventing dimer formation or by competition. One modified oligopeptide inhibitor, pepstatin A, has been reported (Grinde et al. 1989), and many others are under study.

HIV-1, like other retroviruses, also contains an enzyme necessary for integration of proviral DNA into cellular DNA (Varmus and Brown, 1989). Retroviral integrases appear to act on linear forms of the provirus (Brown et al. 1989). The initial step is probably a cleavage that removes the terminal two bases from the 3'end of each viral DNA strand, exposing the 3'-0H group to be joined to the target host DNA. The target DNA is then cut and this cleavage is coupled with joining of the viral 3'-0H ends to the target 5'-P ends, resulting in a gapped intermediate which is then repaired. Mutants defective for integration have been reported not to replicate in T cell lines in culture (Haseltine, 1989), suggesting that at least in some cell types integration may be a requirement for replication and, hence, a good target for inhibition. Integrase inhibitors have not yet been described.

LATER EVENTS IN HIV-1 REPLICATION

Once provirus becomes integrated within host cell DNA, it's subsequent expression is regulated by a number of cellular and viral factors. Certain stimuli (antigens, other viruses) may lead to activation, transcription of proviral DNA to RNA, translation to viral proteins, assembly, and release of HIV-1. Although host cell machinery is utilized for this replication, several viral proteins (e.g., tat, rev, nef) help regulate these processes. Modulators of these viral proteins are under development. Modifications of DNA in an antisense configuration (complementary to sequences of HIV-1 RNA) may bind to viral RNA and block viral proteins from being formed. Several forms of modified oligonucleotides (e.g., methylphosphonates, phosphorothioates) have demonstrated anti-HIV-1 activity *in vitro* (Agrawal et al. 1988; Sarin et al. 1988). If problems of stability, nuclease resistance, solubility and specificity can be satisfactorily addressed, these agents may become useful clinical agents.

Glycoprotein processing occurs late in the virus replication cycle under the control of cellular enzymes. The gp120 of HIV-l is heavily glycosylated and is dependent upon the processing action of trimming glucosidase enzymes which remove three glucose residues while the protein is in the rough endoplasmic reticulum. If such processing is inhibited, mature complex oligosaccharide structures are not formed and the immature gp120 may be functionally disabled. Inhibitors of α-glucosidase 1, such as castanospermine (CAS) and N-butyl deoxynojirimycin (N-butyl DNJ) interfere with HIV-l replication *in vitro* (Gruters et al. 1987; Karpas et al. 1989; Walker et al. 1987). Viral glycoproteins are produced with higher molecular weights than expected, resulting in decreased viral infectivity and reduced syncytium formation. Antiviral activity has been demonstrated for CAS in a murine leukemia virus model system (Ruprecht et al. 1989). Animal toxicology studies of N-butyl-DNJ have been performed and phase I clinical trials are in progress.

Interferon-α (IFN-α) was shown to inhibit HIV-l replication *in vitro* in 1985 (Ho et al. 1985). Subsequent studies also demonstrated anti-HIV-1 activity of IFNs-β and γ (Hartshorn et al. 1987a). Although the mechanisms of IFN inhibition of HIV-l have not been completely established, both acute and chronic infection can be inhibited *in vitro* (Ho et al. 1985; Poli et al. 1989). The mechanisms of inhibition may vary for acute and chronic infections, as previously shown for murine retroviral systems. In chronic infection, IFN-α appears to act late in the virus replicative cycle, perhaps on the release of formed virus particles from the cell (Poli et al. 1989).

High dose IFN-α regimens have induced major responses in AIDS-associated Kaposi's sarcoma in approximately 30% of treated patients (Krown, 1988). The mechanisms underlying these responses are unclear. Anti-HIV-1 activity, as measured by reductions in serum p24 antigen or negative blood cultures on therapy, as well as increases in circulating CD4 cells, have been observed in responders (deWit et al. 1988; Lane et al. 1988), although the role of HIV-l in Kaposi's sarcoma remains unclear. However, the results in patients with advanced AIDS following OIs have not been so promising. In a double-blind, placebo-controlled trial of recombinant IFN-α 2a, 67 patients were enrolled at 5 centers over a 3 year period (Interferon Alpha Study Group, 1988). Groups received either placebo, 3 million units or 36 million units 3 times a week for 12 weeks. There were no significant differences in median survival, OIs, CD4 cell counts, or serum p24 antigen levels during therapy among the three groups. These studies, taken together, would suggest that IFN-α may be more beneficial in patients in earlier stages of infection, where immune function is more intact, than in advanced disease. However, controlled studies need to be done in early infection, given IFN's considerable toxicity including fever, fatigue, influenza-like syndromes, gastrointestinal disturbances, leukopenia, thrombocytopenia, liver function abnormalities, and transient neurologic disturbances.

OTHER ANTIVIRAL AGENTS

The list of experimental agents under investigation increases with each passing week. Some, such as suramin, HPA-23, and AL721 have been largely abandoned because of excessive toxicity or lack of benefit. Others remain under study including agents such as ribavirin, ampligen, dextran sulfate, peptide T, and tricosanthin (GLQ223, Compound Q).

COMBINATION THERAPY

The short-lived clinical benefits observed with AZT monotherapy, together with the emergence of AZT resistance during chronic treatment, suggest that combination antiviral chemotherapy will be required for prolonged control of HIV-1 infection (Johnson and Hirsch, 1989). Combination therapies have proven useful in the approach to a variety of bacterial, fungal, and neoplastic diseases. Agents that attack different sites in the HIV-1 replicative cycle offer the greatest promise for synergistic interactions, which may allow the use of individual agents below their toxic concentrations. In addition, combination regimens should be broadly reactive in a variety of cell types, should not display additional toxicity, and may help prevent the emergence of drug resistance.

Not all combinations will be synergistic (combined effects greater than the sum of the independent activities of the components when measured separately). Some combinations will be merely additive and others frankly antagonistic. It is important that drug interactions be evaluated by mathematical analysis, rather than by often misleading visual inspection of data. Several mathematical analyses (multiple drug effect analysis based on the median-effect principle, isobologram technique, fractional product method) are available to address drug interactions (Chou and Talalay, 1984).

A number of anti-HIV-1 drug combinations have shown promise *in vitro*, i.e., synergistic virus inhibitory activity without enhanced toxicity (Johnson and Hirsch, 1989). Some of these are now in clinical trials. Among the most promising are the following:

Interferon-α Plus Zidovudine (Hartshorn et al. 1987b)

Recombinant IFN α-A (rIFNα-A) showed strong synergistic antiviral interactions with AZT against HIV-1 in peripheral blood mononuclear cells (PBMC). Synergism was observed over a broad range of drug concentrations (rIFNα-A, 8-128 U/ml; AZT, 0.01-0.16 μM) and multiplicities of infection, without additive antiproliferative effects against cultured cells. In a separate study (Berman et al. 1989), it was shown that the drug concentrations in combination that inhibit bone marrow progenitors *in vitro* are much higher than those required to inhibit HIV-1 replication, suggesting that a good therapeutic/toxic ratio of this combination might be achieved *in vivo*.

A number of Phase I clinical trials of IFN-α and AZT have been conducted, or are underway. In the most promising results reported (Kovacs et al. 1989), 39 patients with AIDS and Kaposi's sarcoma received AZT 250, 100 or 50 mg orally every 4 h; 6 weeks later IFN-α was begun at 5 million U/day and increased every 2 weeks until a maximum tolerated dose was achieved. Patients received the maximum tolerated combination dose for a minimum of 12 weeks thereafter. The optimal regimen appeared to be 100 mg AZT every four hours combined with 5-10 million U/day of IFN-α. At these doses, peak serum levels of IFN-α (32-250 U/ml) and AZT (0.4-3.85 μM) were in ranges previously shown to be synergistic *in vitro* (Hartshorn et al. 1987b). Of 22 patients who received a stable dose of both drugs for 12 weeks, 11 had a complete or partial response and 8 showed antiviral effects (loss of culture positivity or serum p24 detection). Dose-limiting toxicities included neutropenia (57%), fatigue (16%), thrombocytopenia (14%) and hepatic dysfunction (10%). Randomized studies comparing AZT vs AZT + IFN-α are needed. Other nucleoside analogs, e.g., ddC, have also shown anti-HIV-1 synergism with IFN-α *in vitro* (Vogt et al. 1988).

AZT Plus rsCD4

Strong anti-HIV-1 synergistic interactions were seen *in vitro* between these compounds using peripheral blood mononuclear cells, a CD4-positive T cell line and a monocyte cell line (Johnson et al. 1989a). Low-dose synergy against HIV-l was found in all 3 cell types. Although the drugs as single agents lost effectiveness over time, the 2 drug regimen showed more complete suppression throughout 14 days in culture. Individual patients have tolerated this combination well and larger studies of this combination are planned.

AZT Plus Inhibitors of Trimming Glucosidases

Combining AZT with either of two glycosylation inhibitors, castanospermine or N-butyl DNJ, results in synergism against HIV-l *in vitro*, predominantly in T cell lines (Johnson et al. 1989b,c). No additive toxicity was observed. If phase I clinical trials of n-butyl-DNJ as monotherapy are promising, this combination may prove useful as well.

Three Drug Combinations

We have evaluated whether more complete *in vitro* HIV-l suppression could be attained with a 3-drug combination, each component of which attacks a different target in the virus replicative cycle (Johnson et al. 1990). AZT, rsCD4, and rIFNα-A inhibited HIV-l synergistically in 2- and 3-drug regimens, both in PBMC and in a CD4-positive T cell line. The 3 drug regimen suppressed virus replication more completely; in acutely infected cells, single drug regimens lost effectiveness in 10-14 days and 2 drug regimens in 14-18 days. In contrast, the 3 drug regimen showed nearly complete virus suppression over 28 days in culture.

A number of other 2 drug combinations have been evaluated *in vitro* (Johnson and Hirsch, 1989). In many of these studies, insufficient data were provided to distinguish between synergism and additivism. Some of these, e.g., the combination of AZT and acyclovir are in Phase II/III clinical trials currently. Antagonism between drugs in combination has also been observed. Ribavirin, which has weak anti-HIV-1 activity, inhibits the antiviral activity of AZT in many cell types *in vitro* (Vogt et al. 1987). This is thought to result from ribavirin-induced elevation of deoxythymidine triphosphate levels, which results in the feed back inhibition of cellular thymidine kinase required for phosphorylation of AZT to its active triphosphate form.

BIOLOGICAL RESPONSE MODIFERS

A detailed discussion of biological response modifiers (BRM) is beyond the scope of this chapter. A variety of approaches have been employed to stimulate host immune responses utilizing hormones, lymphokines (e.g., interleukin-α, IFN), and non-specific inhibitors (e.g., IMREG-l, diethyldithiocarbamate, isoprinosine). Since stimulation of host immune responses may be a double-edged sword, activating HIV-1 replication at the same time, it is likely that BRM will be most useful in combination with effective antivirals. Some such clinical trials (e.g., AZT and interleukin-2) are underway currently.

ISSUES IN THE CONDUCT OF CLINICAL TRIALS

One of the most contentious issues in AIDS research is the proper conduct of clinical trials designed to evaluate safe and effective treatments for HIV-1 infection. Although the process by which new drugs are tested has been criticized in the past, never has public scrutiny been so intense as it is now. The parties to these disputes often include patients and their advocates, primary care physicians, physicians conducting the trials, pharmaceutical companies, governmental sponsors (usually the NIH) and federal or state regulatory agencies (usually the FDA). Although all would agree on the goal, many differences arise concerning drug availability, trial entry criteria, proper controls, endpoints, and other elements of study design.

The initial step in the development of an anti-HIV-1 drug is demonstration of its activity *in vitro* against the virus. Equating *in vitro* inhibitory activity against a virus with clinical efficacy is an assumption that is made far too often. In fact, most drugs that demonstrate antiviral activity *in vitro* fail somewhere along the long trail between discovery and licensure. Nevertheless, demonstration of anti-HIV-1 activity is usually the first step along that path. Initial *in vitro* tests should employ a variety of cell types, various virus preparations including fresh clinical isolates, and a variety of treatment regimens (e.g., continuous vs. intermittent). Standardized criteria should be utilized for evaluating drug effects on viral replication (e.g., 50% reduction in virus yield) and cell toxicity.

Once sufficient *in vitro* data suggest that further testing is warranted, appropriate animal models should be studied to evaluate various routes of administration, absorption and bioavailability following oral administration, pharmacokinetics and excretion mechanisms, capacity to reach infected tissues in adequate concentrations, stability in body fluids and toxicity. The maximum tolerated dose should be determined in several species, including primates if possible. Predictions concerning starting doses and toxicity for humans are often possible from animal studies, although with some agents (e.g., dideoxycytidine), unexpected human toxicity may be observed despite detailed animal studies. Unfortunately, good animal models for infection with HIV-1 have not been clearly defined. Although HIV-1 replicates in chimpanzees, and a closely related virus, Simian immunodeficiency virus (SIV), causes disease in macaque monkeys, these are costly and impractical models for routine testing of antiviral drugs. It is hoped that some of the recent small animal models described (Mosier et al. 1988; Ruprecht, 1989) will prove useful in facilitating the study of anti-HIV-1 drugs.

When sufficient *in vitro* and animal studies have been performed to suggest the promise of clinical utility, human trials may begin. Usually pilot or Phase I trials are conducted initially, in which gradually escalating doses are administered to a limited number of individuals in order to establish pharmacokinetic and excretion patterns and maximum tolerated doses. Candidates for these Phase I studies of new anti-HIV-1 drugs generally include patients with advanced infection, since the disease itself may alter the pharmacokinetics or toxicity of the drug being studied. However, one must not make conclusions about drug efficacy from such early Phase I trials or from anecdotal case reports, particularly in disorders with a variable natural course such as HIV-1 infection. History is replete with incorrect conclusions based on preliminary uncontrolled studies. Examples of treatments that gained wide acceptance only to be proven inappropriate after carefully controlled trials include idoxuridine for herpes simplex encephalitis (Boston Interhospital Virus Study Group, 1975), cytosine arabinoside for herpes zoster (Stevens et al. 1973), and photodynamic treatment of recurrent herpes simplex infection (Myers et al. 1975).

If promising results are obtained regarding the bioavailability and toxicity of an antiviral agent, one can then proceed to the most difficult and complex arena, that of large scale controlled clinical trials. Several factors must be considered in planning such trials in order to achieve satisfactory results, since a poorly planned clinical trial is worse than no trial at all. Valuable time and resources can be wasted on obtaining an inconclusive or even incorrect answer. A controlled clinical trial is a substantial undertaking, requiring both time and resources. Prior to beginning clinical trials care must be taken to adequately define entry requirements for target populations, proper endpoints, adequate mechanisms for randomization and analysis (Peto et al. 1976). Many a clinical trial has failed because randomization was inadequate (treatment groups differed as to certain covariates and were not stratified according to these confounding variables), endpoints were vague or controversial, sample sizes were too small, assessment was subjective, loss to follow-up was too great, or codes were broken prematurely. Institutional Review Boards and patients or their advocates must participate in the risk-benefit analysis, because if the patient does not understand the trial and cannot give true informed consent, it may be doomed to fail. Withdrawal from existing trials is one of the key elements limiting their successful completion. Thus, the patient or patient advocate must be a willing and informed participant.

Individuals must never be denied clearly appropriate treatment even if trial protocols are thereby disrupted. In the study of new AIDS therapies, the need to alter protocols on the basis of new information is a constant concern. On one hand, premature disruption of a protocol on the basis of faulty information can do irreparable harm to the study, making it impossible to compare data obtained before and after the change. On the other hand, delay in incorporating important new data in the study may adversely affect the study subject's health or lead to subject withdrawal from the protocol. The situation of aerosolized pentamidine is a case in point. Once sufficient information was accumulated to convince the FDA and other health officials that aerosolized pentamidine was an appropriate form of prophylaxis for PCP in patients also receiving AZT for advanced HIV-1 infection, its use (or the use of equivalent forms of prophylaxis, e.g., trimethoprim-sulfamethoxazole) was incorporated into all relevant NIH-sponsored AIDS trials.

One of the most criticized and misunderstood aspects of clinical trials is the use of blinding procedures. Although a double blind trial may be the most logical and rapid way to answer a question, it is often inherently unsatisfying both to the patient and his or her physician. Once an agent is established as an effective form of treatment (e.g., AZT in AIDS), it should be used as the standard against which newer agents are compared. In general, the larger the sample size, the more likely one is to achieve a satisfactory and reproducible answer. Often, this requires multi-institutional collaboration and adequate mechanisms to obtain comparable data that can be analyzed centrally.

The climate surrounding clinical testing of experimental AIDS drugs, as well as regulation practices for investigational drugs by the FDA has also changed markedly in the past few years, in part due to intense public pressure from advocacy groups. In the past, the FDA prided itself on its cautious approach to introduction of new drugs, and this approach allowed the USA to largely escape the thalidomide disaster that befell Europe several decades ago

(McBride, 1977). Although this approach may be useful in some situations, the urgency of the AIDS epidemic has necessitated a more expeditious program of drug evaluation.

The current drug testing process begins with submission by a sponsor, usually a pharmaceutical company, of an Investigational New Drug application (IND) to the FDA. An IND allows only the sponsor and clinical investigators listed in the application to study the drug (Kessler, 1989). Traditionally, Phase 1 trials of clinical pharmacology are followed by Phase 2 controlled clinical trials in several hundred patients and Phase 3 expanded clinical trials in several thousand patients. However, AZT was approved in record time after only one controlled clinical trial in 282 patients. Prior to official licensure, it was made available to several thousand desperately ill patients with AIDS under what is called a Treatment IND program (Creagh-Kirk et al. 1988), which was jointly administered by Burroughs Wellcome Company and governmental agencies (NIH and FDA). During the period of a Treatment IND, patients are permitted access to the drug in question, while the FDA reviews data required for licensure.

It is likely that similar fast track approval will occur for other agents used to treat HIV-1-infected patients. The FDA has promulgated interim procedures that indicate that Phase 2 trials may sometimes be sufficiently conclusive to make a break-through drug available quickly (Department of Health and Human Services, 1987, 1988). In addition, the FDA has taken a more active role in the planning of both Phase 1 and Phase 2 trials in order to ensure that the testing that will be done will optimize the opportunity for subsequent drug approval. Conferences are held with sponsors early and often to plan studies and review data. It remains essential that evidence of efficacy or promise of therapeutic benefit be obtained before a Treatment IND is issued. Premature issuance of a Treatment IND would compromise the ability to conduct properly controlled trials. Few patients would choose to participate in a controlled trial designed to demonstrate efficacy if the drug is readily available to the same population by a Treatment IND mechanism.

In response to growing public pressure for expanded access to selected promising anti-HIV-1 drugs, a "parallel track system" has been developed for promising new agents. This system would be available largely for individuals who for various reasons (e.g., geographical distance, eligibility criteria) are unable to participate in a Phase 2 clinical trial, but might reasonably benefit from the drug. At the time of this writing, ddI is being evaluated under such an expanded access program. Details concerning eligibility criteria, drugs to be covered, liability, informed consents, monitoring for benefit and toxicity, and profit to pharmaceutical companies will probably change as experience with this program is gained. Great care must be taken to avoid compromising the conduct of controlled clinical trials which will continue to provide the most likely mechanism for acquiring valid data on the utility of individual drugs. Issues of enrollment and compliance with clinical trials could be greatly compromised by too easy access to the latest faddish drug ("drug of the month"). At the same time, creativity and flexibility must be maintained to provide the best possible medical care to those unable to participate in available clinical trials.

The final step prior to licensure is submission by the sponsor and approval by the FDA of a New Drug Application (NDA). During this period, clinical data are reviewed and drug labeling is negotiated. In 1987, the mean time required for the FDA to approve NDAs was 32.4 months (Kessler, 1989).

In the future, dramatic breakthroughs such as AZT development may be less common than incremental advances. These advances will undoubtedly be difficult to evaluate, since small differences among drugs require large numbers of patients to demonstrate. Developments such as these will pose new problems for investigators, patients and regulators. Nevertheless, the goals for all parties are the same, i.e., the most expeditious control of HIV-1 infection and its complications with the safest possible drugs. Although debates and disagreements on the best way to achieve this goal will still occur, we would be well-advised to put rhetoric aside and develop programs that will allow the proper conduct of clinical trials and at the same time permit the widest possible access to promising drugs for those not engaged in these studies. This task will challenge investigators, patients, and regulators in the years ahead.

REFERENCES

Agrawal S, Goodchild J, Civeira MP, Thornton AH, Sarin PS, Zamecnik P (1988) Oligodeoxynucleoside phosphoramidates and phosphorothioates as inhibitors of human immunodeficiency virus. Proc Natl Acad Sci USA 85:7079-7083.

Berman E, Duigou-Osterndorf R, Krown SE, Fanucchi MP, Chou J, Hirsch MS, Clarkson BD, Chou TC (1989) Synergistic cytotoxic effect of azidothymidine and recombinant interferon alpha on normal human bone marrow progenitor cells. Blood 74:1281-1286.

Boston Interhospital Hospital Virus Study Group and the NIAID Sponsored Cooperative Antiviral Clinical Study (1975) Failure of high does 5-iodo-2'-deoxyuridine in the therapy of herpes simplex virus encephalitis. N Engl J Med 292:599-603.

Brown PO, Bowerman B, Varmus HE, Bishop JM (1989) Retroviral integration: structure of the initial covalent product and its precursor, and a role for the viral IN protein. Proc Natl Acad Sci USA 86:2525-2529.

Capon DJ, Chamow SM, Mordenti J, Marsters SA, Gregory T, Mitsuya H, Byrn RA, Lucas C, Wurm FM, Groopman JE, Broder S, Smith DH (1989) Designing CD4 immunoadhesins for AIDS therapy. Nature 337:525-531.

Chaudhary VK, Mizukami T, Ruerst TR, Fitzgerald DJ, Moss B, Pastan I, Berger EA (1989) Selective killing of HIV-infected cells by recombinant human CD4-pseudomonas exotoxin hybrid protein. Nature 335:369-372.

Chou TC, Talalay P (1984) Quantitative analysis of dose-effect relationships: the combined effects of multiple drugs or enzyme inhibitors. Adv Enzyme Regul 22:27-55.

Clapham PR, Weber JN, Whitby D, McIntosh K, Dalgleish AG, Maddon PJ, Deen KC, Sweet RW, Weiss RA (1989) Soluble CD4 blocks the infectivity of diverse strains of HIV and SIV for T cells and monocytes but not for brain and muscle cells. Nature 337:368-370.

Creagh-Kirk T, Doi P, Andrews E, Nusinoff-Lehrman S, Tilson H, Hoth D, Barry DW (1988) Survival experience among patients with AIDS receiving zidovudine. Follow-up of patients in a compassionate plea program. JAMA 260:3009-3015.

Department of Health and Human Services (1987) New drug, antibodies and biologic drug product regulations. Fed Regist 52:8796-8847.

Department of Health and Human Services (1988) Investigational new drug, antibiotic, and biological drug products regulations: procedures for drugs intended to treat life-threatening and severely debilitating illnesses. Fed Regist 53:41516-41524.

Deray G, Diquet B, Martinez F, Vidal AM, Petitclerc T, Ben Hmida M, Land G, Jacobs C (1988) Pharmacokinetics of zidovudine in a patient on maintenance hemodialysis. N Engl J Med 319:1606-1607.

deWit R, Schattenkerk JK, Boucher CA, Bakker PJ, Veenhof KH, Danner SA (1988) Clinical and virological effects of high-dose recombinant interferon-a in disseminated AIDS-related Kaposi's sarcoma. Lancet 2:1214-1217.

Dournon E, Matheron S, Rozenbaum W, Gharakhanian S, Michan C, Girard PM, Perronne C, Salmon D, de Truchis P, Leporti C, Bouvet E, Dazza MC, Levacher M, Regnier B (1988) Effects of zidovudine in 365 consecutive patients with AIDS or AIDS-related complex. Lancet 2:1297-1302.

Fischl MA, Richman DD, Grieco MH, Gottlieb MS, Volberding PA, Laskin OL, Leedom JM, Groopman JE, Mildvan D, Schooley RT, Jackson GG, Durack DT, King D, the AZT Collaborative Working Group (1987) The efficacy of azidothymidine (AZT) in the treatment of patients with AIDS and AIDS-related complex. A double-blind, placebo-controlled trial. N Engl J Med 317:185-191.

Fischl MA, Richman.DD, Causey DM, Grieco MH, Bryson Y, Mildvan D, Laskin OL, Groopman JE, Volberding PA, Schooley RT, Jackson GG, Durack D, Andrews JC, Nusinoff-Lehrman S, Barry DW, and the AZT Collaborative Working Group (1989) Prolonged zidovudine therapy in patients with AIDS and advanced AIDS-related complex. JAMA 262:2405-2410.

Fisher RA, Bertonis JM, Meier W, Johnson VA, Costopoulos DS, Liu T, Tizard R, Walker BD, Hirsch MS, Schooley RT, Flavell RT (1988) HIV infection is blocked *in vitro* by recombinant soluble CD4. Nature 331:76-78.

Grinde B, Hungnes 0, Tjotta E (1989) The proteinase inhibitor pepstatin A inhibits formation of reverse transcriptase in H9 cells infected with human immunodeficiency virus-l. AIDS Res Human Retrov 5:269-274.

Gruters RA, Neefjes JJ, Tersmette M, de Goede RE, Tulp A, Huisman HG, Miedema F, Ploegh HL (1987) Interference with HIV-induced syncytium formation and viral infectivity by inhibitors of trimming glucosidase. Nature 330:74-77.

Hartshorn KL, Neumeyer D, Vogt MW, Schooley RT, Hirsch MS (1987a) Activity of interferons alpha, beta, and gamma against human immunodeficiency virus replication *in vitro*. AIDS Res Human Retrovirus 3:125-133.

Hartshorn KL, Vogt MW, Chou TC, Blumberg RS, Byington RE, Schooley RT, Hirsch MS (1987b) Synergistic inhibition of human immunodeficiency virus *in vitro* by azidothymidine and recombinant alpha A interferon. Antimicrob Ag Chemother 31:168-172.

Haseltine WA (1989) Development of antiviral drugs for the treatment of AIDS: strategies and prospects. J AIDS 2:311-334.

Helbert M, Robinson D, Peddle B, Forster S, Kocsis A, Jeffries D, Pinching AJ (1988) Acute meningoencephalitis on dose reduction of zidovudine. Lancet 1:1249-1252.

Ho DD, Hartshorn KL, Rota TR, Andrews CA, Kaplan JC, Schooley RT, Hirsch MS (1985) Recombinant human interferon alpha-A suppresses HTLV-III replication *in vitro*. Lancet 1:602-604.

Horwitz JP, Chua J, Noel M (1969) The monomesylates of 1-(2'-deoxy-b-D-lyxo-furanosyl)thymine. J Organ Chem 29:2076-8.

Hussey RE, Richardson NE, Kowalski N, Brown NR, Chang HC, Siliciano RF, Dorfman T, Walker B, Sodroski J, Reinherz EL (1988) A soluble CD4 protein selectively inhibits HIV replication and syncytium formation. Nature 331:78-81.

Interferon Alpha Study Group (1988) A randomized placebo-controlled trial of recombinant human interferon alpha 2a in patients with AIDS. J AIDS 1:111-118.

Jacks T, Power MD, Masiarz FR, Luciw PA, Barr PJ, Varmus HE (1988) Characterization of ribosomal frameshifting in HIV gag-pol expression. Nature 331:280-283.

Jackson GG, Paul DA, Falk LA, Rubenis M, Despotes JC, Mack D, Knigge M, Emeson EE (1988) Human immunodeficiency virus (HIV) antigenemia (p24) in the acquired immunodeficiency syndrome (AIDS) and the effects of treatment with zidovudine (AZT). Ann Intern Med 108:175-180.

Johnson VA, Hirsch MS (1989) Combination therapy for HIV infection. In: Mills J, Corey L, (eds) Antiviral Chemotherapy: New Directions for Clinical Application and Research. Vol 2 pp 275-302. New York: Elsevier.

Johnson VA, Barlow MA, Chou T-C, Fisher RA, Walker BD, Hirsch MS, Schooley RT (1989a) Synergistic inhibition of human immunodeficiency virus type 1 (HIV-l) replication in vitro by recombinant soluble CD4 and 3'-azido-3'-deoxythymidine. J Infect Dis 159:837-844.

Johnson VA, Merrill DP, Chou TC, Hirsch MS (1989b) Synergistic inhibition of HIV-l replication by N-butyl deoxynojirimycin (N-butyl-DNJ) and zidovudine (AZT) (abstr. #504). Proceedings of the Twenty-Ninth Interscience Conference on Antimicrobial Agents and Chemotherapy. Houston, Texas, September 17-20, 1989;185.

Johnson VA, Walker BD, Barlow MA, Paradis TJ, Chou T-C, Hirsch MS (1989c) Synergistic inhibition of human immunodeficiency virus type 1 and 2 replication in vitro by castanospermine and 3'-azido-3'-deoxythymidine. Antimicrob Ag Chemother 33:53-57.

Johnson VA, Barlow MA, Merrill DP, Chou TC, Hirsch MS (1990) Three drug synergistic inhibition of HIV-l replication in vitro by zidovudine, recombinant soluble CD4, and recombinant interferon alpha-A. J Infect Dis (In press).

Kahn J, Davis AJ, Groopman JE, Kaplan L, Sherwin S (1989) Pharmacokinetics studies of recombinant soluble CD4 in patients with AIDS and AIDS related complex. In: "Program and Abstracts of the Fifth International Conference on AIDS". Montreal, Canada June 4-9. Abstract ThB05.

Karpas A, Fleet GWJ, Dwek RA, Peturson S, Namgoong SK, Ramsden NG, Jacob GS, Rademacher TW (1988) Aminosugar derivatives as potential anti-human immunodeficiency virus agents. Proc Natl Acad Sci USA 85:9229-9233.

Kessler DA (1989) The regulation of investigational drugs. N Engl J Med 320:281-288.

Kornhauser DM, Petty BG, Hendrix CW, Woods AS, Nerhood LJ, Bartlett JG, Lietman PS (1989) Probenicid and zidovudine metabolism. Lancet 2:473-475.

Kovacs JA, Deyton L, Davey R, Falloon J, Zunich K, Lee D, Metcalf JA, Bigley JW, Sawyer LA, Zoon KC, Masur H, Fauci AS, Lane HC (1989) Combined zidovudine and interferon-a therapy in patients with Kaposi's sarcoma and the acquired immunodeficiency syndrome (AIDS). Ann Intern Med 111:280-287.

Krown S (1988) AIDS-associated Kaposi's sarcoma: pathogenesis, clinical course and treatment. AIDS 2:71-80.

Lamarre D, Ashkenazi A, Fleury S, Smith DH, Sekaly RP, Capon DJ (1989) The MHC-binding and gp120 binding functions of CD4 are separable. Science 245:743-746.

Lane HC, Kovacs JA, Feinberg J, Herpin B, Davey V, Walker R, Deyton L, Metcalf JFA, Baseler M, Salzman N, Manischewitz J, Quinnan G, Masur H, Fauci AC (1988) Anti-retroviral effects of interferon-a in AIDS-associated Kaposi's sarcoma. Lancet 2:1218-1222.

Lane HC, Falloon J, Walker RE, Deyton L, Kovacs JA, Masur H, Banks S, Kirk LE, Baseler MW, Salzman NP, Fauci AS (1989) Zidovudine in patients with human immunodeficiency virus (HIV) infection and Kaposi's sarcoma. A Phase II randomized, placebo-controlled trial. Ann Intern Med 111:41-50.

Larder BA, Darby G, Richman DD (1989) HIV with reduced sensitivity to zidovudine (AZT) isolated during prolonged therapy. Science 243:1731-1734.

Lifson JD, Hwang KM, Nara PL, Fraser B, Padgett M, Dunlop NM, Eiden LE (1988) Synthetic CD4 peptide derivatives that inhibit HIV infection and cytopathicity. Science 241:712-716.

McBride WG (1977) Thalidomide embryopathy. Teratology 16:79-82.

Merigan TC, Skowron G, Bozzette SA, Richman D, Uttamchandani R, Fischl M, Schooley R, Hirsch M, Soo W, Pettinelli C, Schaumburg H, and the ddC Study Group of the AIDS Clinical Trials Group (1989) Circulating p24 antigen levels and responses to dideoxycytidine in human immunodeficiency virus (HIV) infections. Ann Intern Med 110:189-194.

Mitsuya H, Broder S (1986) Inhibition of the in vitro infectivity and cytopathic effect of human T-lymphotropic virus type III/lymphadenopathy-associated virus (HTLV-III/LAV) by 2',3'-dideoxynucleosides. Proc Natl Acad Sci USA 83:1911-5.

Mitsuya H, Weinhold KJ, Furman PA, St. Clair MH, Lehrman SN, Gallo RC, Bolognesi D, Barry DW, Broder S (1985) 3'-azido-3'-deoxythymidine (BWA509U): an antiviral agent that inhibits the infectivity and cytopathic effect of human T-lymphotropic virus type III/lymphadenopathy-associated virus in vitro. Proc Natl Acad Sci USA 82:7096-7100.

Mosier DE, Gulizia RJ, Baird SM, Wilson DB (1988) Transfer of a functional human immune system to mice with severe combined immunodeficiency. Nature 335:256-259.

Myers MG, Oxman MN, Clark JE, Arndt KA (1975) Failure of neutral-red photodynamic inactivation in recurrent herpes simplex infections. N Engl J Med 293:945-949.

Navia MA, Fitzgerald PM, McKeever BM, Leu CT, Heimbach JC, Herber WK, Sigal IS, Darke PL, Springer JP (1989) Three-dimensional structure of aspartyl protease from human immunodeficiency virus HIV-l. Nature 337:615-620.

Ostertag W, Roesler G, Kreig CJ, Kind J, Cole T, Crozier T, Gaedicke G, Steinheider G, Kluge N, Dube S (1974) Induction of endogenous virus and of thymidine kinase by bromodeoxyuridine in cell cultures transformed by Friend virus. Proc Natl Acad Sci USA 71:4980-5.

Peto R, Pike MC, Armitage P, Breslow NE, Cox DR, Howard SV, Mantel N, McPherson K, Peto J, Smith PG (1976) Design and analysis of randomized clinical trials requiring prolonged observation of each patient. I. Introduction and design. Brit J Cancer 34:585-612.

Pizzo PA, Eddy J, Falloon J, Balis FM, Murphy RF, Moss H, Wolters P, Brouwers P, Jarosinski P, Rubin M, Broder S, Yarchoan R, Brunetti A, Maha M, Nusinoff-Lehrman S, Poplack DG (1988) Effect of continuous intravenous infusion of zidovudine (AZT) in children with symptomatic HIV infection. N Engl J Med 319:889-96.

Poli G, Orenstein JM, Kinter A, Folks TM, Fauci AS (1989) Interferon-a but not AZT suppresses HIV expression in chronically infected cell lines. Science 244:575-577.

Richman DD, Fischl MA, Grieco MH, Gottlieb MS, Volberding PA, Laskin OL, Leedom JM, Groopman JE, Mildvan D, Hirsch MS, Jackson GG, Durack DT, Nusinoff-Lehrman S, the AZT Collaborative Working Group (1987) The toxicity of azidothymidine (AZT) in the treatment of patients with AIDS a AIDS-related complex. A double-blind, placebo-controlled trial. N Engl J Med 317:192-7.

Ruprecht RM (1989) Murine models for antiretroviral therapy. Intervirol 305:2-11.

Ruprecht RM, Mullaney S, Anderson J, Bronson R (1989) *In vivo* analysis of castanospermine, a candidate antiretroviral agent. J AIDS 2:149-157.

Sarin PS, Agrawal S, Civeira MP, Goodchild J, Ikeuchi T, Zamecnik PC (1988) Inhibition of acquired immunodeficiency syndrome virus by oligodeoxynucleoside methylphosphonates. Proc Natl Acad Sci USA 85:7448-7451.

Schmitt FA, Bigley JW, McKinnis R, Logue PE, Evans RW, Drucker JL, and the AZT Collaborative Working Group (1988) Neuropsychological outcome of zidovudine (AZT) treatment of patients with AIDS and AIDS-related complex. N Engl J Med 319:1573-8.

Schooley RT, Merrigan TC Jr, Gaut P, Hirsch MS, Holodniy M, Flynn T, Liu S, Byington RE, Henochowicz S, Gubish E, Spriggs D, Kute D, Schindler J, Dawson A, Thomas D, Hanson DG, Letwin B, Liu T, Gulinello J, Kennedy S, Fisher R, Ho DD (1990) A phase I (II escalating dose trial of recombinant soluble CD4 therapy in patients with AIDS or AIDS related complex. Ann Intern Med (in press).

Smith DH, Byon RA, Marsters SA, Gregory T, Groopman JE, Capon DJ (1987) Blocking of HIV-1 infectivity by a soluble, secreted form of the CD4 antigen. Science 238:1704-07.

Stevens DA, Jordan GW, Waddell TF, Merigan TC (1973) Adverse effect of cytosine arabinoside on disseminated zoster in a controlled trial. N Engl J Med 189:873-878.

Swiss Group for Clinical Studies on the Acquired Immunodeficiency Syndrome (AIDS) (1988) Zidovudine for the treatment of thrombocytopenia associated with human immunodeficiency virus (HIV). Ann Intern Med 109:718-721.

Till MA, Ghetie V, Gregory T, Patzer EJ, Porter JP, Uhr JW, Capon DJ, Vitetta ES (1988) HIV-infected cells are killed by rCD4-ricin A chain. Science 242:1166-1168.

Traunecker A, Luke W, Karjalainen K (1988) Soluble CD4 molecules neutralize human immunodeficiency virus type 1. Nature 331:84-86.

Varmus H, Brown P (1989) Retroviruses In: Howe M, and Berg P (eds) Mobile DNA. ASM, Metals Park, Ohio, pp. 53-107.

Vogt MW, Hartshorn KL, Furman PA, Chou TC, Fyfe JA, Coleman LA, Crumpacker C, Schooley RT, Hirsch MS (1987) Ribavirin antagonizes the effects of azidothymidine on HIV replication. Science 235:1376-1379.

Vogt MW, Durno AG, Chou T-C, Coleman LA, Paradis TJ, Schooley RT, Kaplan JC, Hirsch MS (1988) Synergistic interaction of 2'-,3'-dideoxycytidine (ddCyd) and recombinant interferon alpha-A (r-IFN-alpha-A) on HIV-1 replication. J Infect Dis 158:378-385.

Walker BD, Kowalski M, Goh WC, Kozarsky K, Krieger M, Rosen C, Rohrschneider L, Haseltine WA, Sodroski J (1987) Inhibition of human immunodeficiency virus syncytium formation and virus replication by castanospermine. Proc Natl Acad Sci USA 84:8120-8124.

Watanabe M, Reimann KA, DeLong PA, Liu T, Fisher RA, Letvin NL (1989) Effect of recombinant soluble CD4 in rhesus monkeys infected with simian immunodeficiency virus of macaques. Nature 337:267-270.

Wlodawer A, Miller M, Jaskolski M, Sathyanarayana BK, Baldwin E, Weber IT, Selk LM, Clawson L, Schneider J, Kent SB (1989) Conserved folding in retroviral proteases: crystal structure of a synthetic HIV-l protease. Science 245:616-622.

Yarchoan R, Perno CF, Thomas RV, Klecker RW, Allain J-P, Wills RJ, McAtee N, Fischl MA, Dubinsky R, McNeeley MC, Mitsuya H, Pluda JM, Lawley TJ, Leuther M, Safai B, Collins JM, Myers CE, Broder S (1988) Phase I studies of 2'-,3'-dideoxycytidine in severe human immunodeficiency virus infection as a single agent and alternating with zidovudine (AZT). Lancet 1:76-81.

Yarchoan R, Thomas RV, Pluda JM, Perno CF, Mitsuya H, Marczyk KS, Sherwin SA, Broder S (1989a) Phase I study of the administration of recombinant soluble CD4 (rCD4) by continuous infusion to patients with AIDS or ARC. In Program and abstracts of the Fifth International Conference on AIDS. Montreal, Canada June 4-9. Abstract MCP 137.

Yarchoan R, Mitsuya H, Thomas RV, Pluda JM, Hartman NR, Perno CF, Marczyk KS, Allain JP, Johns DG, Broder S (1989b) *In vitro* activity against HIV and favorable toxicity profile of 2'-3'-dideoxyinosine. Science 245:412-415.

Yarchoan R, Mitsuya H, Myers CE, Broder S (1989c) Clinical pharmacology of 3'-azido-2',3'-dideoxythymidine (zidovudine) and related dideoxynucleosides. N Engl J Med 321:726-738.

ABSTRACTS

RAPID DETECTION OF POLYOMAVIRUSES BK AND JC BY A SHELL VIAL CELL CULTURE ASSAY

A. Telenti, W.F. Marshall, J. Proper, A.J. Aksamit and T.F. Smith

Mayo Clinic
Rochester, Minnesota USA

Polyomaviruses BK and JC (BKV, JCV) characteristically exhibit slow growth in cell culture, with a late and subtle cytopathic effect. Improved culture culture methods for detection of polyomavirus would have a significant impact on clinical studies of BKV and JCV disease. One dram shell vials were seeded with four different cell types: MRC-5, HEK and rhabdomyosarcoma cells for infection with BKV and an adult human brain cells for JCV. After inoculation, cultures were subjected to low speed (700x g) centrifugation for 45 min or to 90 min adsorption without centrifugation. Following 16 or 36 hr incubation, the cell monolayers were fixed, reacted with a monoclonal antibody to the T antigen of polyomavirus SV40 and stained with anti-mouse fluorescein-labeled globulin. Specific nuclear fluorescence could be observed at 16 hr although optimal detection was accomplished at 36 hr. Centrifugation enhanced both BKV and JCV infectivity by a factor of 1.8 to 5.2-fold when compared with un-centrifuged vials. Rapid detection of BKV and JCV can be accomplished with a shell vial assay inoculated under low speed centrifugation.

DIRECT ANTIGEN DETECTION OF RESPIRATORY VIRUSES USING BARTELS MONOCLONAL INDIRECT FLUORESCENT REAGENTS

S. Matthey, D. Nicholson, S. Ruhs, B. Alden and M. Knock

The University of Iowa
Iowa City, Iowa, USA

This study evaluated the effectiveness of indirect fluorescent antibody (IFA) reagents (Bartels) for the direct detection of seven respiratory viruses. Tissue culture isolation with confirmation of Bartels IFA reagents and direct antigen detection with Bartels pooled and individual antisera were performed on frozen (84/255) and fresh (171/255) specimens and slides. Pooled antisera proved to be of limited value due to nonspecific staining. Thirty (11.7%) of the direct stains had inadequate numbers of cells for testing. Direct staining detected 91.5% (43/47) respiratory syncytial virus (RSV) isolates and 44.4% (4/9) influenza A isolates. By direct staining, no culture isolates of influenza B (4), parainfluenza 3 (2), and parainfluenza 1 (1) were detected. All isolates other than RSV were confirmed by IFA reagents from the Centers for Disease Control. The sensitivity, specificity, and positive and negative predictive values were 74.6%, 85.2%, 66.2% and 89.6%, respectively. The small number of positive specimens necessitates further testing, but individual reagents may provide a useful rapid antigen screen provided cultures are not performed.

INFLUENZA IN ELDERLY VACCINEES

D.M. McLean

University of British Columbia
Vancouver, Canada

Acute respiratory illness, often superimposed on chronic pulmonary disease, affected 26 of 74 elderly patients on one floor of an Extended Care Unit in Vancouver between 25 and 31 March, 1989. Symptoms were generally mild and remitted after 2-3 days, afflicting 12 cases from March 25 through 28, 5 cases March 29, 4 cases March 30, 5 cases March 31. Influenza A (HINI) antigenically resembling A/Taiwan/1/86 was isolated from throat swabs from 1 of 14 subjects tested between 28 and 31 March. Trivalent influenza vaccine was administered to 16 of 26 cases including the virus excreter who was 1 of 11 vaccinees among 14 patients examined virologically. Influenza vaccine was administered to 52 of 77 (67.5%) patients of this ward on 9 and 10 November 1988. Assessment of vaccination status on 20 April 1989 revealed that 37 of 69 (53.6%) current patients received influenza vaccine in 1988, 19 of whom, including the excreter, had also received vaccine in 1987 and all except the excreter were vaccinated in 1986. Previous experience with elderly patients during winter 1983-84 revealed cumulative seroconversion frequencies >70% to all components of the 1983-84 trivalent influenza vaccine, with geometric mean antibody titers >40 (protective level) 2-12 weeks after vaccination, declining <40 by 24 weeks when only 36% had HI antibody titers to influenza A (HINI). The present outbreak occurred 20 weeks after vaccination, when antibody titers were expected to decline below protective levels.

SPASTIC PARAPARESIS DUE TO HUMAN T-LYMPHOTROPIC VIRUS TYPE 1 (HTLV-1) IN THE UNITED STATES

R.R. McKendall[1], M.D. Lairmore[2] and J. Oas[1]

University of Texas Medical Branch
[1]Galveston, Texas, and
[2]CDC, Atlanta, Georgia, USA

We report preliminary findings in two American-born patients with spastic paraparesis and well documented infection with HTLV-1. Clinically, the patients primary problems were weakness, clumsiness and spasticity of the legs, typical of spastic paraparesis. Urinary retention without incontinence was also present. Cerebrospinal fluid (CSF) analysis showed moderate pleocytosis, protein elevation, elevated IgG index in both patients and oligoclonal bands in one. Flower lymphocytes were not present in blood or CSF. Radiographically, the spinal cord was normal. However, on magnetic resonance imaging of brain, one patient had multiple frontal lobe lesions. ELISA to HTLV-1 antigens revealed reactivity of both serum and CSF. Western immunoblot studies of serum and CSF revealed reactivity to p19, p24, gp46, p53 and gp68. Western immunoblot assays comparing serum and CSF at equal IgG concentrations showed more intense reactivity in CSF than serum for p24 indicating an intrathecal virus specific response. Using an HTLV-1 specific *pol* primer pair, polymerase chain reaction studies of cells obtained from peripheral blood and from CSF demonstrated virus-specific sequences for HTLV-1 but not HTLV-2. Virus was also demonstrated in PBL and CSF by antigen capture. To our knowledge, these two cases are the only U.S.A. cases in whom virus isolation from the CSF has been accomplished. Of importance is the possibility that one of the patients may represent a case of endemic infection. Risk factors of the two will be presented.

INDUCTION OF AN ANTI-HEPATITIS A VIRUS (HAV) PRIMING RESPONSE BY POLYHEDRIN INCORPORATING AN HAV VP3 PEPTIDE

J. McLinden[1], J. Stapelton[2], V. Ploplis[1] and E. Rosen[1]

[1]American Biogenetic Sciences
Notre Dame, Indiana and
[2]University of Iowa Hospital,
Iowa City, Iowa, USA

HAV causes approximately 100,000 cases of reported hepatitis in the U.S.A. each year. HAV, a member of the picornaviridae family, is a nonenveloped icosahedron virus that contains 4 structural proteins VP1, VP2, VP3 and VP4. Analysis of neutralizing anti-HAV antibodies and neutralization escape mutants suggests that major antigenic sites are located on VP1 and VP3 (Ping et al., *PNAS 85*:8281, 1988). We have incorporated peptide sequences of putative hepatitis A epitopes into the polyhedrin protein made by the baculovirus *Autographa californica* (AcNPV). AcNPV produces a proteinaceous paracrystalline occlusion body (OB) late after infection of susceptible insect cells. The OB is composed primarily of a single protein, polyhedrin. The sequence encoding amino acids 11-24 of VP1 was inserted after aa43 of polyhedrin. This HAV peptide has been shown to generate low titre anti-HAV antibodies in rabbits, (Emini et al., *J Virol 55*:836, 1985). In addition, amino acids 65-75 of VP3 were inserted after amino acid 1 of polyhedrin. Mutation of VP3 amino acid 70 results in resistance to several anti-HAV neutralization antibodies. Rabbits were inoculated with wild type polyhedrin, recombinant polyhedrin containing the VP1 sequence or recombinant polyhedrin containing the VP3 sequence. Animals received an initial inoculation followed by 2 boosts at two week intervals. These animals did not produce anti-HAV antibodies by immunoassays, Western blots and virus neutralization assays. Following immunization with gradient purified HM175 strain of HAV, the animals inoculated with the VP3 polyhedrin protein, but not the animals inoculated with the VP1 polyhedrin, rapidly generated anti-HAV antibodies that recognized denatured HAV proteins by Western blots, suggesting the induction of a priming response by the VP3 polyhedrin protein. The ability of the recombinant polyhedrin protein expressing HAV epitopes to stimulate a priming response may be valuable for the production of a HAV vaccine.

HIGH BACKGROUND IN POLYMERASE CHAIN REACTION ASSAY DUE TO CONTAMINATION FROM OVERLAPPING DNA FRAGMENTS OVERCOME BY NESTING TECHNIQUE

K. Porter-Jordan, E. Rosenberg, J. Keiser and C.T. Garrett

The George Washington Medical Center
Washington, District of Columbia, USA

The polymerase chain reaction (PCR) assay, a promising technique for the rapid and sensitive detection of viral infections, has frequently shown high levels of background in samples containing no template DNA. This background reduces the sensitivity of the assay, makes interpretation of borderline positive samples extremely difficult, and has raised questions about the usefulness of the assay as a diagnostic tool. We have studied this high background in our PCR assay for CMV. Despite extensive precautions to avoid carryover, we observed faint bands on liquid hybridization at a frequency ranging from 100% for a 72 bp sequence to 79% for a 167 bp sequence. In one experiment in an entirely new region of the CMV genome, we found 42/70 water-only samples positive (60%). Positive signal can be eliminated by DNase treatment of water or positive samples, but not by filtration through an isotropic membrane, autoclaving or ultraviolet irradiation. Concentrating water by speedvac evaporation produces a stronger signal. A lag time of 10 to 12 cycles is observed before the reactions with water show amplicon by liquid hybridization. Nested PCR overcomes the background, and in serial dilutions of a positive sample shows 1,000 fold increase in sensitivity by liquid hybridization detection. Our results suggest that the background signal is arising from small fragments of DNA, which may be produced by autoclaving viral culture material or cloned sequences. Nested PCR, appropriately controlled for the number of cycles at each step, should successfully overcome such false positives due to fragmented DNA, no matter whether the contamination occurs at the collection site, in processing, or in the laboratory.

CLINICAL AND LABORATORY FEATURES OF HUMAN IMMUNO-DEFICIENCY VIRUS TYPE 1 (HIV-1) POSITIVE POSTPARTUM WOMEN

M.P. Reyes, C. Meriwether, E.C. Moore, D. Philpot and F. Cohen

Hutzel Hospital and
Children's Hospital of Michigan
Wayne State University
Detroit, Michigan, USA

Thirty-three mothers with positive ELISA and Western blot antibody tests for HIV-1 were seen at the Maternal Infant Center for HIV Infection (MICH) for 3 years from August 1986 to August 1989. Ages ranged from 16 to 41 years. There were 28 black women and 5 white women. Twenty were intravenous drug abusers (IVDA), 4 were sexual partners of IVDAs, and 7 had sexual partners who were HIV-1 positive. Two patients denied belonging to any high risk group for HIV-1 infection. Three patients had positive serology for syphilis. Eighty-three percent of patients had antibody titers to cytomegalovirus, 16% to toxoplasma and 95% to Epstein-Barr virus. There were 2 antigen carriers of hepatitis B virus. Absolute T-helper subpopulation (T4) counts ranged from 251 to 1,708. Four patients have received azidothymidine (AZT). None of the 33 patients had acquired immunodeficiency syndrome (AIDS) (CDC Classification Group IV) at the time of delivery. One patient developed AIDS (esophageal candidiasis) twenty-four months later, 6 months after induced abortion of a next pregnancy. Although the exact date of onset of HIV-1 infection is not known, it appears that the rate of progression to AIDS in this group of postpartum women is similar to the Brooklyn experience.

ACUTE AND CHRONIC HEPATITIS: A NOVEL ROTAVIRUS-INDUCED DISEASE IN MICE

M. Riepenhoff-Talty[1], I. Uhnoo[1], H.B. Greenberg[2], J.E. Fisher[1], P. Chegas[1], H. Barrett[1], P.K. Li[1] and P.L. Ogra[1]

[1]State University of New York at Buffalo
School of Medicine and
The Children's Hospital
Buffalo, New York,
[2]V.A. Hospital
Stanford University
Palo Alto, CA, USA

Rotaviruses (RV) are the most important cause of severe diarrhea in young children. Murine rotavirus (MRV) produces diarrheal disease in suckling mice and, similar to human infection, MRV replication is limited to the small intestinal villi. In the present report, we describe the outcome of heterologous RV infection with rhesus rotavirus (RRV) strain in normal Balb/c mice and mice with severe combined immunodeficiency (SCID). Oral feeding of RRV to 1 to 3 day old mice caused diarrhea between two and five days post inoculation (pi). Dark urine, clay-colored feces and icterus were observed in 82% of SCID mice and 18% of BALB/c mice 7 to 15 days pi. Immunocompetent Balb/c mice recovered with 0% mortality while 30% of SCID mice died and the surviving animals developed chronic liver disease. Histopathological examination of livers from affected animals revealed a diffuse hepatitis with focal areas of parenchymal necrosis. Infectious virus was isolated from the livers and RV was demonstrated by specific immunofluorescent assay and by electron microscopy. Alanine aminotransferase and aspartate aminotransferase were significantly elevated in the serum of symptomatic mice. These data represent the first evidence for systematic dissemination and extraintestinal spread of RV to the liver producing hepatitis. The implications of these findings may be applicable to the vaccine strategy in humans that utilizes live heterologous RV.

COST-EFFECTIVE RAPID CULTURE (RC) FOR HERPES SIMPLEX VIRUS (HSV) USING 96-WELL PLATES AND POLYCLONAL FLUORESCENT ANTIBODY

R.A. Littlewood, L.E. Sibau and C.H. Sherlock

University of British Columbia
Vancouver, British Columbia
Canada

Shell vial cultures with centrifugation and 24-hr antigen detection have provided speed and accuracy for the diagnosis of HSV infection, but technical time and monoclonal reagent costs are high for such screening tests. We adapted the RC system to reduce the time and reagent costs. Clinical specimens from multiple sites were tested. Two 96-well cell-culture plates were seeded with human foreskin fibroblasts (HFF) and 0.1 ml of each specimen was inoculated into 1 well on one plate, 2 wells on the other. The plates were centrifuged at 700x g for 40 min, incubated for 16-24 hr and the cells were washed and fixed with 1:1 acetone/methanol. The single-well plate was stained with rabbit anti-HSV polyclonal IgG and anti-rabbit IgG FITC conjugate (Dako). The plate was inverted, scanned with a 10x objective by fluorescence microscopy and graded 1-4+. If any well stained positively, the 2-well plate was prepared similarly and the 2 wells corresponding to the Dako-positive specimen were stained with monoclonal anti-HSV-1 and anti-HSV-2 FITC conjugates (Syva) in separate wells. Each specimen was cultured in parallel by a conventional 10-day 2-tube cell culture (TC); positives were typed with the Syva reagents. Two thousand ninety-six specimens were cultured; 1,697 (79%) were negative by both methods. Of 409 (20%) positive by TC, 353 (85%) were positive by Dako and, of these, 322 (79%) were typeable (positive) by Syva. Thirty-nine were positive by Dako, but negative by TC; of these, 11 were typeable by Syva. Dako fluorescence (DKF) was $\geq$3+ in 230 RCs; of these TC was positive in 221 (96%). Therefore, a positive RC was defined as DKF positive, Syva positive or DKF $\geq$3+, Syva negative. Compared with TC, the performance of RC was: sensitivity 82%, specificity 99%, positive and negative predictive values 95% and 96%. The false-negative RCs correlated with increased time to CPE in TC: of the 56 Dako-negative TC positive specimens, 42 (75%) took $\geq$3 days to CPE compared with only 40% of all TC-positive specimens. There was no correlation with HSV type or specimen site. Total costs were 30% of the 2 shell-vial/monoclonal antibody system reported by others. We conclude that the microplate/polyclonal antibody RC system is highly cost effective with good accuracy but, as with other RC systems, should be used together with some form of conventional TC.

CHANGES IN INFECTIVITY AND ULTRASTRUCTURE OF HERPES SIMPLEX VIRUS TYPE 2 (HSV-2)

M. Shimozuma, R.C. Miner, W.L. Epstein, W.L. Drew and K. Fukuyama

University of California
San Francisco and
Mount Zion Hospital
San Francisco, California, USA

We purified neutral pH soluble proteins from rat epidermis and investigated their effects on HSV-2. Corneocytes were separated from 2-day old Sprague Dawley rats and proteins extracted in 0.1% acetic acid were dialyzed against 50 mM sodium phosphate buffer, pH 7.0. Fractions purified by gel filtration on a Sephacryl S-300 column followed by chromatography on a Mono S column (FPLC) were tested for inhibition of wild type HSV-2 on human fibroblasts. We found that 20-30 Kda proteins demonstrate dose- and time-dependent inhibition of plaque formation. Exposure of a virus suspension to 1.1 µg/ml for 4 hr decreased viral titer by 50%. Amino acid analysis of this anti-HSV protein shows characteristics for histidine-rich proteins of rat epidermis containing about 7% histidine and traces of leucine and lysine. HSV-2 with or without epidermal protein incubation was fixed with osmium tetroxide in Dalton's chrome buffer and processed for electron microscopy. We observed that the treated HSV-2 was surrounded by an electron-dense material and gave an appearance of thickening of the envelope. The zone between the capsid and envelope became electron-lucent. We conclude that these epidermal proteins most likely bind the viral particle to alter its infectivity and ultrastructure. The presence of this inhibition in epidermis may partially explain the relative resistance of skin to HSV-2.

ISOLATION OF MEASLES VIRUS IN PRIMARY RHESUS MONKEY KIDNEY (RMK) CELLS FROM A CHILD WITH ACUTE INTERSTITIAL PNEUMONIA WHO CYTOLOGICALLY HAD GIANT CELL PNEUMONIA WITHOUT A RASH

C. Siegel and S. Johnston

Bellin Memorial Hospital
Green Bay, Wisconsin, USA

The isolation of measles virus in primary RMK cells in patients with documented giant cell pneumonia who have presented without a rash is limited. The diagnosis is usually made by cytologic examination of nasal of bronchial secretions in which characteristic multinucleated giant cells with intranuclear and intracytoplasmic inclusion bodies are observed. The diagnosis of giant cell pneumonia has been associated with measles virus, but not exclusively. Canine distemper, herpes viruses, and adenoviruses have been associated with these multinucleated giant cells. In addition, vitamin A deficiency has also been cytologically associated with multinucleated giant cells. We describe the isolation of measles virus from bronchial washings and sputum in RMK cells at four days from an 11-year old child with acute interstitial pneumonia who was in remission for acute lymphocytic leukemia. Classical cytopathic effect, consisting of syncytial and hole formation on the RMK monolayer was apparent. In addition, a foamy appearance of the monolayer in certain specified areas was noted in otherwise clean RMK cells of that specific lot. Confirmatory testing with measles antibody of the infected monolayer by immunofluorescence was positive for measles antigen and negative for mumps, parainfluenza, influenza, adenovirus, and herpes virus. Serological studies revealed a rise in titer to $\geq$10,240. Cytologic examination of the same bronchial fluid revealed the typical multinucleated giant cells, with the characteristic inclusions associated with measles virus. Since this disease is usually severe and often fatal, prompt recognition of this virus in tissue culture is essential no only to the patient who can then be treated with immunoglobulin, but also to prevent the spread of the virus to other patients and medical personnel. These findings also support direct evidence for the etiologic role of measles virus in giant cell pneumonia that has been previously detected only either histologically or cytologically.

A CONFORMATIONAL EPITOPE ON THE FUSION (F) PROTEIN OF RESPIRATORY SYNCYTIAL VIRUS (RSV) DETECTED BY A NEUTRALIZING FUSION-INHIBITING MONOCLONAL ANTIBODY

E.K. Subbarao and J.L. Waner

Case Western Reserve University
Cleveland, Ohio and
The University of Oklahoma Health Sciences Center
Oklahoma City, Oklahoma, USA

A murine monoclonal antibody (MAb) (2D8) prepared against a clinical isolate of RSV that reacted with purified F protein in an ELISA and revealed a unique immunoblotting pattern, allowed us to characterize a conformational epitope of the RSV fusion protein. A lysate of HEp-2 cells infected with RSV Long (Ag) was subjected to PAGE under reducing and non-reducing conditions, transferred to nitrocellulose paper and reacted with 2D8. No reaction was seen when the Ag was electrophoresed under reducing conditions (treatment with 2% SDS, 5% ME and heat to 100ºC). Under non-reducing conditions (treatment with 0.1% SDS without heat), a band was seen at 140-170 Kd which corresponds to previous descriptions of the fusion protein 'dimer'. The epitope detected by 2D8, however, was not affected by SDS or ME, but was destroyed by heating to 100ºC, and 2D8 did not react with any of the subunits of the F dimer. Autoradiography of ^{14}C-glucosamine labeled Ag revealed that the protein detected by 2D8 was glycosylated. This was confirmed by a marked diminution in intensity of immunofluorescence (IF) when 2D8 was used to stain HEp-2 cell monolayers that were infected with RSV in the presence of 10 µg/ml of tunicamycin or 10 mM deoxy-D-glucose, which interfere with normal glycosylation. Inclusion of 2D8 ascites in media used to refeed HEp-2 cell monolayers infected with RSV resulted in inhibition of cell-to-cell fusion. 2D8 ascites neutralized virus by >50% at dilutions of >1:512. When used as a diagnostic reagent for IF of cells from nasopharyngeal washings of children with symptoms of respiratory tract disease, RSV antigen was detected by 2D8 in 63 of 66 specimens that tested positive using a commercial anti-RSV antibody (Bartels). We conclude that our MAb detects a unique conformational neutralizing epitope on the F protein of RSV that is necessary for cell-to-cell fusion, is present and detectable in clinical specimens and could be of importance as a diagnostic reagent and as a possible vaccine.

POLYMERASE CHAIN REACTION (PCR) DETECTION OF JC VIRUS DNA IN PATIENTS WITH PROGRESSIVE MULTIFOCAL LEUKOENCEPHALOPATHY (PML)

A. Telenti, A.J. Aksamit, J. Proper and T.F. Smith

Mayo Clinic
Rochester, Minnesota, USA

PCR was used in the detection of polyomavirus JC (JCV) T antigen gene sequences in 52 sections of paraffin-embedded brain tissue from 39 patients. Tissue sections of 6 mm size were deparaffinized with xylene and absolute alcohol, suspended in 100 mL of water and freeze-thawed to release viral genome. After a denaturing step and 30 cycles of amplification, PCR products were detected by agarose gel electrophoresis and dot blot hybridization. The dot blot hybridization analysis was simplified with the use of a PCR-made specific radiolabeled probe. Specific amplified products were detected in 26 of 31 sections from patients with PML, diagnosed by *in situ* hybridization, and in 4 of 5 specimens from demyelinating lesions that were negative by *in situ* hybridization. Three of 16 sections from patients without demyelinating lesions had positive results by PCR; however, only one specimen remained positive upon retesting. The four specimens of PCR-negative PML tissue were also negative when investigated with oligonucleotide primers specific for the late antigen gene of JCV and the T antigen gene of polyomavirus SV40. PCR serves as a more rapid and simple technique for the diagnosis of PML in brain tissue than *in situ* hybridization.

PREVALENCE OF *CHLAMYDIA TRACHOMATIS* IN TWO POPULATIONS BY CULTURE IN MITOMYCIN-C TREATED McCOY CELLS AND BY A DIRECT FLUORESCENT ANTIBODY (DFA) METHOD

P. Walpita[1], A. Jalowayski[1], V. Mason[1], D. Chadwick[1] and J.D. Connor[2]

[1]University of California San Diego and
[2]Children's Hospital
San Diego, California, USA

The prevalence rate of *C. trachomatis* was compared in two populations: (1) sexually active symptomatic adolescent females presenting to the Department of Adolescent Medicine (DAM), and (2) children between 10-17 years, brought to the Child Protection Clinic (CPC) with alleged sexual abuse. All specimens (vaginal or endocervical) were examined for *C. trachomatis* by a cell culture method (CCM) involving mitomycin-C treated McCoy cells. The usefulness of a DFA method (Kallestad) for the detection of Chlamydia in these two populations was also evaluated. *C. trachomatis* was isolated by the CCM from 26 (12.5%) of the 287 adolescents. This isolation rate was significantly higher than the 9.4% (27 of 287) by the DFA method (McNemar's test, p=0.003). The agreement between the two methods in the CPC group was also poor. Chlamydia could be isolated by CCM from three of the 186 girls; only one of these was positive by DFA. Two additional girls were Chlamydia positive by DFA, but could not be culture confirmed. The sensitivity, specificity, positive predictive value and the negative predictive value for the DFA compared to the CCM in the adolescent group were, respectively, 67%, 99%, 89% and 95%. The corresponding figures for the CPC group were 33%, 99%, 33% and 99%. Many factors may influence the DFA results, including the choice of 'gold standard' against which it is evaluated. Mitomycin-C treated McCoy cells used in this study have the sensitivity equal to that or irradiated cells and better than that of cyclohexamide treated McCoy cells. Mitomycin-C treatment has the additional advantage in that it produces large inclusions similar to those in irradiated cells, but with the ease of drug treatment. The data confirm that culture for Chlamydia using a sensitive procedure is essential, particularly in the low prevalence groups, although DFA may be useful as a supplementary tool.

ADDITION OF CALCIUM TO CELL TREATMENT MEDIUM CONTAINING DEXAMETHASONE AND DIMETHYL SULFOXIDE FURTHER ENHANCES THE DETECTION OF CYTOMEGALOVIRUS FROM CLINICAL SPECIMENS

P.G. West[1] and W.W. Baker[2]

[1]SmithKline BioScience Laboratories
Norristown, PA and
[2]Villanova University
Villanova, Pennsylvania, USA

Extensive study of the molecular biology of human cytomegaloviruses (HCMV) has revealed the fact that there is a unique interdependence between the virus and the physiological state of the host cell. The virus replicates most efficiently in young, actively metabolizing human embryonic fibroblasts, however, the only cultures available to most diagnostic virology labs are commercial cultures which are confluent and quiescent. In previous studies it has been shown that the addition of various combinations of dexamethasone and dimethyl sulfoxide to commercial MRC-5 cells pre- and post-inoculation has resulted in increased growth of both HCMV AD 169 and HCMV from clinical specimens. By the addition of 2mM calcium to the post-inoculation medium, we have now seen increases in viral growth on the order of 7-10 fold over that seen in untreated cells and 2-5 fold over that seen in cells treated with the above reagents. We tested 30 low-titer clinical specimens on (a) untreated cells, (b) cells treated with dexamethasone and dimethyl sulfoxide and (c) cells treated with dexamethasone, dimethyl sulfoxide and 2mM Ca. Of the 30 specimens, 27 (77%) tested positive. Of these, 8 (35%) were positive only on (c); 7 (30%) were positive only on (b) and (c) and the remaining 8 (35%) were positive on all three systems, but in every case had more inclusions on (b) and (c). We have investigated the possibility that the combination of reagents acts as a growth factor which stimulates cell division in the static cultures.

EVALUATION OF A NEW CELL CULTURE ISOLATION SYSTEM

D.L. Wiedbrauk and K.M. Welch

Difco Laboratories Research and Development Center
Ann Arbor, Michigan, USA

During the past several years, many traditional virus isolation methods have been replaced by centrifugation-enhanced shell vial procedures. These procedures have resulted in significantly reduced virus isolation times while maintaining the high levels of sensitivity associated with older methods. Although glass vials have traditionally been used for centrifugation-enhanced isolations, many laboratories have begun to use multiwell cluster trays for this purpose. Recently, a new shatter-resistant cell culture isolation system was developed for use with the centrifugation-enhanced isolation procedures. The isolation efficiency of this system was compared with the standard shell vial method and the cluster tray procedure. In this evaluation, standardized herpes simplex virus inocula were used to infect replicate MRC-5, human foreskin fibroblast, primary rabbit kidney, mink lung, and Vero cell monolayers established on 12 mm glass coverslips in glass vials, in 15 mm diameter cluster trays, and in 14 mm diameter CentriVials. The results indicated that the shell vial method gave an average of 2.62×10^6 focus-forming units (FFU)/0.25 ml while the cluster tray and CentriVial methods yielded 3.96×10^6 and 3.02×10^6 FFU/0.25 ml, respectively at 16 hr post inoculation. At 48 hr post-infection, the same inoculum yielded an average 11.52×10^6 FFU/0.25 ml by the shell vial method, 12.7×10^6 FFU/0.25 ml using cluster trays, and 14.94×10^6 FFU/0.25 ml after 48 hr. In our hands, the shell vial, cluster tray, and CentriVial methods had equivalent isolation efficiencies with five different cell lines. Therefore, the CentriVial isolation system may provide an important alternative to glass shell vials and cluster trays for virus isolations.

HUMAN IMMUNODEFICIENCY VIRUS TYPE 1 (HIV-1) AND HEPATITIS B VIRUS (HBV) SEROPREVALENCE IN AN EMERGENCY ROOM SETTING

S.J. Coffee, A.F. Taylor and R.C. Alexander

San Bernadino County Public Health Laboratory
San Bernadino, California, USA

In 1988, the San Bernadino County Public Health Department initiated a blinded study to measure the HIV-1 and the HBV seroprevalence among emergency medical patients at a local trauma center. Seropositivity rates of 1.5% for HIV-1 and 1.6% for HBV were found. Ninety percent of the HIV-1 and 80% of the HBV seropositive individuals were male. A comparison of seropositivity rates between trauma and non-trauma patients showed no statistically significant difference for HIV-1 or HBV. Although these findings suggest that the blood of trauma patients poses no increased risk compared to that of non-trauma patients, exposure to blood and other body fluids should be minimized to prevent possible infection.

PATHOGENICITY OF CHALCONE R0-0410 RESISTANT AND DEPENDENT HUMAN RHINOVIRUS TYPE 2 (HRV-2) MUTANTS FOR MAN

W. Al-Nakib, S.R. Yasin and D.A.J. Tyrrell

Kuwait University
Safat, Kuwait and
Harvard Hospital
United Kingdom

Mutants of human rhinovirus type 2 HRV-2, resistant to and dependent on the antirhinoviral compound, chalcone Ro 09-0410, have been selected in cell culture. A total of 42 volunteers were challenged with either the drug resistant mutant [SR2-410(r)] (15 volunteers), drug dependent mutant [SR2-410(d)] (15 volunteers) or a wild type HRV-2 which had undergone the same passage level and prepared in exactly the same way as the resistant mutant (12 volunteers). Thirty-three percent, 66.6% and 72.7% of volunteers challenged with the wild-type HRV-2 developed cold symptoms, shed virus and showed serological evidence of infection, respectively. In contrast, only 13.3%, 26.6% and 23.0% of volunteers challenged with the drug resistant mutants developed colds, shed virus or showed serological evidence of infection. Furthermore, none of the volunteers challenged with the drug dependent mutant became infected or had symptoms of colds. These results, therefore clearly demonstrate for the first time that rhinovirus drug resistant mutants show a greatly reduced infectivity and virulence in man when compared with the wild-type. However, the results, at the same time, show that they are capable of infecting man and producing disease. In contrast, drug dependent mutants clearly lost their ability to infect man in the absence of the drug.

EVALUATION OF A DIRECT, RAPID EIA FOR THE DIAGNOSIS OF HERPES-VIRUS INFECTIONS

S.J. Zimmerman, E. Moses, N. Sofat, W.R. Bartholomew
and D. Amsterdam

Erie County Medical Center and
State University of New York
Buffalo, New York, USA

Laboratory confirmation of clinically suspect herpesvirus infection is medically important because the presentation is often uncharacteristic and virus specific chemotherapy is available. We evaluated a 12 minute, direct, monoclonal antibody-based enzyme immunoassay (SureCell, Kodak, Rochester, NY) which aids in the detection of herpes simplex virus (HSV) infection; the assay system is also approved for culture confirmation. The test was evaluated from direct clinical samples and compared to conventional culture methodology. A total of 206 specimens from uncharacterized lesions; 137 female cervical/urogenital; 48 male urogenital; 3 rectal; 3 eye; 5 oral; and 10 colposcopy were collected on dacron/cotton swabs and placed in viral transport medium (VTM) consisting of MEM plus bovine albumin and antibiotics. Within 6 hr of receipt, 0.2 ml of the vortexed VTM was inoculated into each of two replicate cell cultures (MRC-5 and A549). Cell monolayers were observed daily for ten days. When positive CPE was detected, it was semiquantitated and HSV confirmed using an indirect immunoperoxidase reagent (Ortho, Raritan, NJ). The procedure for the SureCell assay conformed to the manufacturer's recommendations. A positive reaction is evidenced by visual inspection of the well with a resultant pink color. When conventional culture was compared to EIA results, the overall sensitivity, specificity, positive predictive value, negative predictive value, and the agreement were respectively 66.2%, 99.3%, 97.7%, 86.4%, and 88.8%. Variables affecting the sensitivity are the stage of lesion and conventional culture methodology. Review of culture results for the 22 EIA false negative tests indicated that 11 positive cultures were detected after 48 hr of incubation. CPE observed at 48, 72, and 96 hr alters the sensitivity for the EIA to 69.8%, 69.6%, 70.0%. To ensure detection of SureCell HSV-negative specimens, it is recommended that an unused aliquot of VTM be tested in cell culture.

EVALUATION OF SURCELL (KODAK) ENZYME IMMUNOASSAY FOR DIRECT DETECTION OF HERPES SIMPLEX VIRUS (HSV) IN CLINICAL SPECIMENS

D. Gall, C. Reed, V. Crespi and T.I. Aquino

St. John's Mercy Medical Center and
Washington University School of Medicine
St. Louis, Missouri, USA

Kodak's Surecell HSV enzyme immunoassay was evaluated for direct detection of HSV in clinical specimens. Two hundred and eighteen clinical specimens were tested following the manufacturer's directions. All specimens were also inoculated into diploid fibroblast and rabbit kidney cell monolayers. Specimens showing cytopathic effect in the monolayers were confirmed by immunofluorescence. Sixty-four specimens were positive by Surecell and 98 were positive by cell culture. Viral typing was done in 75 of the culture positive specimens. The sensitivity, specificity, and positive and negative predictive value for Surecell was 57.1, 93.3, 87.5, and 72.7%. The sensitivity of the test was similar for HSV 1 and HSV 2 (55.8 and 53.1%, respectively). The sensitivity of Surecell in 29 specimens showing CPE in 1 day was 86%, while sensitivity dropped to only 44% for 61 specimens showing CPE after 2 or more days. This is most likely related to differences in the viral titer among specimens. False positive Surecell tests were found in 8 specimens (1 each of throat, sputum, mouth, lip, penis, vagina, cervix and rectum). Ninety-six percent of 18 HSV 2 specimens were isolated from lesions below the waist, while 79% of 43 HSV 1 specimens were obtained from lesions above the waist. Surecell is capable of direct detection of HSV in clinical specimens. Test sensitivity appears to be highly dependent on the viral titer of the specimen.

CHANGES IN LABORATORY ASPECTS OF RESPIRATORY SYNCYTIAL VIRUS

J. Clark

Humana Hospital Sunrise
Las Vegas, Nevada, USA

In the past five years, our laboratory has noticed changes in four aspects of RSV infection. First, our diagnostic rate has increased. Second, we are isolating more respiratory syncytial virus (RSV). Third, the virus appears to be growing faster (faster isolation time), and fourth, the "RSV season" appears to have shifted from the usual November to May pattern to beginning later in December and lasting longer into June. The diagnosis of RSV in the laboratory usually depends on direct fluorescent antibody detection of virus in nasal washings or upon the usual culture methods. In the past five years, we have shifted from a procedure of bedside inoculation and started using a method of nasal washings which yielded a 39% rapid recovery rate of specimen tested. All negative specimens were subjected to culture and the virus isolation rate of cultures was increased too. We feel the faster growth rate and higher isolation rate is also a consequence of a better inoculum because of the nasal washing method for obtaining specimens. This method has also aided in the rapid recovery of other viruses, such as adenovirus, rhinovirus, herpes simplex, enterovirus, influenza and para-influenza. Why the virus seasonality has apparently shifted from the usual beginning of November to December is not apparent at this time. It may be that the extension of the November recovery season into June may be a consequence of improved recovery and diagnosis in the lab. Because of the improved accuracy of the test, physicians have been sending specimens now as late as July. After four years of testing in duplicate, we found that fluorescent antibody determination and culture would detect 91% of the RSVs present. There were 3% false positives and 6% false negatives. If the specimen tests negative, it is then cultured. The availability of ribavirin as a treatment for RSV makes a quick report of a positive result extremely valuable. Our high pediatric and neonatal population makes it essential for rapid isolation and treatment of the virus.

IN VITRO AND IN VIVO ANTIVIRAL ACTIVITY OF HYPERICIN

J. Tang, J.M. Colacino and S.H. Larsen

Lilly Research Laboratories
Eli Lilly and Company,
Indianapolis, Indiana, USA

Hypericin is an aromatic polycyclic anthrone first isolated from the plant St. Johnswort (*Hypericum triquetrifolium* Turra) and shown to have dramatic anti-retroviral activity against Friend leukemia virus (FLV) and radiation leukemia virus in mice (Meruelo, D., G. Lavie, and D. Lavie, Proc. Natl. Acad. Sci. USA, 85:5230-5234, 1988). *In vitro*, using the XC-cell assay, hypericin was tested against Moloney leukemia virus (MoLV) and displayed only marginal activity (IC_{50} = 6 µg/ml). In other cell types, hypericin did not display selective antiviral activity against herpes simplex virus type 1 (HSV-1), influenza A/Brazil, adenovirus type 2, or poliovirus. The 50% cytotoxic concentration for various cells was approximately 25 µg/ml. When virus was incubated with serial dilutions of hypericin before infecting cells, the drug was virucidal to all enveloped viruses tested (HSV-1, influenza virus A/Brazil, and MoLV) at concentrations of 1.56 µg/ml to 25 µg/ml. Hypericin was not virucidal to the non-enveloped viruses tested (adenovirus and poliovirus). Other known antiviral agents, including ribavirin, acyclovir, and enviroxime, with the exception of azidothymidine, were not virucidal at concentrations of 0.78 µg/ml to 50 µg/ml. These data indicate that the mechanisms of viral inactivation for hypericin is dependent upon the presence of a viral lipid envelope. *In vivo*, hypericin (50 µg/ml) was effective against FLV or HSV-1 only if incubated with the virus for 1 h at 37°C before infecting CD-1 mice. Under these conditions, hypericin protected mice from FLV-induced splenomegaly, dramatically reduced FLV viremia and rescued mice from HSV-1 induced encephalitis and death. Hypericin was not effective if pre-incubated with virus for 1 hr at 4°C or if administered to mice beginning at 1 h post-infection. The various components of hypericin, as isolated from plant material, include pseudohypericin and protohypericin which differ slightly from hypericin. These components are now undergoing isolation and characterization and will be studied for their antiviral activity.

DETECTION OF MOLLUSCUM CONTAGIOSUM VIRUS (MCV) BY *IN SITU* HYBRIDIZATION

B. Forghani[1], M. Hurst[1], C. Chan[1], J. Dennis[1], and G. Darai[2]

[1]California State Department of Health Services
Berkeley, California, USA, and
[2]Institüt für Medizinische Virologie der Universität
Heidelberg, Federal Republic of Germany

MCV is a member of the Poxviridae family which has recently became recognized as a major sexually transmitted disease. The clinical syndromes of MCV are often confused with herpes simplex virus infections, as both produce inflamed and ulcerated lesions in the genital area. Precise epidemiological data are not available to assess the frequency of MCV infections in the general public, but it is considered that 20-30% of patients attending sexually transmitted disease clinics are infected by MCV. At present, no reliable laboratory diagnostic assay is available for MCV infection because the virus cannot be serially propagated in cell culture. However, MCV produces an abortive limited growth with some cytopathic effect (CPE) in certain cell lines of human origin. We have developed an *in situ* hybridization assay for detection and identification of MCV genome in clinical specimens. The assay is based on a molecular cloned (pMCV-1-B-C1), biotinylated DNA probe and alkaline phosphatase-labeled biotin as detector signal and 5-bromo-4-chloro-3-indolyl phosphate-nitro-blue tetrazolium (BCIP/NBT) as substrate. Briefly, human fetal diploid lung (HFDL) cell monolayers were prepared in Lab-Tek slides (Miles Labs, IN) and infected with the clinical specimens. After 24-48 hr post-infection, depending on the extent of CPE, the culture medium was removed, the monolayer was rinsed with saline, air-dried and fixed with 4% para-formaldehyde. The fixed cells are ready for immediate use or stored at -70°C for future analysis. The DNA probe was titrated and used under full stringency condition with 50% formamide. Only MCV infected cells showed homology to the MCV probe with intense purple-brown staining. Probe specificity was demonstrated by *in situ* hybridization of cells infected with different viruses and by Southern blot hybridization of different regions of the MCV genome and other viral DNA. Dot blot hybridization with known quantities of MCV and other viral DNA indicate that our probe is capable of specifically detecting 1 pg of MCV DNA. The *in situ* hybridization procedure described here is the first successful identification of an MCV genome in clinical samples by molecular hybridization.

INFECTIVITY FROM HERPES SIMPLEX VIRUS (HSV) ON HEP2 CELLS TREATED WITH SATURATED AND UNSATURATED FATTY ACIDS

F. Galdiero, A. Folgore and M.A. Tufano

Instituto di Microbiologia
I Facolta' di Medicina e Chirugia Universita' de Napoli
Napoli, Italy

The dynamic behavior of biological membranes is determined to a large extent by their content of saturated and unsaturated fatty acids with short or long chains. For this reason, adsorption and penetration of viruses may be conditioned by the type of fatty acids in the membranes themselves. We have verified the consequences of alterations of the lipid phase of HEp2 cells on infectivity of HSV. With this aim, we have treated HEp2 cells with saturated and unsaturated fatty acids with chain lengths from 8 to 18 carbon atoms. The results showed that in HEp2 cells treated with saturated fatty acids with 8 or 10 carbon atoms infectivity from HSV increased, while it was reduced if HEp2 cells were treated with saturated fatty acids and 18 carbon atoms. As to unsaturated acids, those with 8 carbon atoms, 2-octenoic acid and 2-octynoic acid, increase infectivity, while low concentrations of monounsaturated acids with 14 and 18 carbon atoms, cis-9-tetradecanoic acid and cis-9-octadecenoic acid, reduce infectivity.

CLINICAL APPLICATIONS OF THE POLYMERASE CHAIN REACTION (PCR) FOR CYTOMEGALOVIRUS (CMV) INFECTIONS IN ACQUIRED IMMUNODEFICIENCY SYNDROME (AIDS) PATIENTS

M. J. Gill[1] and S.A. Cassol[2]

[1]Faculty of Medicine
University of Calgary
[2]The Canadian Red Cross Blood Transfusion Service
Calgary Alberta Canada

Primer mediated enzymatic gene amplification (polymerase chain reaction, PCR) offers a novel and potent new tool for the detection of CMV. We have undertaken an evaluation of this methodology in the diagnosis and in monitoring the efficacy of antiviral therapy in 3 AIDS patients with CMV infection. For amplification we use a 368 bp segment of the fourth exon of the major CMV IE gene. This region is highly conserved with no homology to human DNA. Two primers, SC-6 and SC-8, 33 and 30 bases long determined from published sequence data were synthesized. Amplified sequences were detected using Southern transfer and hybridization with [^{32}P] SC-2, a 30 base long synthetic probe. Blood from patients with AIDS has been routinely stored for PCR evaluation. Following red blood cell lysis using NH_4C1 lysis buffer, the DNA was extracted from the total leukocyte fraction. One 45-year old male receiving Zidovudine, showed a low level of CMV DNA in his lymphocytes prior to the development of CMV retinitis. The high levels of the CMV DNA found at the onset of the disease declined significantly with ganciclovir therapy. A 42-year old male developed progressive pneumonia. Repeated bronchoscopy revealed only inclusion bodies diagnostic of CMV. Therapy with IV ganciclovir was unsuccessful and the patient died. Retrospective PCR analysis detected only small quantities of CMV DNA, despite a wide variety of hybridization conditions and different primers. A 26-year old male with AIDS developed CMV retinitis. Despite ganciclovir therapy, he deteriorated. PCR studies showed a persistently high level of CMV DNA raising the possibility of drug resistant virus. Our experience suggests that PCR is a promising technique in the diagnosis and monitoring therapy of CMV infection.

COMPARISON OF HERPCHEK WITH CULTURE FOR DETECTION OF HERPES SIMPLEX VIRUS (HSV) FROM CLINICAL SPECIMENS IN VIRAL TRANSPORT MEDIA

C.A. Gleaves, C.F. Lee and D.H. Rice

Fred Hutchinson Cancer Research Center
Seattle, Washington, USA.

The Dupont Herpchek assay is a forward sandwich enzyme immunoassay (EIA) for the rapid direct detection of HSV antigen in clinical specimens. We retrospectively assayed 301 clinical specimens (229 culture positive and 72 culture negative) with this system. Specimens were transported to the virus lab in viral transport media (VTM) and included throat (172), genital (56), other oral (55), skin (17), and eye (1). Specimens were inoculated into human fibroblasts (HF) and A549 cell culture tubes for viral isolation, and the remainder of the sample was placed at -70^0C. Specimens were thawed, vortexed and resuspended in 10x Herptran concentrate prior to testing. Each sample was then added in duplicate to designated wells of a microtiter plate for the EIA assay. Herpchek detected 147/150 (98.0%) culture positive specimens from symptomatic patients, 63/79 (79.7%) culture positive specimens from patients considered asymptomatic, and 210/229 (91.7%) culture positive specimens overall. Of the 19 discrepant samples, 18 were from throat specimens from 16 asymptomatic patients. The additional sample was from an oral lesion and required 7 days to grow in culture. Of the EIA negative/culture positive samples, an average of 3.4 days was required before CPE was observed in culture (range 2 to 10 days) as compared to an average of 2.1 days for the EIA positive/culture positive samples (range 1 to 4 days). Herpchek reacted with all 19 viral isolates, suggesting that virus concentration and/or sampling variability and/or technical variabilities may have affected results. Herpchek was negative for 70/72 (97.2%) culture negative specimens. Subsequent clinical data on the 2 EIA positive/culture negative specimens suggests that these 2 samples were true EIA positives. These data suggest that the Herpchek assay can be used with clinical specimens submitted in conventional VTM. However, VTM samples which are EIA negative, particularly with EIA values close to the EIA positive cut off value, need to be cultured.

EVALUATION OF THE CYTOMEGALOVIRUS (CMV)-CUBE ASSAY FOR THE DETECTION OF CMV IMMUNE STATUS IN MARROW TRANSPLANT PATIENTS

C.A. Gleaves, S.F. Wendt, D.R. Dobbs and J.D. Meyers

Fred Hutchinson Cancer Research Center
Seattle, Washington, USA

We evaluated a new membrane dot immunobinding assay (CMV-CUBE; Difco, Detroit, Michigan) for the detection of CMV antibody in marrow transplant patients and donors. The CMV-CUBE assay was compared to a commercially available EIA (CMV STAT; Whittaker Bioproducts) and latex agglutination (LA) (CMVScan; Becton Dickinson) tests. Sera were collected from 311 transplant patients and donors prior to transplantation. Sera were tested by EIA at a 1:21 dilution in wells coated with CMV antigen and read at a 550-nm wavelength for a positive or negative reaction. Sera were screened by LA, undiluted and mixed with CMV antigen-coated latex beads on a card slide and positive or negative reactions were indicated by the presence or absence of agglutination. Sera were tested by the CMV-CUBE tube assay at a 1:5 dilution and added to a membrane cassette with CMV antigen bound to the membrane. The membrane was washed with buffer before being reacted with a chromogenic enzyme-linked conjugate for 1 minute, washed again and reacted with a chromogenic reagent. The reaction was stopped after 2 minutes. A positive reaction exhibits a blue spot in the center of the membrane with a white background. A negative reaction exhibits a white membrane. A total of 164 sera were positive for CMV antibody by one or more of the three assays with 153/164 (93.3%) samples positive by all three tests. A total of 147 sera were CMV antibody negative. CMV-CUBE detected 157/164 (95.7%) CMV positive samples. As compared to EIA, CMV-CUBE has a sensitivity of 97.6% (six sera were EIA positive/CUBE negative) with a specificity of 99.4% (one serum was CUBE positive/LA negative). The CMV-CUBE assay is a simple and rapid visual assay which can be used for the qualitative detection of CMV antibody in patient serum. The data show that the CMV-CUBE assay is a useful test for the determination of patient and donor immune status to CMV in the marrow transplant patient population.

TREATMENT OF INFLUENZA A VIRUS INFECTION IN MICE WITH RIMANTADINE, AMANTADINE, RIBAVIRIN, OR GAMMA INTERFERON, AND THE EFFECTS OF THESE AGENTS ON IMMUNE RESPONSES

K. West, J.E. Herrmann, M. Bruns and F.A. Ennis

University of Massachusetts Medical School
Worcester, Massachusetts, USA

Rimantadine, amantadine, ribavirin and recombinant gamma interferon were evaluated for treatment of influenza A virus infection in mice. Evaluation included determination of antiviral activity, measured by titration of virus in lung tissue, and assessment of the effect of the antiviral agents on immune responses *in vivo* and *in vitro*. The immune responses measured were virus-specific cytotoxic T lymphocyte (CTL) responses, natural killer (NK) cell activity, lymphocyte proliferation *in vitro*, the production of virus-neutralizing serum antibodies, and the resistance of mice to re-infection after recovery from primary infection. Rimantadine was found to be the most effective agent tested in reducing pulmonary virus titers. Rimantadine also suppressed the CTL and neutralizing antibody responses, but the suppression was found to be virus-specific and not a general phenomenon. The decreased antibody response did not lower resistance to re-challenge at a dose equivalent to that used for the primary infection (50 infectious units), but did permit infection at very high doses (1 X 10^5 infectious units). Recombinant gamma interferon did not reduce pulmonary virus concentration, nor did it enhance or suppress either the CLT response or the NK cell response. Ribavirin significantly inhibited *in vitro* lymphocyte proliferation responses to B and T-cell mitogens (concanavalin A, phytohaemagglutinin, lipopolysaccharide) and influenza virus antigen at drug concentrations 8-fold lower than any other drug tested. None of the drugs, however, suppressed the proliferative responses to the mitogens or influenza A virus with lymphocytes obtained from mice treated *in vivo*.

THE PRESUMPTIVE IDENTIFICATION OF ENTEROVIRUSES USING RHABDOMOSARCOMA (RD), HEP-2, AND PRIMARY RHESUS MONKEY KIDNEY (RMK) CELL LINES

S. Johnston and C. Siegel

Bellin Memorial Hospital
Green Bay, Wisconsin, USA

The limited supply of the Lim Benyesh-Melnick antisera pools for the typing of enteroviruses has made routine neutralization testing inappropriate in most clinical laboratories. Moreover, few clinical virology laboratories have the expertise or can afford the time or expense of running neutralizations. This study indicates that a presumptive identification of enterovirus groups can be made on the basis of characteristic cytopathic effect displayed in RD, Hep-2, and RMK cell lines. Echoviruses and Coxsackie A viruses could be isolated in RD and RMK cells, but not in Hep-2 cells. Coxsackie B virus could be isolated in all three cell lines. We recommend the use of these cell lines to make presumptive enterovirus group identifications for routine viral isolates.

A COMPARISON OF THE USE OF A SERUM SUPPLEMENT (OMNI SERUM) AND FETAL BOVINE SERUM (FBS) IN CELL CULTURE USED TO ISOLATE VIRAL AGENTS FROM CLINICAL SPECIMENS

S. Johnston and C. Siegel

Bellin Memorial Hospital
Green Bay Wisconsin, USA

Traditionally, FBS has been the principle component in media used in the growth and maintenance of cell cultures. Recent shortages have affected the cost and availability of FBS to the clinical laboratory. Furthermore, lot to lot variability can affect cell culture performance and growth. We evaluated a commercially available serum supplement (Omni Serum) for use in the growth of cell cultures and for use in maintenance media used in the isolation of viruses from clinical specimens. Rhabdomyosarcoma (RD) and mink lung (ML) cells raised on 5% Omni Serum displayed the same sensitivity and integrity in tubes (RD and ML) and vials (MRC-5) as those grown in 10% FBS and maintained with 5% FBS. Our trials indicate that Omni Serum is a viable substitute for FBS used in maintenance media for cell culture tubes and vials used in viral isolation from clinical specimens. Furthermore, the cost of Omni Serum is less than that of FBS.

ENHANCEMENT OF COXSACKIEVIRUS B_3 (CB_3) REPLICATION IN VERO CELLS BY INDOMETHACIN (IND)

R. Khatib, M.P. Reyes and F. Smith

Grace Hospital and
Wayne State University School of Medicine
Detroit, Michigan, USA

In a previous study using a murine model, we showed that IND increased CB_3 titers in the heart, worsened animal mortality and histopathologic changes. In this study, we elected to examine the effect of IND on CB_3 replication *in vitro*. IND was diluted in distilled water and sodium carbonate, filtered and added to Vero cell monolayers in microtiter plates (final concentration of IND was 2×10^{-5} - 2×10^{-7} m/well) after 32 hr incubation, 2×10^5 $TCID_{50}$ of CB_3 was added and reincubated for 18 hr. Cells were scraped, sonicated and virus titers were compared. Titers in control wells were $10^{2.75}$ $TCID_{50}$ whereas they were $10^{3.2}$, $10^{4.2}$ and $10^{5.5}$ $TCID_{50}$ in wells supplemented with 1×10^{-6}, 1×10^{-5}, or 1×10^{-4} M of IND, respectively. In three additional triplicate experiments, a trend toward enhancement of CB_3 titers by IND was apparent. (Log of control titers were 5.25 ± 0.5, 5.93 ± 0.81, 5.87 ± 0.63, compared to 7.16 ± 0.76, 6.4 ± 0.35, 7.5 ± 0.10 or 6.05 ± 0.42, 5.11 ± 1.44, 7.02 ± 1.0 in the presence of IND at 1×10^{-6} or 1×10^{-4} m/0.1 ml, respectively). We repeated the experiment using a plaque assay method. Vero cells were planted on 6-well microtiter plates in the presence of increasing concentrations of IND incubated for 72 hr, then CB_3 was added at MOI of 0.001 or 0.005. After 2 hr, media was discarded and cells were overlayed with media supplemented with 0.5% methylcellulose, incubated for 72 hr, then plaques were counted. IND significantly increased plaque formation as shown:

DRUG CONCENTRATION

MOI	0	10^{-8}	10^{-7}	10^{-6}	10^{-5}	10^{-4}
0.001	1.76 ± 0.58	ND	2.67 ± 2.51	$9.34 + 1.53$	6.34 ± 0.58	13.67 ± 2.08
0.005	7.25 ± 2.63	9 ± 2.45	10 ± 1.42	11 ± 0	16.25 ± 4.35	32.75 ± 7.14

These results demonstrate a dose dependent enhancement of CB replication *in vitro* by IND. These findings may explain (at least partially) the basis for IND deleterious effects in CB_3 murine infection.

STRUCTURAL AND PRELIMINARY SEQUENCE ANALYSIS OF THE HUMAN HERPESVIRUS 6 (HHV-6) GENOME

G.J. Lindquester[1], T. Dambaugh[2], R. Allen[1] and P.E. Pellett[3]

[1]Rhodes College
 Memphis, Tennessee
[2]E.I. DuPont
 Wilmington, Delaware
[3]Centers for Disease Control
 Atlanta, Georgia, USA

HHV-6, the probable cause of *roseola infantum* (*exanthum subitum*, sixth disease) can be isolated from peripheral blood lymphocytes and grows in T cells *in vitro*. We are studying the genome of an isolate of HHV-6 from Zaire, HHV-6(Z29). HHV-6(Z29) genome has no strong DNA homology to the other 5 known human herpesvirus under stringent hybridization conditions. The G+C content of the genome is approximately 44% as determined by isopycnic density gradient centrifugation. We have constructed a plasmid clone library of the HHV-6(Z29) genome and are determining restriction endonuclease maps for BamHI and KpnI. The genome consists of a 141 kilobase unique segment bracketed by a pair of directly repeated sequences that vary from 10 to 13 kb in length in viral DNA isolated at different passages. Sequence analysis of HHV-6(Z29) has identified blocks of genes conserved in several other human herpesviruses.

EVALUATION OF THE RECOMBIGEN TEST FOR THE DETECTION OF ANTIBODIES TO HIV-1

S.L. Aarnaes, R.L. MacDonald, L.M. de la Maza and E.M. Peterson

University of California Irvine Medical Center
Orange, California, USA

A 5 minute latex agglutination test (Recombigen®-HIV LA, Worchester, MA) for HIV-1 antibody status was compared to an EIA (Abbott Laboratories, N. Chicago, IL). The LA test uses a recombinant gp41 and gp120 protein as antigen, whereas, the EIA uses a viral lysate. The LA was tested according to manufacturers instructions where cards are rotated at 60 rpm for 5 minutes on a horizontal shaker and agglutination is read with a fluorescent magnified lamp. The EIA was also tested according to manufacturers instructions where a positive sample needed to be repeatably positive. All positive specimens by either method were further tested by Western blot (E.I. DuPont De Nemours and Co., Inc., DuPont, Wilmington, DE). Of the 121 specimens examined there were 40 positive and 68 negative by all three methods. Of the 13 discrepant samples all were positive by EIA and all were negative by LA. Western blots were negative for 4 of these 13 and positive for the remaining 9. All of these latter nine had a gp41 and a gp120 band. The majority of these 9 discrepant samples had OD values by EIA <2, whereas, all 40 samples that showed agreement between EIA and LA had an OD >2. The overall sensitivity and specificity of the latex agglutination compared to EIA was 75% and 100% and when compared to Western blot, 82% and 100%, respectively. Upon contacting the manufacturer of the LA about these results, they suggested retesting false negative samples by LA, however, with the modifications of rotating cards by hand and using a halogen magnified lamp to read the end result. Using this method of the 9 EIA positive, LA negative samples, 7 were now also positive by LA. Therefore, we recommend modification of the package insert for the LA product to eliminate a large percentage of the false negative samples.

CONTRIBUTORS

LARRY J. ANDERSON   Division of Viral Diseases, Center for Infectious Diseases, Department of Health and Human Services, Centers for Disease Control, Atlanta, Georgia, USA

NAN ANDERSON   Department of Microbiology, Faculty of Medicine, University of Toronto, Toronto, Ontario, Canada

GRAEME BARNES   Department of Gastroenterology, Royal Children's Hospital, Melbourne, Parkville, Victoria, Australia

RUTH F. BISHOP   Department of Gastroenterology, Royal Children's Hospital, Melbourne, Parkville, Victoria, Australia

NEIL R. BLACKLOW   Division of Infectious Diseases, Department of Medicine, University of Massachusetts Medical School, Worcester, Massachusetts, USA

BRUCE C. BYRNE   State University of New York, Division of Hematology/Oncology, Department of Medicine, Syracuse, New York, USA

ELIZABETH CIPRIANI   Department of Gastroenterology, Royal Children's Hospital, Melbourne, Parkville, Victoria, Australia

FRANCES W. DOANE   Department of Microbiology, Faculty of Medicine, University of Toronto, Toronto, Ontario, Canada

GARTH D. EHRLICH   State University of New York, Division of Hematology/Oncology, Department of Medicine, Syracuse, New York, USA

RICHARD W. EMMONS   Viral and Rickettsial Disease Laboratory, State of California Health and Welfare Agency, Department of Health Services, Berkeley, California, USA

M.A. EPSTEIN   University of Oxford, Nuffield Department of Clinical Medicine, John Radcliffe Hospital, Headington, Oxford, United Kingdom

PEKKA E. HALONEN   Department of Virology, University of Turku, SF-20520, Turku 52, Finland

JOHN C. HIERHOLZER   Respiratory Virus Laboratory, Division of Viral Diseases, Center for Infectious Diseases, Department of Health and Human Services, Centers for Disease Control, Atlanta, Georgia, USA

YORIO HINUMA   Shionogi Institute for Medical Science, Mishima, Settsu-shi, Osaka 566, Japan

MARTIN S. HIRSCH   Massachusetts General Hospital, Harvard Medical School, Infectious Disease Unit, Boston, Massachusetts, USA

JOHN HOPLEY   Department of Microbiology, Faculty of Medicine, University of Toronto, Toronto, Ontario, Canada

SHIRLEY KWOK   Cetus Corporation, Emeryville, California, USA

FRANCIS LEE   Department of Microbiology, Faculty of Medicine, University of Toronto, Toronto, Ontario, Canada

JENNIFER LUND   Department of Gastroenterology, Royal Children's Hospital, Melbourne, Parkville, Victoria, Australia

KATHRYN PEGG-FEIGE   Department of Microbiology, Faculty of Medicine, University of Toronto, Toronto, Ontario, Canada

BERNARD J. POIESZ   State University of New York, Division of Hematology/Oncology, Department of Medicine, Syracuse, New York, USA

JOHN SNINSKY   Cetus Corporation, Emeryville, California, USA

JOHN A. STEWART   Herpes Virus Laboratory, Division of Viral Diseases, Centers for Disease Control, Atlanta, Georgia, USA

LEANNE UNICOMB   Department of Gastroenterology, Royal Children's Hospital, Melbourne, Parkville, Victoria, Australia

KEITH WELLS   State University of New York, Division of Hematology/Oncology, Department of Medicine, Syracuse, New York, USA

AUTHOR INDEX

Aarnaes, S. L., 271
Aquino, T. I., 258
Aksamit, A. J., 239, 251
Allen, R., 270
Al-Nakib, W., 256
Amsterdam, D., 257
Alden, B., 240
Alexander, R. C., 255
Anderson, L. J., 17, 187
Anderson, N., 1

Baker, W. W., 253
Barnes, G., 85
Barrett, H., 246
Bartholomew, W. R., 257
Burns, M., 266
Bishop, R., 85
Blacklow, N. R., 111
Byrne, B. C., 47

Cassol, S. A., 263
Chan, C., 261
Chegas, P., 246
Chadwick, D., 252
Cipriani, E., 85
Clark, J., 259
Cohen, F., 245
Coffee, S. J., 255
Colacino, J. M., 260
Conner, J. D., 252
Crespi, V., 258

Dambaugh, T., 270
Darai, G., 261
de la Maza, L. M., 271
Dennis, J., 261
Doane, F. W., 1
Dobbs, D. R., 265
Drew, W. L., 248

Ehrlich, G. D., 47
Emmons, R. W., 129
Ennis, F. A., 266
Epstein, M. A., 207
Epstein, W. L., 248

Fisher, J. E., 246
Folgore, A., 262
Forghani, B., 261
Fukuyama, K., 248

Galdiero, F., 262
Gall, D., 258
Garret, C. T., 244
Gill, M. J., 263
Gleaves, C. A., 264, 265
Greenberg, H. B., 246

Halonen, P. E., 17
Hierholzer, J. C., 17
Herrmann, J. E., 266
Hurst, M., 261
Hinuma, Y., 147
Hirsch, M. S., 217
Hopley, J., 1

Jalowayski, A., 252
Johnston, S., 249, 267, 268

Keiser, J., 244
Khatib, R., 269
Knock, M., 240
Kwok, S., 47

Lairmore, M. D., 242
Larsen, S. H., 260
Lee, C. F., 264
Lee, F., 1
Li, P. K., 246
Lindquester, G. J., 270

Littlewood, R. A., 247
Lund, J., 85

MacDonald, R. L., 271
McKendall, R. R., 242
McLean, D. M., 241
McLinden, J. M., 243
Marshall, W. F., 239
Mason, V., 252
Matthey, S., 240
Meriwether, C., 245
Meyer, J. D., 265
Miner, R. C., 248
Moore, E. C., 245
Moses, E., 257

Nicholson, D., 240

Oas, J., 242
Ogra, P. L., 246

Pegg-Feige, K., 1
Pellett, P. E., 270
Peterson, E. M., 271
Philpot, D., 245
Ploplis, V., 243
Poiesz, B. J., 47
Porter-Jordan, K., 244
Proper, J., 239, 251

Reed, C., 258
Reyes, M. P., 245, 269
Rice, D. H., 264
Riepenhoff-Talty, M., 246
Rosen, E., 243

Rosenberg, E., 244
Ruhs, S., 240

Sherlock, C. H., 247
Shimozuma, M., 248
Sibau, L. E., 247
Siegel, C., 249, 267, 268
Smith, F., 269
Smith, T. F., 239, 251
Sninsky, J., 47
Sofat, N., 257
Stapelton, J., 243
Stewart, J. A., 163
Subbarao, E. K., 250

Tang, J., 260
Taylor, A. F., 255
Telenti, A., 239, 251
Tufano, M. A., 262
Tyrell, D. A. J., 256

Uhnoo, I., 246
Unicomb, L., 85

Walpita, P., 252
Waner, J. L., 250
Welch, K. M., 254
Wells, K., 47
Wendt, S. F., 265
West, K., 266
West, P. G., 253
Wiedbrauk, D. L., 254

Yasin, S. R., 256

Zimmerman, S. J., 257

SUBJECT INDEX

Acetylcholine receptor, 132
Acrylamide gel, 80
Acyclovir, 166, 167, 260
    AZT combined therapy, 227
Adenosine deaminase, 223
Adenovirus, 4, 6, 18, 22, 25-35, 45, 169,
        259
    enteric, 111, 112
    serotype *40*, 111
    serotype *41*, 111
Agent, antiviral, *see* separate
        compounds
Agglutination test, 271
AIDS patient, 48, 53, 73, 151, 163, 217-
        236, 263
    Clinical Trials Group (USA)
        studies, 221-222
    drug-resistant virus, 263
    -related complex, *see* ARC
    see HIV-1
Airfuge ultracentrifuge, 8
AL-721, 225
Alanine aminotransferase, 246
Amantadine, 266
Ampligen, 225
Anemia, macrocytic
    AZT, 220
    retrovirus, 47
Antibody
    -antigen complex, 6
    ELISA, 153
    fluorescing, 163, 240, 252, 259
    fusion-inhibiting, 250
    labeled, 24
    monoclonal (MAB), 8, 9, 18, 24, 26,
        250
    neutralizing, 250, 266
    polyclonal, 2, 247
    problems, 154
Antigen-antibody complex, 6
Antiviral agent, *see* separate compounds
ARC patient, 73, 219, 220
Arthritis and retrovirus, 47
Aspartate aminotransferase, 246
Aspartic protease, microbial, 224
Aspirate, nasopharyngeal, 19, 42

Astrovirus, 111-114, 119, 122-123
    animal, domestic, 122
    detection methods, 122
    diarrhea, watery, 122
    gastroenteritis, 122-123
    Marin County agent, 114, 122, 123
    removed from Picornavirus group,
        122
    type *5*, 114
ATL, *see* T-cell leukemia virus in adult
Autographa californica, 243
Avidin, 22
    -biotin peroxidase staining, 137
AZdU, *see* 3'-Azido-2',3'-deoxyuridine
3'-Azido-2',3'-deoxyuridine (AZdU),
        222
Azidothymidine (AZT), *see* Zidovudine
AZT, *see* Zidovudine

Baculovirus, 243
    occlusion body, paracrystalline, 243
    polyhedrin, 243
Bat virus, *see* Lyssavirus serotypes
B-cell, 165
    EBV-infected, 165
    HBLV-infected, 163
Biotin, 20, 22
    -avidin peroxidase staining, 137
BK strain of polyomavirus, 239
Blood, virus-infected
    blood bank, 154
B-lymphotropic virus, human, *see*
        Herpesvirus-6 (HBLV)
    B-cell specific, 163
Bovine serum, fetal, 268
Bronchiolitis, 19
Burkitt's lymphoma, endemic, 207, 208
N-Butyl deoxynojirimycin, 225
    AZT combined therapy, 227

Calcium, 253
Calicivirus, 112, 114, 116, 120-121
    animal, domestic, 120
    cold food, 121
    gastroenteritis, 120-121
    sea lion, 120

Calicivirus (continued)
    shellfish, 121
    strains, 121
    water, 121
Calithrix jacchus, see Marmoset
Candidiasis, esophageal, 245
Carcinoma
    bronchogenic, 207
        cigarette smoking, 207
    nasopharyngeal, 207, 208
    retrovirus, 47
Castanospermine, 255
    AZT combined therapy, 227
CD4 receptor for HIV-1, 217
    autoantibody against, 218
    AZT combined therapy, 227
    recombinant, soluble, 218
    toxin hybrid preparations, 218
Cell
    culture in shell vial, 239
    cytotoxicity, 218
    line
        A-549, 257
        B, 163, 165
    fibroblast, diploid, 258
        foreskin fibroblast, human, 254
        HEK, 239
        HEp-2, 250, 262, 267
        HUT-78, 50, 51, 57, 58
        JJHAN, 163
        Killer, natural, 266
        McCoy, 252
        mink lung, 254
        MRC-5, 239, 253, 254, 257
        rabbit kidney, 254, 258
        rhabdosarcoma, 267
        rhesus monkey kidney, 267
        T, 163, 165, 166
        Vero, 254
    tropism of HHV6, 165
Chalcone R.09.0410, 256
Chicken
    herpesvirus, 208
    lymphoma, 208
    Marek's disease, *see* Marek's
        disease
Chlamydia trachomatis
    abuse, sexual, 252
    antibody, fluorescing direct method,
        252
    culture on McCoy cells, 252
Cluster tray, multiwell, cultivation, 254
Cocklevirus, 114
Complement, 2
Compound Q, 255
Concanavalin A, 266
Conjunctivitis, hemorrhagic, 19
    coxsackievirus A-24, 18
Cord blood lymphocyte and HHV6, 165
Coronavirus, 18, 19, 22, 24, 26-32, 35, 44,
        45, 111, 112
    enteric, 111, 112

Cotton rat, 187
Cottontop tamarin, *see Saguinus oedipus*
Coxsackievirus, 13
    A, 267
    A-24, 18, 31, 32
    B, 267
    B-3, 269
Cultivation, viral
    cluster tray, multiwell, 254
    shell vial centrifugation, 254
Cyclohexamide, 252
Cynomolgus monkey, 155
Cytomegalovirus, 28, 30, 167, 168, 171,
        172, 175-179, 184-185,
        244, 245, 253
    African green monkey, 168
    AIDS patient, 263
    assay, commercial, 265
    detection, 253
        fibroblast, embryonic, human,
            253
            calcium, 253
    marrow transplant patient, 265
    membrane dot immunobinding assay,
        265
    pneumonia, 263
    polymerase chain reaction, 263
    retinitis, 263
Cystosine arabinoside and herpes
        zoster, 229
Cytotoxicity, cellular, 218

Deoxycholate, 168
2'-Deoxy-conformycin, 154
    for ATL patient, 154
Dexamethasone, 253
Dextran sulfate, 225
Diarrhea and rotavirus
    calf, 86
    infant, 85
        deadly, 86
        discovered in *1973*, 86
    mouse, neonate, 86
Diarrhea, viral, 111-128 *see* separate
        viruses
Dideoxyadenosine (ddA), 223
    reverse transcriptase inhibited, 223
    triphosphate, 223
Dideoxycytidine (ddC), 223-224
    anti-HIV activity, 223
    AZT-alternating regimen, 223
    reverse transcriptase inhibition, 223
    toxicity, 223
        neuropathy, peripheral, 223
2',3'-Dideoxy-2',3'didehydrothymi-
        dine, 58
Dideoxyinosine (ddI), 223-224, 230
    anti-HIV activity, 223
    toxicity, 223
        fatigue, 223
Diethyldithiocarbamate, 227
Dimethyl sulfoxide, 20, 253

Ditchlingvirus, 114
DNA, 77, 81-84
    amplification, 50 *see* Polymerase
            chain reaction
    hybridization, 163
    polymerase I, 80
        DNA-dependent, 49
        Klenow fragment, 48
        Taq, 48, 50
    viral, 169
Dog bites in the USA, 130
    rabies treatment, *see* Rabies
Dot immunoblot
    assay, 169
    hybridization, 137, 261
Duncan syndrome, 211
    EBV infection, 211
Duvenhagevirus, 132

EBV, *see* Epstein-Barr virus
Echovirus, 13, 28, 267
EIA, *see* Enzyme immunoassay
Electron microscopy, 1-16, 164, 246, 248
    antigen-antibody complex, 1, 2
    film, 2, 5, 6
    HHV6, 164-167
    immunoassay, 1-16
    lymphocyte, 164
    methods, comparison of, 5-9
ELISA, *see* Enzyme immunoassay
Encephalitis, 185
Enterovirus, 4, 7, 14, 18, 19, 22, 24, 26-29,
            31, 32, 259
    identification, 267
    Lim Benyesh antisera pool, 267
Enviroxime, 260
Enzotin reagent, 19
Enzyme immunoassay (EIA) test, 8, 9, 17,
            74, 137, 171, 172, 245,
            250, 271
    capture test
        all-monoclonal, 20
        polyclonal, 21
    commercial, 257, 258, 264, 265
    formatting, 22-27
        listed, 23
    for herpes simplex virus, 257, 258
    improved in *1984*, 17
    one-incubation TR-FIA test, 21-22
    problems, 154
Enzyme-linked immunosorbent assay,
            see Enzyme
            immunoassay
Epitope, 189-194, 196-198, 243
    definition, 190
Epstein-Barr virus (EBV), 168, 172, 184,
            185, 245
    infection
        B-cell, 165
        Burkitt's lymphoma, endemic,
            207

Epstein-Barr virus (EBV)(continued)
        carcinoma, nasopharyngeal,
            207
        Duncan syndrome, 211
        recombinant, 209
        tumor, human, 207
        vaccine, 207-215
            human use, 207-208
            gp340, 209-210
            rationale for, 207-208
            subunit prototype, 208-209
            phase I, 210-212
            protocol, 211
Escherichia coli DNA polymerase I
    Klenow fragment, 48
Europium, 18, 20, 21, 23, 25, 34, 36
Exanthem subitum, *see* Roseola

Fatigue, chronic, syndrome, 184
Fibroblast, human, 258, 264
    calcium useful, 253
    embryonic, 22, 253
    foreskin, human 247
    lung, embryonic, diploid, 22
Ficoll gradient, 170
Fluorescein isothiocyanate, 18
3'-Fluorodideoxythymidine, 58
Fluoroimmunoassay, monoclonal, 17-45
    history, 18
Formatting, 22-27
Friend leukemia virus, murine
    hypericin, 260

Gancyclovir, 166, 167, 263
Gastroenteritis virus, human *see*
            separate viruses
    electron microscopy, 111
    pediatric, 121
    small, round, 111-128
    virology, medical, comparative,
            113-115
Gene amplification, enzymatic
    primer-mediated, *see* Polymerase
            chain reaction
Giant cell pneumonia
    adenovirus infection, 249
    canine distemper, 249
    herpesvirus infection, 249
    measles infection, 249
    vitamin A deficiency, 249
Glial cell, 169
Glioblastoma, 165
β-Globin, 52
γ-Globulin, hyperimmune, 134, 135
GLQ223, 225
α-Glucosidase-1, 225
    inhibitors, 225
        castanospermine, 225
        N-butyl deoxynojirimycin, 225
Glycoprotein, 35
Gold, colloidal, 1, 6, 7, 14, 15

Granulocytopenia and AZT, 221

HAM, see Myelopathy, HTLV-I-
    associated
Hawaii virus, Norwalk-like, 114
HBLV, see B-lymphotropic virus,
    human
HDCV, 135
HEF, see Fibroblast of human foreskin
HEp-2 cell line, 22, 267
Hepatitis virus, 55
    A, 4, 243
    B, 212, 245, 255
    murine, 246
Herpesvirus
    animal, 168-169, 208
    human, 164-186 see Herpesvirus-6
    latency, 178
    simian, 164
Herpesvirus saimiri, 168
Herpesvirus-6, human (HHV6), 163-
    186, 270
    ACIF test, 171
    AIDS patient, 163, 169
    antibody, 168
    antigen, 168-170
        analysis by, 168-170
    associations, clinical, 172-175
    B-cell infected, 163
    cell
        culture, 170
        tropism, 164-165
    demonstrated by
        electron microscopy, 163, 166-168
        immunofluorescent antibody,
            163,170-171
    discovery in 1986, 163
    diseases, 175
    DNA, double-stranded, 169
        primer sequence
    ecology
        AIDS patient, 163, 169
        roseola patient, 163
    electron microscopy, 166-168
    genome analysis, 270
    growth characteristics, 164, 166
        T-cell line, 171
    history, 163-164
    immunofluorescence assay, 163, 170-
        171
    laboratory diagnosis of infection,
        170-172
    latency, 178-180
    leukemia patient, 163
    lymphadenopathy patient, 163
    lymphoma patient, 163
    properties, physical, 166-167
    proteins, 168
    reactivation, 178-180
        reinfection, 179
        transplant patient, 178

Herpesvirus-6 (continued)
    roseola infantum, 270
    sensitivity to antiviral drug, 166
    seroepidemiology, 176-180
    seroprevalence, 176-177
    symptoms, 175
Herpes simplex virus, 168, 171, 172, 176,
        185, 247, 248, 259, 261
    antibody, fluorescent, polyclonal,
        247
    assay, commercial, 264
    culture, rapid, 247
    encephalitis and idoxuridine, 228
    enzyme immunoassay (EIA), 257, 258
    fibroblast cells, 264
        human, 248
    HEp-2 cell line, 262
        fatty acid-treated, 262
    idoxuridine and encephalitis, 228
    infectivity and fatty acids, 262
    photodynamic treatment, 229
    rat epidermis protein, 248
    sandwich enzyme immunoassay in
        fibroblast cells, 264
    skin resistance to, 248
    specimen, clinical, 264
    type-1, 258
    type-2, 258
Herpes zoster and cytosine arabinoside,
        229
HHV-6, see Herpesvirus-6, human
HIV, see Human immunodeficiency
        virus
Horse virus and "staggers" in Nigeria,
        132
HPA-23, 225
HRIG, see Human rabies
        immunoglobulin
HTLV, see T-cell lymphoma-leukemia
        virus, human
Human immunodeficiency virus (HIV)
    type-1, 48-54, 57, 58, 73, 77-79, 83,
        84,170, 271
        agent, antiviral, 225
        agglutination test, 271
        AIDS patient, 149 see AIDS
        antibody detection, 271
        attachment inhibitor, 217
        AZT, 219-222 see Zidovudine
            combination therapy, 226-
            227
            resistance is emerging, 222
        binding and entry, 217-218
        cell membrane receptor CD4,
        217
            rsCD4, 218
        combination therapy with AZT,
        226-227
            analysis, mathematical,
            226
            three drugs, 227

Human immunodeficiency virus (continued)
  dideoxycytidine, 223-224
  dideoxyinosine, 223-224
  DNase-treated, 58
  drug abusers, 245
  emergency room, 255
    seroprevalence, 255
  gag gene, 53, 54, 57, 58
  glycoprotein
    gp120 for attachment, 217
    processing, 225
    inhibition as a target, 225
  human herpesivirus-6, 177, 186
  immunoadhesin, 218
  infection, 217-236 *see* AIDS
  integrase, 224
  interferon, 225
  lymphadenopathy
    -associated virus (LAV)
      HTLV-II/ARV, 149
    syndrome, 178
  nucelotide sequence of certain
      regions, 69-70
  oligonucleotides, 224
  postpartum women, 245
  progression to AIDS, 245
  protease, 219, 224
  receptor on cell membrane CD4,
      217
  receptor preparation rsCD4, 218
  replication, late events, 224-225
  response modifiers, biological,
      227
  reverse transcriptase, 149, 179,
      219
  risk groups for, 177, 186
  sex partner, 245
  syphilitic, 245
  syncytium formation by, 149
  T-cell specific, 163
  trials, clinical, 228-231
  type-2 in West Africa, 48, 163
Human rabies immunoglobulin (HRIG),
      135
Hydrocortisone, 166
Hypericin, 260
Hyperplasia, lymphoid, 175

Idoxuridine, 228
IF, *see* Immunofluorescence
Immune
  complex, 1-3
  deficiency and retrovirus, 47
      ˙*see* HIV, AIDS
Immunity
  deficiency, 47, 175, *see* HIV, AIDS
  suppression, 175
Immunoadhesin hybrid protein, 218
Immunoassay, 1-16, 243
  electron microscopy, 1-16

Immunoelectron microscopy, 1-16
  gold, colloidal, 1
  history, 1-2
  methods, 3-8
Immunofluorescence (IF), 258
  antibody, 153, 154, 163
  assay, 246, 249, 250
  rabies, 137
Immunoperoxidase reagent, 257
Immunoprecipitation, 171
Immunosorbent electron microscopy, 4
Immunostimulation, 227
Indomethacin, 269
Infection, viral, *see* separate viruses
  polymerase chain reaction
      assay, 244
Influenza virus, 14, 15, 18, 259
  A, 30, 35, 43, 240, 241
  B, 240
  infection in human elderly
      vaccinees, 241
  infection, murine, 266
    drugs, antiviral, 266
  vaccine, trivalent, 241
Integrase of HIV-1, 219, 224
  integration of proviral DNA into
      cellular DNA, 224
Interferons, 134, 135, 225, 226, 266
  toxicity , 225
Interleukin
  alpha, 227
  -2, 152, 166, 170
Iododeoxyuridine, 147
Iscom, 210
Isoprinosine, 227

JC strain of polyoma virus, 251
  DNA, 251
JJHAN cell line, 163, 170
Jurkat cell line, 174

Kaposi's sarcoma, 222, 225, 226
  interferon, 225
Kawasaki disease, 186
Keratoconjunctivitis, epidemic, 32
Kotonkanvirus, 132
  Nigeria from *Culicoides* gnats, 132

Lagos bat virus, 132
Lanthanide, 18
Latex agglutination, 265
Lectin, 6
Lentivirus, 149
Leukemia, adult
  human herpesvirus-6, 163
  retrovirus, 47
  T-cell form (ALT), 155 *see* ATL
Leukoencephaly, multifocal,
      progressive, 251
  JC polyoma virus, 251
  laboratory diagnosis, 251

Ligase T4, 56
Lim Benyesh antisera pool for
            enterovirus typing, 267
Lipopolysaccharide, 266
Lung cell, human
      diploid, fetal, 261
Lupus erythematosus, systemic, 154
      retrovirus, 47
Lymphadenopathy syndrome
      HIV-1, 178
      HHV-6, 163
Lymphocyte, 165, 167, 266
Lymphoma, 175
      HHV-6, 163
      retrovirus, 47
Lyssavirus serotypes
      bat virus, 132
      Duvenhage, 132
      Lagos bat, 132
      Mokola, 132
      rabies virus, 132

MAB, see Antibody, monoclonal
Marek's disease, 208
      herpesvirus, 208
      vaccine against, 208
      virus, 169
Marin County virus, 114, 122, 123
      astrovirus type 5, 114
Marmoset (Calithrix jacchus), 209
      Oka varicella virus, 209
Marrow transplant patient, 265
      cytomegalovirus, 265
      immunity status, 265
McCoy cell line, 252
      cyclohexamide-treated, 252
Measles virus, 184
      pneumonia, 249
Megakaryocyte, 165
Methylphosphonate, 224
Mink lung cell line, 254, 268
Mitomycin-C, 252
Mokola virus, 132
Molluscum contagiosum virus, 261
Moloney murine leukemia virus, 48, 50
Monkey B virus, 169 see Rhesus monkey
Mononucleosis, infectious, 151, 175
Montgomery County virus (Norwalk-
            like), 114
Mouse, 246
      leukemia virus, 225
      rotavirus infection, 246
MRC-5 cell line, 268
Mumps virus, 30
Murine leukemia virus, 225
Myelopathy, HTLV-I-associated, 47,
            148
      non-lethal, 160
      U.S.A., 242

Nasopharyngeal carcinoma, see
Carcinoma

Natural killer cell activity, 266
Neoplasm in animal, 147
      retrovirus, 147
Neutralization test, viral, 171, 243, 250
Neutropenia and AZT, 220
Norwalk virus
      characteristics, biological, 115-116
      diagnosis, 117
      epidemiology, 117
      gastroenteritis, 115, 116
      immunity, clinical, 118
      immunoassay, 115
      nursing home outbreaks, 118
      pathology, clinical, 116
      relatedness to calicivirus, 118
      transmission by
            air, 117
            fecal-oral route, 117
            vomitus, 117
      winter vomiting disease in
                  Norwalk, Ohio (1968),
                  116
Norwalk-like viruses, 112,118-120
      discovery in 1972, 111
      gastroenteritis, 118-120
      Hawaii agent, 119
      Montgomery County (Maryland)
                  agent, 119
      Otofuke agent, 120
      Sapporo agent, 120
      Snow Mountain agent (Colorado
                  resort camp), 119, 127
      Taunton (Great Britain) agent, 120
NP-40, 168
Nycodenz gradient, 168

Obodhiang virus in Sudan
      from mosquitoes, 132
cis-9-octadecenoic acid, 262
      herpes simplex virus, 262
2-Octenoic acid, 262
      herpes simplex virus, 262
2-Octynoic acid, 262
      herpes simplex virus, 262
cis-9-tetra decenoid acid, 262
      herpes simplex virus, 262
Oka virus, see Varicella virus
Oligomer restriction hybridization, 53
Oligonucleotide with anti-HIV-1
            activity, 224
      see Methylphosphonate,
                  Phosphorothioate
Oncovirus, 149
Otofuke virus (Norwalk-like), 114

PA, see Particle agglutination test
Parainfluenza virus, 18, 19, 25-35, 45,
            240, 259
Paramatta virus, 114
Paramyxovirus, 31
Paraparesis, spastic, 148
      see Myelopathy

Particle agglutination test for antibody, 153-154, 160
  problems, 154
Parvovirus, 115
Pentamidine, 229
Pepstatin A, 224
Peptide T, 225
Phosphonoformic acid, 166, 167
Phosphorothioate, 224
9-12,2-Phosphorylmethoxye-
        thyladenine,58
Phosphotungstic acid stain, 2
Phytohemagglutinin, 170, 266
Picornaviridae, 243
Picornavirus, 24, 31
Pneumocystis carinii pneumonia, 220
Pneumonia, 19, 220, 249
Poliovirus, 5, 84
Poxviridae, 261
'   sexually transmitted, 261
Polyhedrin (protein), 243
  recombinant, 243
  VP3, 243
Polymerase chain reaction, 244, 263
  AIDS, 263
  amplification protocol, 59-61
  assay, 244, 251
  carry-over problem, 58-59, 77-82
  cytomegalovirus, 263
  HHV-*6*, genome, 170
  inverse, 55, 56
  ligation-mediated, 55
  limitations, 244
  liquid hybridization, 63-64
    spot blot method, 64-66
  reactions, 61-63
  round-table discussion, 77-84
  spot blot method of hybridization,
        64-66
Polyomavirus, 239
Protease of HIV-*1*, 219, 224
  target of antiviral drug, 224
  see Aspartic protease, microbial
Protein A of *Staphylococcus aureus*, 3-6
Proteinase, *see* Protease
Pseudomonas aeruginosa exotoxin A, 218
Pseudorabies virus of swine, 168

Rabbit kidney cell line, 254, 258
Rabies, 129-145
  acquisition, indigenous, 130
  animal reservoir, 131, 132
  antibody
    assay, 137
    immunofluorescent, 137
    monoclonal, 131
    neutralizing, 131
  carrier state, 133
  cell culture, 137
  chronic, 133
  clinical, 133

Rabies (continued)
  costs in the U.S.A., 130
  diagnosis, 134, 136, 138
  dog bites in the U.S.A., 130
  ecology, 130-133
  genes, structural, five, 130
  global, 129
    humans, 129
  history, 129
  human, 129, 130, 133-135
    non-fatal, 133
  immunofluorescence stain, 137
  infection, specificity of, 133
  laboratory diagnosis, 136-138
    methods described, 137
  mouse inoculation test, 137
  non-fatal human cases, 133
  prevention, 129, 135-136
  probe, molecular, 137
  reservoir in animals, 131, 132
  Semple-type vaccine, 135, 136
  stability, genetic, 131
  strains, 131
  symptoms, 133
  taxonomy, 131
  T-cell response, 133
  transplant, corneal, 135
  treatment is a big challenge, 134-135
  vaccine
    anti-idiotypic, 136
    duck embryo, 135
    horse serum, 135
    human serum, 135
    oral, 136
    peptide, synthetic, 136
    Semple, 135, 136
    single, 131
    sources of, 136
  vaccinia vaccine, recombinant, 136
  virus, 130-133
    morphology, 130
    pathogenicity, loss of, 131
    proteins, 131
    RNA, single-strand, 130
    RNA transcriptase, RNA-
        dependent, 130
    wildlife form, 136
Racoon rabies along U.S. East Coast, 132
Radiation leukemia virus, murine, 260
  hypericin, 260
Rat epidermis protein, 248
  herpes simplex virus type 2, 248
Reoviridae, 86 *see* Rotavirus
Reovirus, 28
Respiratory syncytial virus (RSV), 18,
        19, 22-29, 31, 32, 35, 44,
        187-205, 240
  antibody fluorescence, direct, 259
  antigens, 189-191
    difference among strains, 187-
        191

Respiratory syncytial virus (continued)
    broncheolitis, 187-190
    detection methods, 190
        enzyme immunoassay (EIA),
            190, 191
        immunofluorescence, 190, 259
        immunoprecipitation, 190
        Western blot analysis, 190
    discovery in 1957, 187
    epidemiology, 200-201
        strains, 200-201
    epitope, 189-198
        conformational, 250
        differences, 191-194
    F protein, 189, 191-194, 198, 250
    glycoprotein, 196, 198
        F, 189, 191-194, 198, 250
        G, 189, 191-194, 197-199
    G protein, 189, 191-194, 197-199
    groups, major, two, 187
        A, 195
        B, 195
    immunity, protective, 194-200
    infection, 188-189
    laboratory aspects, changes in, 259
    neutralization of glycoprotein, 193-
        197
    pneumonia, 187-190
    proteins listed, 191
    respiratory tract, lower, illness
        global in children, 187-189
    ribavirin, 259
    site, antigenic, on virus, 189, 190, 196
    strain
        antigen differences, 187-191
        differences are important, 187-
            205
        epidemiology, 200-201
        immunity, protective, 194-200
    vaccine
        development, 187-205
        unsuccessful, 187
    worldwide lower respiratory tract
        illness
        children, 187-189
Respiratory virus infections, 14, 17-45
    detection method, 240
    diagnosis, rapid, 17-45
    enzyme immunoassay (EIA), 17
    fluoroimmunoassay (FIA), 17
        time-resolved, 17-45
    immunofluorescence (IF), 17
    viruses involved, 17
Response modifier, biological, 227
    immunostimulation, 227
Restriction endonuclease, 53, 56
Retrovirus (RNA tumor virus), human,
        47-75
    adult T-cell lymphoma-
        leukemia (ATLL), 47
    autoimmune disease, 47

Retrovirus (RNA tumor virus)(continued)
    cytopathic disease, 47
    diagnosis, problems of, 48
    diseases associated with, 47-48, 147
        autoimmune, 47
        cytopathic, 47
        T-cell lymphoma-leukemia, 47
    endogenous, 148
        transmission is vertical, genetic,
            150
    lentivirus group, 149
    malignancy, 47
    neoplasm in animal, 147
    oncovirus group, 149
    polymerase chain reaction, 47-75
    primer pairs, 49
    screening for, 80
    sequence endogenous with human
            genome, 51-52
    transmission
        blood transfusion, 150-151
        germ cells, 148
        horizontal, 150
        vertical, 150
    type
        B, 148
        C, 147
        D, 148
Reverse transcriptase, 50, 77, 84
    of HIV-1, 219
    target of antiviral drug, 154
Rhabdosarcoma, human
    cell line, 22, 239, 267, 268
Rhabdoviridae, 131-132
Rhabdovirus group, see Rabies virus
Rhesus monkey
    kidney cell line, 249, 267
        measles virus, 249
    rotavirus, 246
Rhinotracheitis virus, bovine, 168
Rhinovirus, 84, 259
    type-2, human (HVR-2), 256
Ribavirin, 134, 225, 227, 259, 260, 266
Ricin, 218
Rimantadine, 266
RNA, 83
    transcriptase, RNA-dependent, 130
    see Retrovirus
Roseola (exanthem subitum)
    antibody titer, 173-174
        neutralizing, 174-175
    atypical, 174-175
    complications, 173
    epidemiology, 172
    features, clinical, 172-173
    HHV6, 163, 172, 173
    Japan, 169, 172, 177
    patient, 163, 171, 185
    symptoms, 173
    transmission, experimental, human,
        172

Rotavirus, 4-7, 13, 28, 85-112
    antibody against, 87, 92-94
        cord blood, 95
        neutralizing, 92-93, 98
        tests listed, 93
    Bangladesh, 85
    blood, 88, 95
    cell-mediated immune response, 94-
        95
    characteristics, 86-87
    coproantibody, 98-99
    detection by electron microscopy, 86
    developing countries, 85
    diarrhea
        calf, 86
        infant, 86*
        mouse, neonate, 86
    dose, infectious, oral, 87-88
    electron microscopy for detection, 86
    feces, shedding by, 88-89
        coproantibodies, 98-99
    hepatitis, murine, 246
    immune response
        to animal rotavirus in human, 93
        cell-mediated, 94-95
    immunity, clinical, 89
    immunocompromised patient, 110
    infants, 85
    infection, neonatal, 95-96
    interferon, 95
    interleukin-2, 95
    intestinal tract, human 85-110
    murine, 246
    killer cell, natural, 95
    non-group A, 111, 112
    pathophysiology, 87-88
    pig strain, 87, 88
    polypeptide, 87, 89, 91
    protection, 99-100, see antibody,
        vaccine
    reinfection, 89-94
        is easy, 99
    rhesus monkey, 246
    RNA, double-stranded, 86
    serology, 91-96
    serotype, 87, 89, 91
    simian strain, 88
    symptoms, clinical, 88-89
    vaccine
        program, 109, 110
        strains, 95
Rubella, virus, 8

Saguinus oedipus (cottontop tamarin),
        208-210
Semple vaccine against rabies, 135
Sapporo virus (Norwalk-like), 114
Sarcoidiosis, 175
Sarcoma and retrovirus, 47
Serum
    agar method, 4
    supplement, commercial, 268

Shell vial culture, 247, 254
Simian immunodeficiency virus, 218, 228
Sindbis virus, 5
Site, antigenic, 189, 190
    definition, 190
SIV, see Simian immunodeficiency virus
Sixth disease, see Roseola by HHV6
Slot blot hybridization, 50, 51, 57, 80,
        164
Small, roun, featureless agents
    of human gastroenteritis, 112-128
Snow Mountain virus (Norwalk-like),
        114
Southern blot analysis, 73, 80, 169, 261,
        263
Specimens, clinical, for virus isolation,
        268
Spot blot hybridization method, 64-66
SRSV (small round-structured viruses),
        see Gastroenteritis
        virus
Staphylococcus aureus protein A, 3, 4
Stevens-Johnson syndrome, 221
Streptavidin, 20
Sucrose gradient, 168
Suramin, 225

Taunton virus (Norwalk-like), 114
T-cell, 169, 170, 270
    antigen, 165
        CD series, 165
    cytoxicity, 266
    helper cell, 245
    line Jurkat, 174
    virus-immune, 131
T-cell lymphoma-leukemia, viral, in
        adult (ATL), 73, 82, 147,
        151-155
    antibody detection, 153-155
    carrier, 147
    control, 155
    features, clinical, 151
    groups, five
        acute, 151
        chronic, 151
        crisis-acute, 151
        lymphoma-type, 151
        smoldering, 151
    hepatomegaly, 151
    immune deficiency is common, 151
    infection, opportunistic, 151
    Japan, endemic, 147, 150
    laboratory diagnosis, 152-155
    lesion, cutaneous, 151
    leukemogenesis, 149-150
    lymphadenopathy, 151
    Martinique, 147, 148
    MT-2 positive cell line, 149
    pathogenesis, 149-150
    prevention, 155
    seroepidemiology, 150-151
    Taiwan, 147
    T-cells, abnormal, 151

T-cell lymphoma-leukemia virus,
     human, 47, 50-54, 66-69,
     72-74, 77, 82, 147-161
    HTLV-I
        antibody, 153-155
        antigen, 148
        blood transfusion,
            transmission by, 150-151
        cerebrospinal fluid, 242
        control, 154
        DNA analysis, 149
        drug abuser, intravenous, 47
        ELISA, 153
        endemic, worldwide, 47
        nucleotide sequences of certain
            regions, 66-69
        paraparesis, spastic, 242
        properties, 148-149
        provirus DNA, 150-152
        structure, 148
        transmission by blood
            transfusion, 150-151
        worldwide endemic, 47
    HTLV-II, 47, 50-54, 72-74, 149
        drug abuser, intravenous, 72-73
        hairy cell leukemia, 47
        nucleotide sequences of certain
            regions, 69
        T-cell leukemia, 47
T-lymphotropic virus, simian (STLV),
     149
cis-9-Tetradecenoic acid, 262
    herpes simplex virus, 262
Thalidomide disaster, 229
Therapy, antiviral, evaluation, 263
Thrombocytopenia, 222
T-lymphocyte, see T-cell
Tobacco mosaic virus, 1
Torovirus, 128
    Berne agent, 128
    Breda agent, 128
    diarrhea, 128
Toxoplasmosis, 175, 245
Tract, intestinal, human
     rotavirus, 85-110
Transplant patient, 178, 179, 180, 184,
     185
Tricosanthin, 225
Trimethoprim-sulfamethoxasole, 229

Vaccinia virus, 198, 199
    hemagglutinin gene, 154

Vaccinia virus (continued)
    recombinant, 154
    WR strain, 209
Varicella virus
    marmoset, 209
    Oka strain, 209
    recombinant, 209
Varicella-zoster virus, 168, 171, 172, 176
Vero cell line, 254, 269
Viruses, see separate viruses

Western blot analysis, 74, 160, 168-171,
     242, 243, 271
Wollan virus, 114, 115
W virus, see Wollan virus

XC cell assay, 260
Xenon light, 18
X-linked lymphoproliferative
            syndrome (Duncan
            syndrome), 211
    Epstein-Barr virus, 211

Zidovudine (AZT), 58, 166, 167, 219-222,
     230, 245, 260, 263
    cerebrospinal fluid, 219
    combined therapy, 226, 227
    Friend murine leukemia virus
            inhibited in 1974, 219
    glucuronidation, 219
    hemodialysis, 219
    infections, opportunistic, reduced by,
        220
    interferon in combined therapy,
        226
    licensed in U.S.A. in 1987, 220
    mortality greatly reduced, 220
    synthesized as anti-cancer drug in
        1974, 219
    therapy, combined, 226, 227
    thymidine kinase, 219
    thymidylate kinase, 219
    toxicity, 220, 221
        anemia, macrocytic, 220
        drug fever, 221
        granulocytopenia, 221
        headache, 220
        insomnia, 220
        nausea, 220
        neutropenia, 220
        rash, 221